Origins of
Anatomically Modern
Humans

Origins of Anatomically Modern Humans

Edited by

MATTHEW H. NITECKI
and
DORIS V. NITECKI
Field Museum of Natural History
Chicago, Illinois

PLENUM PRESS • NEW YORK AND LONDON

Library of Congress Cataloging in Publication Data

Nitecki, Matthew H.
Origins of anatomically modern humans / edited by Matthew H. Nitecki and Doris V. Nitecki.
p. cm.—(Interdisciplinary contributions to archaeology)
Includes bibliographical references and index.
ISBN 0-306-44675-8
1. Human evolution—Congresses. 2. Paleolithic period—Congresses. I. Nitecki, Doris V. II. Title. III. Series.
GN281.N565 1993 93-40104
573.2—dc20 CIP

ISBN 0-306-44675-8

A Division of Plenum Publishing Corporation
233 Spring Street, New York, N.Y. 10013

Printed in the United States of America

Contributors

Paul G. Bahn • 428 Anlaby Road, Hull, HU3 6QP England

Ofer Bar-Yosef • Department of Anthropology, Peabody Museum, Harvard University, Cambridge, Massachusetts 02138

Rebecca L. Cann • Department of Genetics and Molecular Biology, John A. Burns School of Medicine, University of Hawaii at Manoa, Honolulu, Hawaii 96822

Catherine Farizy • Laboratoire d'Ethnologie Préhistorique, Université de Paris, 1 Pantheon–Sorbonne, 44, Rue de l'Admiral Moucher, 75014 Paris, France

David W. Frayer • Department of Anthropology, University of Kansas, Lawrence, Kansas 66045

F. Clark Howell • Department of Anthropology, University of California, Berkeley, California 94720

Arthur J. Jelinek • Department of Anthropology, University of Arizona, Tucson, Arizona 85721

Richard G. Klein • Department of Anthropology, University of Chicago, 1126 East 59th Street, Chicago, Illinois 60637

J. Koji Lum • Department of Genetics and Molecular Biology, John A. Burns School of Medicine, University of Hawaii at Manoa, Honolulu, Hawaii 96822

Doris V. Nitecki • Department of Geology, Field Museum of Natural History, Chicago, Illinois 60605

Matthew H. Nitecki • Department of Geology, Field Museum of Natural History, Chicago, Illinois 60605

Geoffrey G. Pope • Department of Anthropology, University of Illinois, Urbana, Illinois 61801

Olga Rickards • Dipartimento di Biologia, II Universita di Roma, Tor Vergata, Roma, Italia

Tal Simmons • Department of Anthropology, Western Michigan University, Kalamazoo, Michigan 49008

Fred H. Smith • Department of Anthropology, Northern Illinois University, DeKalb, Illinois 60115

Olga Soffer • Department of Anthropology, University of Illinois, 607 S. Mathews, Urbana, Illinois 61801

Christopher Stringer • Department of Palaeontology, The Natural History Museum, Cromwell Road, London, SW7 5BD England

Alan G. Thorne • Department of Prehistory, Australian National University, Canberra, ACT 2601 Australia

Milford H. Wolpoff • Department of Anthropology, University of Michigan, Ann Arbor, Michigan 48109

Preface

This volume is based on the Field Museum of Natural History *Spring Systematics Symposium* held in Chicago on May 11, 1991. The financial support of Ray and Jean Auel and of the Field Museum is gratefully acknowledged.

When we teach or write, we present only those elements that support our arguments. We avoid all weak points of our debate and all the uncertainties of our models. Thus, we offer hypotheses as facts. Multiauthored books like ours, which simultaneously advocate and question diverse views, avoid the pitfalls and lessen the impact of indoctrination. In this volume we analyze the anthropological and biological disagreements and the positions taken on the origins of modern humans, point out difficulties with the interpretations, and suggest that the concept of the human origin can be explained only when we first attempt to define *Homo sapiens sapiens*.

One of the major controversies in physical anthropology concerns the geographic origin of anatomically modern humans. It is undisputed, due to the extensive research of the Leakeys and their colleagues, that the family Hominidae originated in Africa, but the geographic origin of *Homo sapiens sapiens* is less concretely accepted. Two schools of thought exist on this topic. The received view, promoting an African origin for *H. sapiens sapiens*, is in direct opposition to a multiregional model of anatomically modern humans, which attempts to illustrate how anatomically modern human populations evolved independently in various geographic areas instead of evolving in Africa and then migrating and displacing other hominid populations elsewhere.

A short step away from the interest in the geographic origin of anatomically modern humans is the concern with the actual biological differences between ourselves and our predecessors. These differences, and the behavioral modifications are reflected in Ice Age art, a characteristic unique to anatomically modern humans.

The commonly hypothesized date for the origin of anatomically modern humans, approximately 200,000 ybp, or more generally between 100,000 and 300,000, is derived not through the archaeological records, but through the molecular clock, based on the human mitochondrial DNA.

We have tried very hard not to favor any viewpoint, but to present a symmetrical shape of the argument. We are neither advocating nor denying the African origin of modern humans. We are aware that the great pendulum of intellectual fashion is now swinging away from the mitochondrial interpretation of human origin. But we are also aware that new questions are being constantly asked. We hope that this approach to the complexities of our origin and who we are, and our attempt to provide both sides of the medal, will be of heuristic value.

Matthew H. Nitecki
Doris V. Nitecki

Contents

PART I. INTRODUCTION

PART II. WHAT ARE MODERN HUMANS?

PART III. AFRICAN CENTER OF ORIGIN

PART IV. MULTIREGIONAL HYPOTHESIS

Part I

Introduction

Paleoanthropology is both history and science – it is, in fact, an *historical science*. It interprets and reconstructs the life of extinct humans from their preserved skeletal anatomy, their assumed behavior, and the processes that shaped them; by comparison with other evolutionary disciplines, the anthropological data are sparse. Nevertheless, the general consensus is that the appearance of modern humans altered the biomass of the earth and that humans subdued and utilized the inert matter and the seemingly unlimited solar energy to divert the flow of water to paint the deserts with chlorophyll, and to cause extinctions selectively. A more extreme view is that the human is an angel with frightful skills to build and to destroy – a sexual superstar outsexing all others, and overpopulating the earth with his own kind. Thus, there appears to be an accord on the results of human activities, and, perhaps, even on the bases for accepting the consensus. However, while the time after the origin is believed to be known, the place(s) and time(s) of origin(s) of humans and the definition(s) of *Homo sapiens* are matters of grave discords.

At present the main quest of human anthropology is *where* did humans originate? *Who* they were, and *when* they appeared are secondary disagreements. Thus, the place of the origin appears to some anthropologists to be the single most important point of contention.

In his introductory chapter Richard Klein surveys the main points of disputes on the human origin. Human anthropology, although based on fragmentary fossil material, advances to broad solutions of definitions and origins of humans. Geology (geochemistry), genetics (molecular biology), and archaeology support these advances. Thus, we have arrived at the point where some certainty may be claimed on the methods and techniques used. Lately, molecular biology offered this great certainty. Klein does not reject the mtDNA information as not useful – it certainly is that; nor can the work of Wilson and Cann be discarded – it is only the beginning of an application of a new method. It will, and is, being refined. It is one of many approaches to the study of the origin of modern people. While Klein, as all of us, seeks certainty, he is aware that the proofs appear in the courts and juries of anthropological opinions and judgments.

Chapter **1**

The Problem of Modern Human Origins

RICHARD G. KLEIN

Like many other scientific fields, paleoanthropology has been revolutionized by the late twentieth-century breakdown and reformulation of traditional disciplinary boundaries. Human paleontologists, who once dominated the field, must now draw increasingly on research by molecular geneticists, geochronologists, archeologists, and other specialists to address major paleoanthropological questions. As illustrated by the contributions to this volume, the multidisciplinary approach has especially influenced the fascinating and highly controversial issue of modern human origins. With these contributions in mind, my purpose here is to summarize the major competing explanations for modern human origins and to outline how genetics, geochronology, and archeology can help to choose between them.

THEORIES OF MODERN HUMAN ORIGINS

Many scientific theories of modern human origins are clearly possible, but specialists and lay people alike now tend to recognize just two major alternatives. The first posits that modern humans evolved more or less simultaneously from earlier, nonmodern populations throughout the occupied world, while the second suggests that modern humans emerged in a circumscribed geographic region and subsequently replaced nonmodern populations elsewhere.

As formulated, especially by Wolpoff (1992; Wolpoff et al. this volume), the idea of geographically widespread, more or less concurrent, evolu-

---

RICHARD G. KLEIN • Department of Anthropology, University of Chicago, Chicago, IL 60637.

Origins of Anatomically Modern Humans, edited by Matthew H. Nitecki and Doris V. Nitecki. Plenum Press, New York, 1994.

tion toward modern humans is now commonly called the Theory of Multiregional Evolution. In essence, multiregionalists argue that modern racial variability in craniofacial form is already detectable in very ancient, nonmodern fossils from widely separated regions of the world and that this implies fundamental genetic continuity between nonmodern and modern populations in each region. They explain the early development of racial features by a combination of gene drift and regionally specific natural selection pressures. However, they believe that persistent gene flow constrained the degree of racial divergence, so that ancient, nonmodern populations as far apart as present-day Cape Town, Paris, Beijing, and Jakarta never approached the status of separate species. An obvious implication is that unlike racial features, which conferred no more than local, natural selective advantage, modern features conferred a significant selective advantage everywhere. It was presumably this universal advantage that allowed modern features to spread so widely, while racial features remained geographically confined.

At least two different variants of the multiregional model are current and represented in the present volume. The first and stronger view (favored by Wolpoff) accepts the possibility that different modern features evolved in different regions and subsequently combined everywhere via gene flow and natural selection. This is essentially the position advocated by Li and Etler (1992), who argue specifically that the modern human face originated in east Asia, while the modern braincase originated far to the west, perhaps in Africa. A weaker (and arguably much more plausible) multiregional variant is that modern morphology originated as a single complex in a geographically restricted region, most probably Africa, but spread from there as much by gene flow as by population replacement (Smith 1992, this volume)

The major alternative to the multiregional model was christened the "Noah's Ark" hypothesis by Howells (1976) and is now commonly referred to as the "Out-of-Africa" theory, since its most prominent contemporary proponents (including Stringer 1992, this volume) favor Africa as the modern human birthplace. Unlike multiregionalists, proponents of the Out-of-Africa theory see no compelling fossil evidence for racial continuity between nonmodern and modern humans in widely separate parts of the world. Instead, they believe that gene drift and natural selection drove disparate archaic populations along divergent evolutionary trajectories from the time when archaic humans first colonized Eurasia (from Africa), between roughly 1.4 and 1 million years ago. The result was the development of separate human lineages that culminated in incipient, if not full-fledged species. Present evidence allows for perhaps three such lineages: a European one that can be called *Homo neanderthalensis* (assuming that the well-known Neanderthals were its terminal product); a Far Eastern one that can be tentatively referred to as *Homo erectus* (based on the conventional name for its intermediate products; its end product is less well known); and an African one that can be called *Homo sapiens* (assuming that it alone introduced the morphology that characterizes living humans).

Like the multiregional hypothesis, the Out-of-Africa theory comprises at least two variants. The stronger version, which has been vigorously encouraged by mitochondrial DNA (= mtDNA) studies summarized by Cann (1992; Cann, Rickards & Lum, this volume), holds that modern humans (the fully modern products of the *Homo sapiens* lineage) replaced their archaic contemporaries essentially without gene exchange. The weaker version (dubbed the "African Hybridization and Replacement Model" by Bräuer 1992) postulates gene exchange or interbreeding during replacement and is distinguished from the weaker version of the multiregional hypothesis mainly by the more limited degree of gene exchange it envisions.

THE RELEVANCE OF GENETICS

The likelihood that modern human genetic diversity can illuminate human evolution has been understood for decades, but genetic studies had only minor impact before 1987, when Cann, Stoneking, and Wilson (1987) published a landmark analysis that attracted wide attention. As recounted by Cann, Rickards, and Lum (this volume), they analyzed the variability in mtDNA from about 140 individuals of different modern geographic or racial origin. They chose mitochondrial (as opposed to nuclear) DNA because it diversifies more rapidly and because it is inherited more simply (individuals receive all or nearly all their mtDNA exclusively from their mothers.) Cann, Stoneking, and Wilson used taxonomically popular parsimony analysis to construct a phylogenetic tree that required the smallest number of mutations to explain the mtDNA variability in their sample, that is, to relate the various mtDNA types according to their degree of similarity. The tree had two main branches – one comprised exclusively of Africans and a second including some Africans and individuals of all other origins. Since the individual branches that required the most mutations to explain (and that had thus probably existed longest) were African, Cann, Stoneking, and Wilson concluded that the most plausible root for the tree was in Africa.

By itself, the mtDNA tree did not imply that the common mitochondrial ancestor was already modern, but the estimated mutation rate behind the mtDNA variability indicated that the common ancestor had lived in Africa about 200,000 years ago. This was a time when archaic Eurasian lineages were already distinct, and the implication seemed clear that archaic Eurasians (perhaps above all, the Neanderthals) contributed little, if any, mtDNA to living humans. An additional, more comprehensive and methodologically sophisticated analysis of mtDNA variation in living humans (Vigilant et al. 1991) reconfirmed the structure of the original tree and continued to suggest that the last shared mitochondrial ancestor of all living humans had lived in Africa roughly 200,000 years ago.

Stringer (this volume) notes that Beaumont, de Villiers, and Vogel (1978), Bräuer (1984), Rightmire (1984), and others had implicitly or explicitly begun to postulate a largely African origin for modern humans before the mtDNA parsimony tree was published, but their conclusions depended heavily on a relatively small number of relevant African fossils, many of which are very fragmentary or poorly dated. The mtDNA results appeared far less equivocal and convinced many observers that the stronger version of the Out-of-Africa theory was essentially correct. All paleoanthropologists, whatever their view of the fossil record, would probably agree that it was the mtDNA tree and not the fossils that prompted the lavish press coverage that modern human origins have received since 1987.

Unfortunately for those who thought that the issue had been largely decided, the mtDNA results proved to be at least as flawed as the fossil record they were meant to complement. The fundamental problem is that Cann, Vigilant, and their coworkers inadvertently misapplied the parsimony analysis behind their mtDNA tree, and remedial analyses of the same data, in fact, produce a plethora of trees, some of which root in Africa and some of which root elsewhere (Hedges et al. 1992; Maddison, Ruvolo & Swofford 1992; Templeton 1992). There are no statistical grounds for preferring African trees to their non-African alternatives, and even a substantial increase in the database will not solve the problem. In short, for now, the detailed mtDNA results that appeared to support the Out-of-Africa hypothesis so conclusively must be discounted.

This is not to say that genetic studies have become irrelevant to the issue of modern human origins or to the Out-of-Africa hypothesis. For example, Cann, Rickards, and Lum (this volume) point out that living humans exhibit only 1/40 to 1/50 as much mtDNA and nuclear DNA diversity as chimpanzees, suggesting that all living humans share a much more recent common ancestor with each other than they share with chimpanzees. Assuming that the rate of genetic divergence has been more or less constant and that the common ancestor of chimpanzees and people lived about 7 million years ago, it follows that the last common ancestor of all living humans existed no more than 125,000 years ago. Cann, Rickards, and Lum also note that ongoing mtDNA and nuclear DNA studies continue to reveal more genetic variability in Africa than on any other continent, implying that modern humans have been evolving in Africa longer than anywhere else. Finally, an analysis of mtDNA variability designed to detect "bottlenecking" in the population ancestral to living humans suggests that a major population expansion occurred in the interval centered on roughly 45,000 years ago (Rogers and Harpending 1992; Rogers, n.d.; Sherry et al., n.d.). This is in keeping with archeological evidence (noted again below) for a dramatic reduction in human numbers in Africa beginning about 60,000 years ago, followed shortly by the appearance of fully modern humans in both Africa and Eurasia and the disappearance of the Neanderthals. However, the demise of the mtDNA parsimony tree clearly undermines the extreme Out-of-Africa position that postulates replacement without interbreeding, and it breathes fresh life into the multiregional hypothesis advocated by Smith (this volume) that posits a substantial interbreeding or gene exchange in the spread of modern morphology.

THE CONTRIBUTION OF GEOCHRONOLOGY

It would be difficult to overstate the importance of radiocarbon dating to the archeology of the last 40,000 years or of radiopotassium dating to our understanding of the earliest phases of human evolution, before 500,000 years ago. Unfortunately, the events surrounding modern human origins occurred in the interval between 200,000 and 40,000 years ago, to which neither radiocarbon nor radiopotassium dating are routinely applicable because of inherent technical limitations, the absence of suitable materials, or both. The most promising methods for filling the breach are surely Thermoluminescence (TL) dating (applied mainly to burnt flint artifacts) and its close cousin, Electron Spin Resonance (ESR) (applied mainly to ancient dental enamel) (Grün and Stringer 1991; Mercier et al. 1993, each with further references). In effect, TL and ESR are techniques for measuring the radiation dose that dated objects have accumulated since burial. The radiation comes mainly from tiny amounts of uranium, thorium, and radiopotassium that occur naturally in most sediments. The average annual radiation dose that a buried object has experienced can be estimated from the annual dose within a site today or from laboratory assay of the sediments. Dividing the total accumulated dose by the annual dosage rate provides the time since burial.

As discussed by Bar-Yosef (this volume), perhaps the most fascinating TL and ESR dates are those suggesting that modern or near-modern people already existed between 120,000 and 80,000 years ago in what is today Israel. This is a time when Neanderthals were clearly the sole occupants of Europe. Since Israel is on the very margin of Africa, and since the animal fossils that accompany the early modern or near-modern fossils come partly from African endemics – including, for example, hartebeest (*Alcelaphus buselaphus*) and warthog (*Phacochoerus aethiopicus*) (Tchernov 1988, 1991) – the TL and ESR dates can clearly be used to support a fundamentally African origin for modern human morphology. Dates that have been obtained for sites providing

Neanderthals in Israel can be used to suggest either that Neanderthals and early moderns/near-moderns cohabited the region for tens of thousands of years or that the Neanderthals replaced early moderns/near-moderns after 80,000 years ago (Bar-Yosef 1989; Tchernov 1991). This might have happened if the Neanderthals were physiologically better adapted to the colder conditions that affected the region beginning about this time. Trinkaus (1981, 1989) has discussed Neanderthal skeletal features that might imply special adaptation to cold.

TL has also provided a widely publicized date of about 36,000 years on burnt flint artifacts associated with the famous Neanderthal skeleton from La Roche à Pierrot shelter, Saint-Césaire, Charente-Maritime, France (Mercier et al. 1991). This is the youngest generally accepted date for a Neanderthal, and taken at face value, it could imply that Neanderthals survived in France for at least four millennia after modern humans (or at least the kinds of artifacts assumed to have been made by modern humans) had appeared in nearby Spain. The Spanish dates were provided by a so-far unusual application of the accelerator mass spectrometry radiocarbon method to specks of ancient charcoal (Bischoff et al. 1989; Cabrera Valdes and Bischoff 1989).

The TL and ESR dates are exciting and important, but their accuracy is clearly limited, mainly because they depend on site-specific variables that are difficult to control (Jelinek, this volume). The most problematic variable is probably the annual dosage rate, which may have varied significantly through time, depending, for example, on largely indeterminate past variation in the moisture content of the burial environment. Problems in estimating the annual dosage rate may explain the relatively weak fit between some of the key Israeli dates and the widely accepted regional Middle Paleolithic sequence (Bar-Yosef, this volume; Jelinek, this volume; Marks 1992). This sequence was first established at Tabun Cave (Mt. Carmel), and as described by Bar-Yosef (this volume), comprises four main stages: the Acheuleo-Yabrudian (or Mugharan) followed by what are conveniently known as the Tabun-D, Tabun-C, and Tabun-B type Mousterian industries, after the successively higher layers in which they occur at Tabun Cave. Based on means and error estimates for ESR and TL determinations summarized by Bar-Yosef (this volume), Bar-Yosef et al. (1992), and Mercier et al. (1993) for Tabun and five other key Middle Paleolithic sites, figure 1 shows that the apparent ages for the successive stages overlap significantly. Advocates of the dating methods may, of course, argue that this implies the stages were partly contemporaneous, but it is at least as likely that some of the dates are inaccurate.

The figure also shows that even within individual sites some of the TL and ESR dates agree only in the loosest sense of the term. The discrepancies may stem partly from a problem that particularly affects the application of

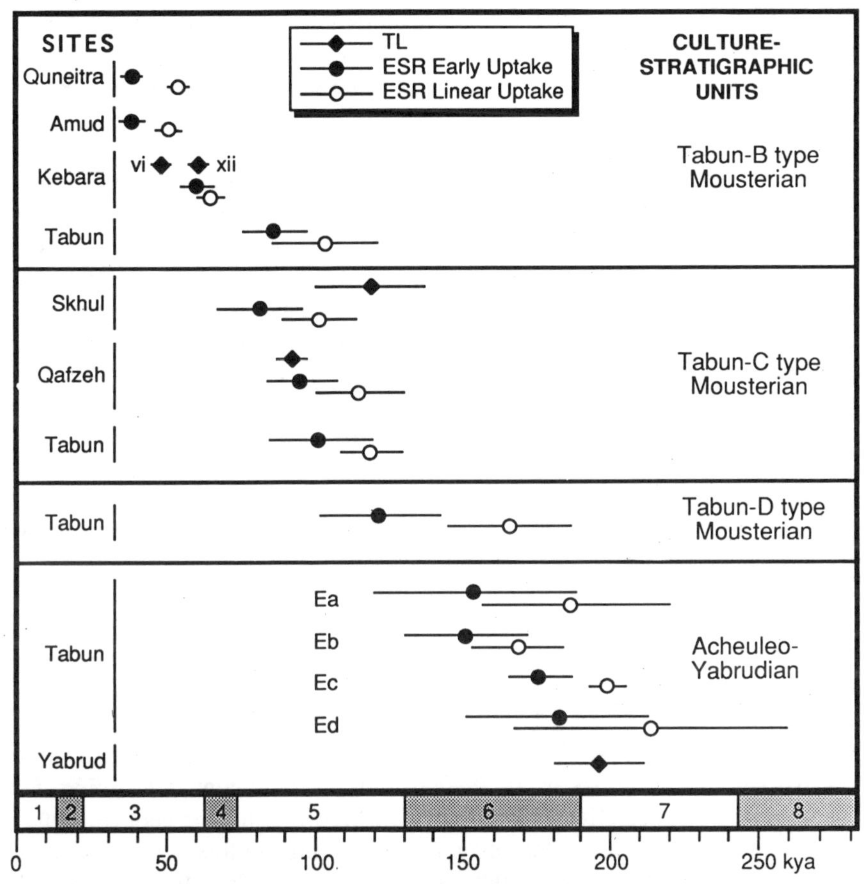

Figure 1. Means and error estimates of Thermoluminescence (TL) and Electron-Spin-Resonance (ESR) dates from southwest Asian Middle Paleolithic sites (data from Bar-Yosef, this volume, Bar-Yosef et al. 1992, and Mercier et al. 1993). The numbered, white and shaded boxes just above the kya scale represent global oxygen-isotope stages. The Roman numerals alongside the Kebara Mousterian dates and the letters alongside the Tabun Acheuleo-Yabrudian dates refer to different levels within the sites.

ESR to dental enamel, which provided all the ESR dates in the figure. Teeth tend to accrue uranium from ground water, and the result can be a significant increase in the annual radiation dosage rate. The precise increase is difficult to determine, but where uranium absorption is an obvious problem, two different age estimates are usually calculated – one assuming that the uranium mainly accumulated shortly after burial (leading to the EU or Early Uptake dates in fig. 1), and the other that it accumulated more or less continuously afterwards (producing the LU or Linear Uptake estimates in the figure). In any given case, the "true" age is presumed to lie somewhere between the two estimates.

In summary, both methodological considerations and some apparently discrepant results suggest that it would be unwise to interpret the TL and ESR dates too precisely, to argue, for example, that Neanderthals and early modern humans overlapped in Western Europe for a few millennia. Clearly, interpretations should be restricted to the relatively gross level (illustrated in fig. 1) at which the dates tend to be mutually consistent. In this regard, it is important to stress that they are also broadly consistent with the conventional sediment-, faunal-, or culture-stratigraphic dating of important archaic and modern or near-modern fossils in both Africa and Eurasia. It follows that even if some of them turn out to be no more reliable than the mtDNA parsimony tree, the relative dating of many key fossils will not have to be reconsidered from scratch. Most notably, it would still be patent that modern or near-modern people were present in Africa and on its Near Eastern periphery long before they appeared in Europe.

Plainly, both the producers and users of TL and ESR dates would benefit if the dates could be checked by independent geochemical means, preferably by a technique based on very different assumptions. The most promising method on the horizon is perhaps the estimation of age from the epimerization of amino acids in ostrich eggshell, which is found at many Asian and African sites. This method has already been used to date some key African sites (Brooks et al. 1990; Brooks, pers. comm.), including, for example, the Klasies River Mouth caves, where it confirms geologic inferences that the principal modern or near-modern human fossils date from the same broad interval (before 80,000 years ago) as their Israeli counterparts.

THE IMPORTANCE OF ARCHEOLOGY

Human paleontologists and geneticists often barely mention the archeological evidence for modern human origins, perhaps because they believe it is too ambiguous or subjective. Even some archeologists effectively discount it, because they feel that archeology lacks sufficiently refined or objective expository tools (Clark and Lindly 1989; Clark 1992). Certainly, some archeological evidence has been poorly collected and analyzed, and some interpretations are plainly flawed by a lack of logical rigor or good sense. However, archeological data have the unquestionable advantage of being relatively abundant. From an epistemological perspective, I believe that reliable patterns can be detected using fundamentally the same principles for evaluating incomplete or contradictory evidence that are successfully employed in judicial cases. If this is accepted, then archeology can contribute to an understanding of modern human origins in at least two major ways. First, it can elucidate behavioral differences between nonmodern and modern humans to help explain why nonmoderns failed to survive. Second, and arguably of greater consequence, it can determine whether modern behavior appeared earlier in some regions than in others and whether its appearance was gradual or abrupt. This, in turn, is clearly relevant to the question of whether modern people originated first in one area or more or less concurrently in many.

There is widespread agreement that nonmodern people, even the latest ones, differed profoundly in their behavior from their fully modern successors. Many authors (e.g., Bar-Yosef, this volume; Farizy, this volume; Freeman 1992; Jelinek, this volume; Klein 1989; Mellars 1989; Soffer, this volume) have outlined the significant behavioral differences between moderns and nonmoderns in Europe, where the last nonmoderns were the well-known Neanderthals. The artifacts they made are commonly assigned to the Mousterian (or Middle Paleolithic) industrial complex, as distinct from the Upper Paleolithic complex that followed. The Mousterians/Neanderthals had much more than an elementary grasp of the mechanics of stone flaking; they apparently built fires at will; they sometimes modified their living floors; they cared for their old and their sick; they buried their dead; and they unquestionably acquired large mammals as food. On the other hand, they manufactured a significantly smaller range of recognizable stone tool types than their successors; their artifact assemblages varied far less through time and space; they obtained stone raw materials from very distant sources much less frequently (suggesting either that their home ranges were relatively small or that their social networks were much simpler); they do not seem to have utilized bone, antler, and similar organic materials to produce formal artifacts; their fireplaces never show the complexity (and enhanced capability for heat production) of many Upper Paleolithic hearths; their sites provide little or no evidence for substantial structures (which, together with the difference in hearths, helps explain why they never occupied especially cold environments in which Upper Paleolithic people flourished); their graves reveal none of the evidence for ritual or ceremony that accompanies many Upper Paleolithic burials; they left no indisputable evidence for art or decoration (which appears unqualifiedly even in very early Upper Paleolithic sites [Bahn, this volume; White 1989]); and they were at least arguably less effective hunters. This list suggests that a very substantial behavioral gulf separated Mousterians from Upper Paleolithic people. Unquestionably, the Mousterians were cognitively human, but only Upper Paleolithic people were cognitively modern in the sense that all living people are.

In Europe, the Upper Paleolithic artifactual evidence for modern behavior seems to have appeared abruptly about 40,000 years ago, and to the extent that human remains occur in Upper Paleolithic sites, they come almost exclusively from fully modern people. This coincidence argues strongly for a largely physical replacement of Neanderthals by modern humans moving in from elsewhere (?Africa), though it would clearly be premature to rule out gene exchange or interbreeding, as hypothesized by Smith (1992, this volume). There is also the problem of what have sometimes been called "transitional cultures" between the Mousterian and the Upper Paleolithic. The most credible of these is unquestionably the Châtelperronian culture of northern Spain and western France. Wherever Châtelperronian assemblages occur in long stratigraphic sequences, they are sandwiched between Mousterian assemblages (below) and Upper Paleolithic ones (above), and they appear to comprise a mix of Mousterian and Upper Paleolithic artifact types. Equally remarkable, the only known associated human remains – isolated teeth from the Grotte du Renne at Arcy-sur-Cure, Yonne, France, and the now-famous partial skeleton from Saint-Césaire referred to above – come from Neanderthals (Stringer, Hublin & Vandermeersch 1984). Perhaps the most plausible interpretation for the association is that the Châtelperronian was produced by Neanderthals in close contact with Upper Paleolithic immigrants (Klein 1973; Harrold 1989), but even this strains credulity if the cognitive gulf between Neanderthals and Upper Paleolithic people was truly profound. The Châtelperronian not only eludes ready explanation, it also presents a puzzle within a puzzle, for among all the Châtelperronians, it is only the remarkable occupants of the Grotte du Renne who seem to have been more or less fully Upper Paleolithic (Farizy, this volume) Those who occupied the Saint-Césaire site, for example, made typical Châtelperronian stone artifacts, but they left no more evidence for the manufacture of formal bone artifacts or art objects than the preceding Mousterians. Nor do they appear to have organized their living area in the same elaborate and typically Upper Paleolithic way as the Châtelperronians of the Grotte du Renne.

Outside of Europe, the relationship between the origins of the modern physical form and the origins of modern behavior is more complicated. Pope (1992; Wolpoff et al. this volume) argues that the Far Eastern archeological record lacks evidence for an intrusive technology or culture until near the end of the Pleistocene and that this implies regional continuity between non-modern and modern populations, as predicted by the multiregional model. In fact, however, Pope's argument is a classic instance of employing an absence of evidence as evidence for an absence. Archeological discoveries in Australia (Jones 1992) and Siberia (Morlan 1987) can be used to argue for a relatively abrupt appearance of modern human behavior in the Far East, but the relevant record in China and southeast Asia is too poorly known for meaningful evaluation, and it may be many years before it can be brought to bear on modern human origins.

The far fuller archeological records of the Near East and Africa broadly resemble the European record, particularly in suggesting the same kind of major behavioral change around 40,000 years ago, the exact date perhaps depending on the place. In the Near East and north Africa, the culture-stratigraphic units that correspond to the European Mousterian (or Middle Paleolithic) (before 40,000 years ago) are commonly also called Mousterian (or Middle Paleolithic) and are characterized by essentially the same artifact types that characterize the European Mousterian. Similarly, in both the Near East and north Africa, the succeeding cultural unit (after about 40,000 years ago) is called the Upper Paleolithic and is marked by basically the same kinds of artifacts that distinguish the Upper Paleolithic in Europe. In sub-Saharan Africa, the unit corresponding broadly to the Mousterian is called the Middle Stone Age, and its successor, corresponding broadly to the Upper Paleolithic, is called the Later Stone Age. The names, however, reflect geographic distance and scholarly tradition more than anything else, and the Middle Stone Age contrasts with the Later Stone Age in basically the same way that the Mousterian contrasts with the Upper Paleolithic elsewhere (Klein 1989; Thackeray 1992).

If archeological evidence were all there were, we might, therefore, be tempted to suggest that the evolution of modern humans followed a similar course throughout western Eurasia and Africa. However, whereas Mousterian artifacts in Europe were produced exclusively by Neanderthals, Mousterian/Middle Stone Age artifacts in Africa were produced by people who were significantly more modern in their morphology, and Mousterian artifacts in the Near East (at least in the part adjoining Africa) were made sometimes by modern or near-modern people and sometimes by Neanderthals. As noted briefly above, relevant TL and ESR dates from the Near East may suggest either that early moderns/near-moderns and Neanderthals cohabited the Near East or that Neanderthals, expanding from Europe under deteriorating climatic conditions, physically replaced early moderns/near-moderns at what was effectively the northeastern periphery of their African range.

The disjunction between the archeological and human fossil records in Africa and the Near East is manifest, but it may be more apparent than real, if the modernity of the relevant African and Near Eastern fossils has commonly been overstated (Simmons, Falsetti & Smith 1991; Smith 1992, this volume; Stringer 1992, this volume; Wolpoff 1992; Wolpoff et al. this volume; Wolpoff and Caspari 1990). In this event, the term near-modern would clearly be more appropriate, and none of the people involved need be directly ancestral to any living humans. Instead, their geographic occurrence in Africa and on its southwest Asian margin need imply only that Africa is the place to look for the earliest appearance of fully modern morphology. The reasons behind its emergence remain obscure, but if, as I believe, the modern cognitive ability implied by Later Stone Age/Upper Paleolithic cultures is biologically rooted, there should be a very close coincidence between the emergence of fully modern humans and the appearance of the Later Stone Age/Upper Paleolithic. And if fully modern humans emerged first in Africa, then the earliest archeological evidence for fully modern behavior should also occur there.

So far, fossil evidence that may bear on the evolution of fully (or truly) modern humans in Africa is totally lacking, and potential archeological evidence for a corresponding shift in behavior is sparse. It is likely to remain so in most of northern and southern Africa, where extremely dry conditions in the relevant interval (between roughly 60,000 and 25,000 years ago) appear to have dramatically reduced human populations. Conditions for continuous human occupation appear to have been far more favorable in equatorial east Africa, and it is perhaps there that we should search for relevant evidence. A transition to modern humans within a relatively small geographic region like east Africa could explain the genetic evidence (referred to above) for a major expansion of the population ancestral to living humans in the interval centered on about 45,000 years ago. In addition, radiocarbon dates indicating that the Later Stone Age began in Kenya before 46,000 years ago (Ambrose, n.d.) suggest that it is in east Africa that the oldest traces of modern human behavior may eventually be documented.

CONCLUSION

The mode of modern human origins remains controversial. The mtDNA parsimony tree that briefly seemed to determine the issue in favor of the Out-of-Africa hypothesis must now be shelved, if not totally abandoned, and proponents of opposing views must refocus on the pertinent fossil record. This is admittedly incomplete and often poorly or questionably dated, but I believe that the morphological evidence, supported by the best available dates and by relevant archeological data, strongly suggests that (1) modern humans indeed emerged first in Africa, and (2) their morphology then spread rapidly to other parts of the world, either by physical replacement, by gene flow, or by a variable combination of the two. The reason for the rapid spread is almost certainly the very significant adaptive advantage conferred by fully modern behavior (or perhaps, more precisely, by the fully modern capacity for culture).

Presently, slim evidence suggests that fully modern morphology and behavior may have emerged first in east Africa, perhaps between 50,000 and

40,000 years ago. This hypothesis can be tested empirically by fresh fieldwork and by the further development of methods that provide reliable absolute dates in the interval between roughly 100,000 and 40,000 years ago. The issue of modern human origins will probably never be solved as conclusively as many molecular biological or physical problems can be, if only because competing theories cannot be tested by experiment and the evidence will always be less complete and more equivocal. However, even if full closure is beyond our grasp, the accumulation of fresh, high-quality evidence will surely allow us to refine alternative explanations and to determine which among them has the greatest probability of being correct.

ACKNOWLEDGMENTS

I thank S. H. Ambrose, D. K. Grayson, H. C. Harpending, R. G. Milo, M. H. Nitecki, and C. B. Stringer for helpful comments on a draft. The National Science Foundation supported my own research relevant to modern human origins.

REFERENCES

Ambrose, S. H. n.d. Chronology of the Later Stone Age in East Africa. Submitted.

Bar-Yosef, O. 1989. Upper Pleistocene cultural stratigraphy in southwest Asia, 154-80. In *The Emergence of Modern Humans: Biocultural Adaptations in the Later Pleistocene*, ed. E. Trinkaus. Cambridge: Cambridge University Press.

Bar-Yosef, O., B. Vandermeersch, B. Arensburg, A. Belfer-Cohen, P. Goldberg, H. Laville, L. Meignen, Y. Rak, J. D. Speth, E. Tchernov, A.-M. Tillier, and S. Weiner. 1992. The excavations in Kebara Cave, Mt. Carmel. *Current Anthropology* 33:497-550.

Beaumont, P. B., H. de Villiers, and J. C. Vogel. 1978. Modern man in sub-Saharan Africa prior to 49000 years BP: A review and evaluation with particular reference to Border Cave. *South African Journal of Science* 74:409-19.

Bischoff, J. L., N. Soler, J. Maroto, and R. Julia. 1989. Abrupt Mousterian/Aurignacian boundary at c. 40 ka BP: Accelerator ^{14}C dates from L'Arbreda Cave (Catalunya, Spain). *Journal of Archaeological Science* 16:563-76.

Bräuer, G. 1984. The "Afro-European *sapiens*-hypothesis" and hominid evolution in East Asia during the Late Middle and Upper Pleistocene. *Courier Forschungsinstitut Senckenberg* 69:145-65.

Bräuer, G. 1992. Africa's place in the evolution of *Homo sapiens*, 83-98. In *Continuity or Replacement: Controversies in* Homo sapiens *Evolution*, ed. G. Bräuer and F. H. Smith. Rotterdam: A. A. Balkema.

Brooks, A. S., P. E. Hare, J. E. Kokis, G. H. Miller, R. D. Ernst, and F. Wendorf. 1990. Dating Pleistocene archeological sites by protein diagenesis in ostrich eggshell. *Science* 248:60-64.

Cabrera Valdes, V., and J. L. Bischoff. 1989. Accelerator ^{14}C dates for early Upper Paleolithic (Basal Aurignacian) at El Castillo Cave (Spain). *Journal of Archaeological Science* 16:577-84.

Cann, R. L. 1992. A mitochondrial DNA perspective on replacement or continuity in human evolution, 65-73. In *Continuity or Replacement: Controversies in* Homo sapiens *Evolution*, ed. G. Bräuer and F. H. Smith. Rotterdam: A. A. Balkema.

Cann, R. L., M. Stoneking, and A. C. Wilson. 1987. Mitochondrial DNA and human evolution. *Nature* 325:31-36.

Clark, G. A. 1992. Continuity or replacement? Putting modern human origins in an evolutionary context, 183-205. In *The Middle Paleolithic: Adaptation, Behavior, and Variability*, ed. H. L. Dibble and P. Mellars. Philadelphia: The University Museum, University of Pennsylvania.

Clark, G. A., and J. M. Lindly. 1989. The case for continuity: Observations on the biocultural transition in Europe and western Asia, 626-76. In *The Human Revolution: Behavioural and Biological Perspectives on the Origins of Modern Humans*, ed. P. Mellars and C. Stringer. Edinburgh: Edinburgh University Press.

Freeman, L. G. 1992. Mousterian facies in space: New data from Morín Level 16, 113-25. In *The Middle Paleolithic: Adaptation, Behavior, and Variability*, ed. H. L. Dibble and P. Mellars. Philadelphia: The University Museum, University of Pennsylvania.

Grün, R., and C. B. Stringer. 1991. Electron spin resonance dating and the evolution of modern humans. *Archaeometry* 33:153-99.

Harrold, F. B. 1989. Mousterian, Châtelperronian and Early Aurignacian in Western Europe: Continuity or discontinuity? 677-713. In *The Human Revolution: Behavioural and Biological Perspectives on the Origins of Modern Humans*, ed. P. Mellars and C. Stringer. Edinburgh: Edinburgh University Press.

Hedges, S. B., S. Kumar, K. Tamura, and M. Stoneking. 1992. Human origins and analysis of mitochondrial DNA sequences. *Science* 255:737-39.

Howells, W. W. 1976. Explaining modern man: Evolutionists versus migrationists. *Journal of Human Evolution* 5:477-95.

Jones, R. 1992. The human colonisation of the Australian continent, 289-301. In *Continuity or Replacement: Controversies in* Homo sapiens *Evolution*, ed. G. Bräuer and F. H. Smith. Rotterdam: A. A. Balkema.

Klein, R. G. 1973. *Ice-Age Hunters of the Ukraine*. Chicago: University of Chicago Press.

Klein, R. G. 1989. *The Human Career: Human Biological and Cultural Origins*. Chicago: University of Chicago Press.

Li Tianyuan, and D. A. Etler. 1992. New Middle Pleistocene hominid crania from Yunxian in China. *Nature* 357:404-7.

Maddison, D. R., M. Ruvolo, and D. L. Swofford. 1992. Geographic origins of human mitochondrial DNA: Phylogenetic evidence from control region sequences. *Systematic Biology* 41:111-24.

Marks, A. E. 1992. Typological variability in the Levantine Middle Paleolithic, 127-42. In *The Middle Paleolithic: Adaptation, Behavior, and Variability*, ed. H. L. Dibble and P. Mellars. Philadelphia: The University Museum, University of Pennsylvania.

Mellars, P. 1989. Technological changes at the Middle-Upper Palaeolithic transition: Economic, social and cognitive perspectives, 338-65. In *The Human Revolution: Behavioural and Biological Perspectives on the Origins of Modern Humans*, ed. P. Mellars and C. Stringer. Edinburgh: Edinburgh University Press.

Mercier, N., H. Valladas, O. Bar-Yosef, B. Vandermeersch, C. Stringer, and J.-L. Joron. 1993. Thermoluminescence date for the Mousterian burial site of Es-Skhul, Mt. Carmel. *Journal of Archaeological Science* 20:169-74.

Mercier, N., H. Valladas, J.-L. Joron, J. L. Reyss, F. Lévêque, and B. Vandermeersch. 1991. Thermoluminescence dating of the late Neanderthal remains from Saint-Césaire. *Nature* 351:737-39.

Morlan, R. E. 1987. The Pleistocene archaeology of Beringia, 267-307. In *The Evolution of Human Hunting*, ed. M. H. Nitecki and D. V. Nitecki. New York: Plenum.

Pope, G. G. 1992. Craniofacial evidence for the origin of modern humans in China. *Yearbook of Physical Anthropology* 35:243-98.

Rightmire, G. P. 1984. *Homo sapiens* in sub-Saharan Africa, 295-325. In *The Origin of Modern Humans: A World Survey of the Fossil Evidence,* ed. F. H. Smith and F. Spencer. New York: Alan R. Liss.

Rogers, A. R. n.d. Genetic evidence for a Pleistocene population explosion. Unpublished manuscript.

Rogers, A. R., and H. C. Harpending. 1992. Population growth makes waves in the distribution of pairwise differences. *Molecular Biology and Evolution* 9:552-69.

Sherry, S. T., A. R. Rogers, H. Harpending, H. Soodyall, T. Jenkins, and M. Stoneking. n.d. Pairwise differences of mtDNA reveal recent human population expansions. Unpublished manuscript.

Simmons, T., A. B. Falsetti, and F. H. Smith. 1991. Frontal bone morphometrics of southwest Asian Pleistocene hominids. *Journal of Human Evolution* 20:249-69.

Smith, F. H. 1992. The role of continuity in modern human origins, 145-56. In *Continuity or Replacement: Controversies in* Homo sapiens *Evolution*, ed. G. Bräuer and F. H. Smith. Rotterdam: A. A. Balkema.

Stringer, C. B. 1992. Replacement, continuity and the origin of *Homo sapiens*, 9-24. In *Continuity or Replacement: Controversies in* Homo sapiens *Evolution*, ed. G. Bräuer and F. H. Smith. Rotterdam: A. A. Balkema.

Stringer, C. B., J. J. Hublin, and B. Vandermeersch. 1984. The origin of anatomically modern humans in Western Europe, 51-135. In *The Origin of Modern Humans: A World Survey of the Fossil Evidence,* ed. F. H. Smith and F. Spencer. New York: Alan R. Liss.

Tchernov, E. 1988. The biogeographical history of the southern Levant, 159-250. In *The Zoogeography of Israel*, ed. Y. Yom-Tov and E. Tchernov. Dordrecht, The Netherlands: Dr. W. Junk.

Tchernov, E. 1991. The Middle Paleolithic mammalian sequence and its bearing on the origin of *Homo sapiens* in the Southern Levant, 77-88. In *Le Squelette Moustérien de Kebara 2*, ed. O. Bar-Yosef and B. Vandermeersch. Paris: Centre National de la Recherche Scientifique.

Templeton, A. R. 1992. Human origins and analysis of mitochondrial DNA sequences. *Science* 255:737.

Thackeray, A. I. 1992. The Middle Stone Age south of the Limpopo River. *Journal of World Prehistory* 6:385-440.

Trinkaus, E. 1981. Neandertal limb proportions and cold adaptation, 187-224. In *Aspects of Human Evolution*, ed. C. B. Stringer. London: Taylor and Francis.

Trinkaus, E. 1989. The Upper Pleistocene transition, 42-66. In *The Emergence of Modern Humans: Biocultural Adaptations in the Later Pleistocene*, ed. E. Trinkaus. Cambridge: Cambridge University Press.

Vigilant, L., M. Stoneking, H. Harpending, K. Hawkes, and A. C. Wilson. 1991. African populations and the evolution of human mitochondrial DNA. *Science* 253:1503-7.

White, R. 1989. Production complexity and standardization in Early Aurignacian bead and pendant manufacture: Evolutionary implications, 366-90. In *The Human Revolution: Behavioural and Biological Perspectives on the Origins of Modern Humans*, ed. P. Mellars and C. Stringer. Edinburgh: Edinburgh University Press.

Wolpoff, M. H. 1992. Theories of modern human origins, 25-64. In *Continuity or Replacement: Controversies in* Homo sapiens *Evolution*, ed. G. Bräuer and F. H. Smith. Rotterdam: A. A. Balkema.

Wolpoff, M. H., and R. Caspari. 1990. On Middle Paleolithic/Middle Stone Age hominid taxonomy. *Current Anthropology* 31:394-95.

Part II

What Are Modern Humans?

The question of human origins requires understanding who the early humans were, the interpretation of the phylogeny of Neanderthals, and the correlation of artifacts with fossils. The derivation of a concept of species from individual fossils, or the reduction (or explanation) of individual bone fragments to the concept of modern humans is very difficult. Questions of who we are, the eternal subjects of myths and speculations of all cultures, are as problematic as the interpretations of behavior of early humans. Attempts to define modern humans from fossils, and to explicate humanness to "biological" traits from morphology are equally difficult. However, humanness must be explicable in more than morphology and behavior, an even more difficult or perhaps impossible task considering the almost incomprehensible human nature and history.

The concept of anatomically modern humans is a complex evolutionary model that can only be explained when simplified. As long as we only describe fossils, we generate little debate. However, as soon as we begin to *reconstruct* and *explain* the past life and the taxa that represent modern humans, we generate controversies.

Ofer Bar-Yosef asks how the hominids from Southwestern Asia contribute to the solution of the origins of modern humans. He utilizes chronostratigraphy, faunal spectra, and lithic assemblages for the environmental reconstructions of the ecozones occupied by Middle Paleolithic hominids. To assess the intensity of human occupations, survival, and competition among various human groups, Bar-Yosef hunts for evidence in stratigraphy, site sizes, and the processes of manufacture of lithic assemblages. Bar-Yosef examines differential survival in marginal zones. He sees the Levant as an optimal region for survival and as a belt for human migrations, as well as a sanctuary from harsh climate. The morphology of the Middle Paleolithic humans is controlled by gene flow between incoming and local population. Human adaptation and technology appear to be a cause of success. Thus, he sees the cultural change to the Upper Paleolithic as a Levantine process preceding similar shifts in Eurasia. Bar-Yosef interprets the movement of humans from the Levant into Europe and central Asia by an analogy to the origins and advent of Neolithic agriculture by diffusion.

Arthur Jelinek deals with the difficult problem of delineating the differences in behavior between modern and archaic humans. While it may be easy to differentiate earlier people by the evidence from the last 200,000 years, it is exceedingly difficult to evaluate. Jelinek examines the archaeological evidence for changes in styles in producing lithic artifacts. According to him we probably cannot demonstrate conclusively when hominids became "fully cultural" (or anatomically "fully modern"), but we must continue to evaluate the anatomical and artifactual information to reconstruct the nature and behavior of our remote ancestors.

Surely the questions that Jelinek asks are very old queries concerning the origin of civilizations. Is culture always produced under similar conditions? Is the origin of culture due to the similarities of human minds? Does culture spread through diffusion? Or, in other words, were the canoe or cave painting invented once or many times? Was culture (= art) invented by Cro-Magnons and not by Neanderthals because of their mental differences, or because they occupied different ecologies? The definition of the modern humans, when based on rigorous anatomical definitions will cause no arguments. However, the definition based on culture and behavior produces problems. Did early humans paint less well? or were they unable to paint at all, as the majority of humans cannot today? Definitions are relative terms.

Catherine Farizy, in her chapter on the Middle Paleolithic culture and behavior, analyzes the data from her recently excavated open-air sites. She discusses processing of meat, procurement of raw material, and technology of lithic industries. She very carefully delineates the postdepositional biases, and correlates the spatial pattern of Middle Paleolithic sites with the age of their occupations. Middle Paleolithic human accumulations appear, however, to be random. Farizy compares these accumulations with the first well-known Early Upper Paleolithic dwellings of the French Châtelperronian. She demonstrates cultural changes through changes in artifact manufacture and through the emergence of ornaments and of other nonutilitarian items.

Olga Soffer concentrates on the Eurasian transition from the Middle to the Upper Paleolithic when modern humans replaced the Neanderthals. To answer why these changes occurred, Soffer unearths evidence from ecology, biology, and archaeology, and argues that Neanderthals and early modern humans utilized ecological resources and solved ecological problems differently. Furthermore, the replacement involved major sociocultural transformations of sexual units of early modern humans. It was these social feminist transformations that facilitated the expansion to and the colonization of the open loessal northern latitudes.

Paul Bahn reviews recent advances in the field of Ice Age art, particularly recent discoveries of open-air Ice Age engravings in southwestern Europe. He also reviews the dating of pigments in the cave art of France and Australia. The art and tools are the main sources of information on the nature, richness, and diversity of prehistoric art and technology, and on their "sudden" appearance and development. While Ice Age art is useless in solving the problem of the modern human origin–because it can support either the Out-of-Africa or the Multiregional hypothesis–it, nevertheless, can offer a hope of better understanding of the behavior of early humans.

The crux of problems of culture is whether culture can be reduced to biology or to physical anthropology. The origin of modern humans is, above all, the problem of the identification, or the definition, of modern humans. While paleontologists can observe and measure fossil skulls, their material is very fragmentary, and their concepts of species are of necessity based on very small samples.

Chapter **2**

The Contributions of Southwest Asia to the Study of the Origin of Modern Humans

OFER BAR-YOSEF

The current debate concerning the origins of modern humans centers around a few questions: (1) How do we classify and date the human fossils of the Late Quaternary, and what are the implications of each classification? (2) Does a "multiregional evolution" model better explain the fossil and cultural records than the "Out-of-Africa" model? and (3) "Why," "when," and "where" did the transition from the Middle to the Upper Paleolithic occur? Each of these questions implies that we would like to know what is their meaning in terms of the evolution of human behavior. Currently these "big" questions end up, in the Western World, in their more mundane, simplified version when colleagues, journalists and learned people, who often implicitly hold a Eurocentric approach, demand to know, Who were the Neanderthals? How different were they from those who inhabited the Near East during the Last Glacial? Is it possible that two human types defined on the basis of a few skeletal remains were contemporary and for how long? Are there any signs of the interbreeding of these two populations? Are the stone tools of the Near Eastern sites evidence the Out-of-Africa movement of humans? And so on. It is quite obvious that we do not have satisfactory answers and even worse a large number of scholars disagree on how to interpret the available fossil and archaeological records. Undoubtedly this is the result of much confusion concerning the taxonomic definition of the fossils, the lack of unambiguous ways of interpreting lithic assemblages, the rarity of clear intra-site spatial organization during the Middle Paleolithic, etc. Under these cir-

---

OFER BAR-YOSEF • Department of Anthropology, Peabody Museum, Harvard University, Cambridge, MA 02138.
Origins of Anatomically Modern Humans, edited by Matthew H. Nitecki and Doris V. Nitecki. Plenum Press, New York, 1994.

cumstances each of the participants in the ongoing debate is forced to articulate his or her interpretation of the documented observations with the hope that one day this intriguing picture will become clearer. Right now it seems that the field is wide open and ready for the most outrageous speculations. As predicted by the law of supply and demand, additional conferences and symposia will increase the production of many more papers and will undoubtedly increase the number of active participants. This chapter falls in the same category. I have already expressed my views elsewhere (Bar-Yosef 1988, 1989a, 1989b, 1992a, 1992b) but with every new version I have tried to incorporate additional data, which unfortunately do not accumulate in step with the organized meetings on this topic.

My chapter begins by enumerating the contributions of Southwest Asia (fig. 2) to the study of the Middle and Upper Paleolithic periods. It is followed by a more elaborate discussion of each topic (fossils, Middle and Upper Paleolithic industries and nonartifactual remains, subsistence and the current chronology). I have also attempted to propose a systematic model which I have often mentioned in my oral presentations, based on the idea that employing a similar approach to the one used for uncovering the origins of agriculture will be fruitful and will enable us to devise future field research (Bar-Yosef 1992a, in press).

SUMMARY OF THE SOUTHWESTERN ASIAN CONTRIBUTIONS

The main contributions of Southwest Asia to the current debate on the origins of modern humans are the following:

1. The morphological and metrical variability among the Middle Paleolithic human fossils in the region is larger than the recorded variability of Middle Paleolithic fossils in Western Europe. The scanty Upper Paleolithic remains demonstrate, as among European fossils, a robust *Homo sapiens sapiens* form in the Early Upper Paleolithic (EUP), and a range of gracile forms in the Late Upper Paleolithic (LUP, also known in the Near East as Epi-Paleolithic).

2. Middle Paleolithic stratified lithic industries fall into two Middle Paleolithic geographic provinces: (a) The Levant and (b) the Taurus-Zagros (including the Caucasus) regions.

3. The earliest Levantine Upper Paleolithic sequence reflects a rapid technological change cautiously dated to around 47/45 ka. From 38/35 ka blade technologies dominate most of the Upper Paleolithic sequence. Assemblages rich in bone and antler objects are mostly related to the Levantine Aurignacian (34/32-28 ka). Before 13 ka the region is extremely poor in art items.

4. Middle and Upper Paleolithic archaeological features such as hearths, burials, bone accumulations, etc., are well recorded and provide some of the needed information for drawing behavioral inferences.

5. Most Middle Paleolithic faunal assemblages reflect hunting of small- and medium-sized animals and the opportunistic scavenging of large bovids. Upper Paleolithic animal bone assemblages were largely the results of hunting and trapping.

6. Middle Paleolithic Levantine chronology, currently based on radiometric measurements (mainly based on TL and ESR with a selection of U/Th dates) coupled with a relative chronology as derived from detailed paleontological analyses, places the Levantine industries and most of the fossils in a known sequential record. Upper Paleolithic sites are dated by an increasing number of ^{14}C dates and can be correlated with other dated Eurasian archaeological entities.

Each of these observations, when viewed in a global context, has far-reaching implications. For example, testing the archaeological ramifications of the Out-of-Africa model can be done by examining whether the earliest Levantine lithic industries or behavioral change, as documented by nonartifactual evidence, is of African origin. This means that, if in a Late Acheulian or Early Mousterian phase humans moved out of Africa, we should expect to find at least one or two assemblages that were produced from local raw materials, but in techniques that prevailed in the original homeland of the newcomers as happened around 0.7-0.5 ma (Bar-Yosef 1987, 1992b). An illustrative example from the Lower Paleolithic is the lowermost assemblage in the Acheulian site of Gesher Benot Ya'aqov which contains high frequencies of flake lava cleavers (Gilead 1973; Bar-Yosef 1975), produced in a typical African technique (Bordes 1961). In spite of large exposures of lava flows in the region, this is the only well-recorded instance where the African "block on block" technique was employed (Stekelis 1960; Bar-Yosef 1987; Goren-Inbar 1992).

Understanding the behavioral messages of the archaeological record depends on the assumption that we know how to "read" and interpret the uncovered remains (Bar-Yosef 1991a, 1992b). Apart from lithic industries there are other archaeological expressions for human behavior such as hearths, spatial distribution of artifacts and bones, burials, the use of ground stone tools, the production of bone and antler tools, the use of red ocher, body decorations such as beads and/or marine shells, art objects, etc. Numerous data sets were retrieved during the excavations of the 1930s that lack the detailed field observations required today. For example, the exact location of the woman from Tabun Cave, known as Tabun I, is of great importance. D. Garrod herself suggested that its attribution to Layer C in Tabun is uncertain, and that it could have been from Layer B (Garrod and Bate

1937:64). However, the increasing number of modern excavations (such as in Tabun, Kebara, Quneitra, Amud, etc.) has reduced the amount of ambiguity in basic excavation records.

In addition, the region furnishes a large body of data relevant for testing hypotheses concerning "prehistoric revolutions." The Levant had served as the core area for the emergence of the "Neolithic or Agricultural Revolution." This period is well radiocarbon-dated and the sequence of pivotal socioeconomic thresholds marking the advent of the new system is known. I, therefore, suggest viewing the Middle to Upper Paleolithic transition by employing as a source of insight the main factors which were behind the processes that led to the agricultural revolution and its early development. In my view, such an approach will cnablc us to formulate improved research strategies. As shown below, one of the many lessons from such an analogy is to locate the core area where the biological and/or cultural change took place. This proposition is made under the assumption that such relatively short-term revolutions did happen and that they occurred in a certain region. Obviously, those who see the biocultural changes as a penecontemporaneous, multiregional process (Wolpoff 1989; Wolpoff et al., this volume) will disagree with this approach, which falls within the current controversy concerning the Out-of-Africa model (Stringer and Andrews 1988; Stringer, this volume; Cann, Rickards & Lum, this volume).

THE REGION AND ITS RECORDS

Southwestern Asia encompasses the area from the Caucasus through the Iranian and Turkish plateaus, the Levant, Mesopotamia, Sinai, and the Arabian peninsula (fig. 1). High mountain ranges with intermountain valleys dominate the landscape of the Caucasus and Transcaucasia, the Pontinian Mountains, the Taurus, and the Zagros Mountains. The Anatolian and Iranian plateaus are high (often over 500 m above sea level) and accommodate a series of lakes. The Iranian plateau is more arid than the Anatolian. The Levant is basically a hilly area, rarely more than 1000 m above sea level, with a narrow coastal plain. The Jordan Rift Valley forms an elongated trough with a few permanent lakes. Mesopotamia is almost a continuous plain from sea level to about 300 m above sea level, dissected by the rivers and their tributaries which descend from the mountains. The Arabian Peninsula has hilly to mountainous ranges along its western edge and a sloping flattish landscape towards the Persian Gulf. Today, only its southern fringes are watered by the monsoons. Most of it, including the Sinai Peninsula, is part of the global Saharo-Arabian desert belt.

Reconstruction of past climates, even if incomplete, indicate that dur-

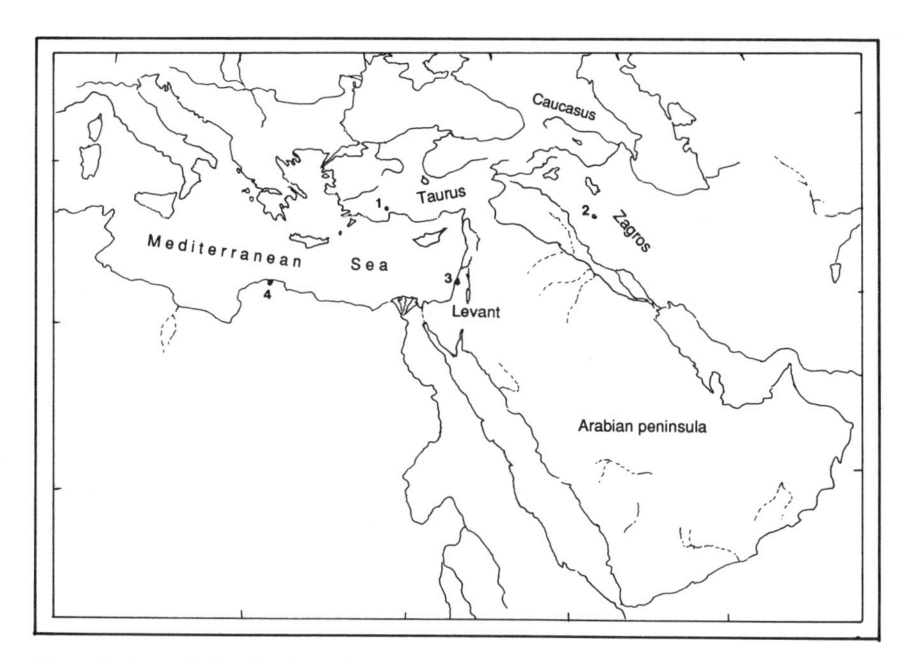

Figure 1. A map indicating the main regions mentioned in the text. 1 - Karain Cave; 2 - Shanidar Cave; 3 - Mount Carmel; 4 - Hauah Fteah Cave.

ing the maximum of the glacial periods the climate of the Iranian and Anatolian plateaus was cold and dry. Caves in high altitudes in the mountains (Caucasus, Armenia, Taurus and Zagros) were uninhabitable for most of the time. The main prehistoric occupations were situated in sites up to 400 m above sea level. Examples of intensely used locales are Tabun Cave (45 m), Kebara (65 m), Hayonim (250 m), Qafzeh (220 m), Shanidar (822 m), Karain (430 m), etc. Many of the high altitude cave sites such as Tor Sabiha (1300 m), Yabroud I (1420 m), and Koudaro (1600 m) were more ephemerally inhabited, and probably only during warm spells. Finally, the broken topography of the northern part of the region was possibly an impediment for close social contacts during colder periods. Interregional access was easiest along the coastal areas which, with the lowered sea levels (Mediterranean, Black and Caspian seas), were far more extensive than today.

The problems in Southwest Asia are the small number and the location of excavated sites. The Paleolithic sequences of both Iran and Turkey are poorly known (Smith 1988; Dibble 1984; Dibble and Holdaway 1990; Yalçinkaya 1981, 1988). A body of information is available from the Caucasus region but, because most of the publications are in Russian or local languages (e.g., Liubin 1989), and radiometrically dated assemblages are few, we unfortunately tend to ignore them. Tentative correlations between the Caucasus, the Taurus-Zagros, and the Levant are achieved only by employing the relative chronologies based on changes in the faunal spectra, and crude sedimentological interpretations which indicate stadial or interstadial conditions.

The advantages of the Levant and in particular the area of the Lebanese mountains, Galilee, and Mt. Carmel (ca. 5,000 km^2), is that the sequence of Middle and Upper Paleolithic is relatively well dated and has already pro-duced many fossils. Thus, it can serve as a cornerstone in building a scheme for global correlations and may provide insights which cannot be gained by examining only the Western European or the South African contexts. While the former and the latter are geographically located in cul-de-sacs (Trinkaus 1989), the Near East and especially the "Levantine corridor" should reflect the interactions between Eurasia and Africa. By comparing the archaeological sequences and fossils from these three regions, despite their size differences, and by taking into account the fragmentary evidence from the regions in between, we should be able to produce testable hypotheses concerning the origin of modern humans and especially the cultural changes attributed to them. We can even try to identify the location of the core area, where it all began. Or inversely, by having well-dated and, hopefully, well-understood archaeological records, we can present a global summary that views the emergence of the Upper Paleolithic as the result of convergent evolution (Lindly and Clark 1990; Straus 1989). In order to clarify what can be learned from the Southwest Asian data sets, I will briefly summarize each of the main subjects, commencing with Middle and Upper Paleolithic archaeology, the nature of the cultural transition, the fossils, and the chronology, and will close with a proposed future research strategy based on the "Neolithic Revolution."

THE ARCHAEOLOGICAL SEQUENCE

General Remarks. Lower and Middle Paleolithic archaeology is generally known through lithic studies of Acheulian and non-Acheulian industries. The apparent lack of major changes in the methods of production and the morphology of tools during more than a million years make this archaeology seem boring. But do we really see the full array of material culture? A short reminder is perhaps required.

Lithic objects were employed by early hominids in Eurasia because chipped hard rocks provide sharp cutting edges that can be used for various tasks. Whether they were scavengers or early hunters, the hominids who have tried, sometimes successfully, to colonize the temperate zone, where animal tissues were an important source of food, needed butchery tools.

They also required sharp objects with which wooden tools could have been made, as shown by such rare finds as the spears from Clacton and Lheringen. It is unfortunate that we do not have even the slightest idea what kind and type of organic objects are missing from the archaeological record. We often pay lip service by stating that early hominids were capable of producing a variety of different items from wood, bark, fibers, etc., as well as working hides and tendons, but the evolution of these artifacts is unknown. Instead of amplifying our efforts to discover waterlogged or frozen sites we fall back on the lithics. But lithics were basically employed in similar ways and for somewhat similar tasks through the ages. Thus, it is not surprising that meticulous microwear and edge damage studies, based on replicatory experiments with the same raw materials, indicate that flakes, blades, and bifacial items were all used for woodworking, butchery, scraping, whittling, etc. (Beyries 1988; Shea 1989; Anderson-Gerfaud 1990). The evolution of hunting techniques during the Middle Paleolithic is generally accepted and, therefore, it is not surprising that a recent study demonstrated that Levallois points were hafted and used for hunting as well as for butchering (Shea 1988). The use of projectile points for butchering is not exclusively Middle Paleolithic. They were employed for the same purpose in the Neolithic and are known from ethnographic observations. Worth mentioning is that there is no evidence for the use of stone points as projectiles in the European Mousterian, but Anderson-Gerfaud (1990) and Schelinski (Plisson 1988) suggest that evidence of woodworking manifest on many artifacts may be the result of the production of wooden spears. In both regions pointed pieces were often hafted (Beyries 1988; Shea 1988).

The transition from scavenging in the Lower Paleolithic to hunting and opportunistic scavenging in the Middle Paleolithic is recorded in animal-bone collections (Chase 1989; Binford 1991) and not in the lithics. Therefore, except for the general knowledge about the functions of stone tools and selected blanks, we have no idea what other objects were produced by the Lower and Middle Paleolithic humans. No unambiguous bone or antler tools are known, and wooden objects, as mentioned above, are very rare due to poor preservation.

The Acheulo-Yabrudian. The study of the lithic industries of this region indicates that there are at least two subregions: (1) the Taurus-Zagros zone, and (2) the Levant, which stretches from southeast Turkey to southern Sinai and includes the Arabian peninsula.

It is unfortunate that little is known about the late Lower Paleolithic and Middle Paleolithic of the region immediately north of the Levant, namely, Turkey, northern Iraq, and western Iran. The Taurus and Zagros Mousterian differ in their operational sequences from the Levantine Mousterian and often end with radial cores (Yalçinkaya 1988; Dibble and Holdaway 1990). Core-reduction strategies of this industry (or industries) are somewhat constrained by the size, availability, and accessibility of raw material, which is predominantly radiolarite. This issue will be further discussed below.

The Levantine sequence commences with the Acheulo-Yabrudian, which in stratified sites such as Yabrud I, Tabun and Zuttiyeh, is always overlain by the Mousterian (fig. 2). The description of this earlier entity illustrates how terminology may bias our interpretations and how regional distribution of sites poses an interesting problem that requires a plausible interpretation.

The stratigraphically pre-Mousterian entity was originally known as the "Acheulo-Yabrudian" (Garrod 1956). In suggesting this term, Garrod accepted the observations of Rust, who excavated in Yabrud Rockshelter I (on the eastern slopes of the Anti-Lebanon mountains) and identified at least three types of assemblages (Rust 1950). The new label replaced what was called "Upper Acheulian (Micoquian)" during the first excavations of Tabun Cave (Garrod and Bate 1937). The original term and the newly modified one indicated the presence of bifaces (hand axes) in the lithic assemblages and were, therefore, considered as part of the Lower Paleolithic. In the 1950s, F. Bordes proposed a dendritic evolutionary model for the Lower and Middle Paleolithic industries. He also republished the Acheulo-Yabrudian and Mousterian assemblages from Yabrud I (Bordes 1955). The Yabrudian "facies" resembles the Charentian Mousterian (Ferrassie and Quina-type Mousterian). Moreover, in the European sequence bifaces of various forms are present throughout the Middle Paleolithic sequence and, therefore, Bordes had no difficulty in assigning all the Levantine industries to this period.

Intensive geochronological studies by Farrand (1971, 1979, 1982) addressed the question of possible contemporaneity of Mousterian and Acheulo-Yabrudian industries, by now included in the Middle Paleolithic. In spite of two alternative chronologies (Farrand 1979), relative dating, geochronological correlations and radiometric readings, there is no shred of evidence that would indicate such a cultural contemporaneity within the northern and central Levant (Bar-Yosef 1989b, and see below).

Currently, the Acheulo-Yabrudian is also known as the "Mugharan Tradition." It differs considerably from the Mousterian in its knapping methods, by which blanks were obtained, as well as in the dominant forms of retouched pieces. The Acheulo-Yabrudian includes three industrial facies (Jelinek 1982a, 1990), categorized on the basis of quantitative and qualitative studies, although they were once considered as independent archaeological entities (Rust 1950; Bordes 1955, but see Bordes 1977).

1. The "Yabrudian facies" contains numerous sidescrapers, often made

on thick flakes, resulting in relatively high frequencies of Quina and semi-Quina retouch, a few Upper Paleolithic tools, rare blades and a few, or total absence of, Levallois products (Skinner 1970; Copeland and Hours 1981, 1983; Jelinek 1982a).

2. The "Acheulian facies" is defined by Jelinek (1982a) as having up to 15% bifaces, with numerous scrapers fashioned in the same way as the Yabrudian ones.

3. The "Amudian facies" (end scrapers, burins, backed knives, and rare bifaces) seems, following the Tabun excavations, to be closer to the Acheulian than the Yabrudian, and contains the evidence for limited practice of Levallois technique (Jelinek 1982a).

Jelinek grouped these industries under one taxon – the Mugharan Tradition. He also observed that the use of Levallois technique in Tabun increased rapidly during the time of the Transitional Unit (X), thus evolving into the fully fledged Mousterian complex (Jelinek 1982a; Copeland and Hours 1981). An alternative interpretation, based on the published geological section (Jelinik 1982b), is that this unit could have been a natural mixture caused by the erosion of Acheulo-Yabrudian deposits and the accumulation of a Mousterian industry.

Resharpening of flake scrapers among the Acheulo-Yabrudian assemblages is much more common than in the succeeding Mousterian. It is rather surprising to see the high level of resharpening of these scrapers, in areas where good quality raw material is in abundance within the immediate vicinity of the sites such as Mt. Carmel and El-Kowm (northeast Syria). Even during harsh winters these outcrops were not covered for many weeks by snow, as could have been the case in European environments.

The known geographic distribution of the Acheulo-Yabrudian sites covers solely the northern and central Levant (fig. 2). Despite intensive surveys, none of the typical and easily identifiable artifacts of this industry were found in the Negev and Sinai or the region of southern Jordan. It seems that the southern boundary of this industrial complex cuts across the various parallel, north-to-south oriented, ecological zones of the Levant. This east-west boundary is not an ecological or climatic boundary unless it represents a considerable northward shift of the desert edge that could have happened under extreme climatic conditions. This boundary resembles the one which marked the expansion of the Aterian versus the Mousterian in northeast Africa. Moreover, the presence of the Acheulo-Yabrudian in the semi-arid, isolated El-Kowm basin in eastern Syria, raises the possibility that this boundary marks the contemporaneity of the Acheulo-Yabrudian with another Upper Acheulian entity, such as is found in Egypt (Caton-Thompson 1952; Wendorf and Schild 1980). A second hypothesis suggests that some of the Negev Mousterian sites are older than currently viewed, and thus contempo-

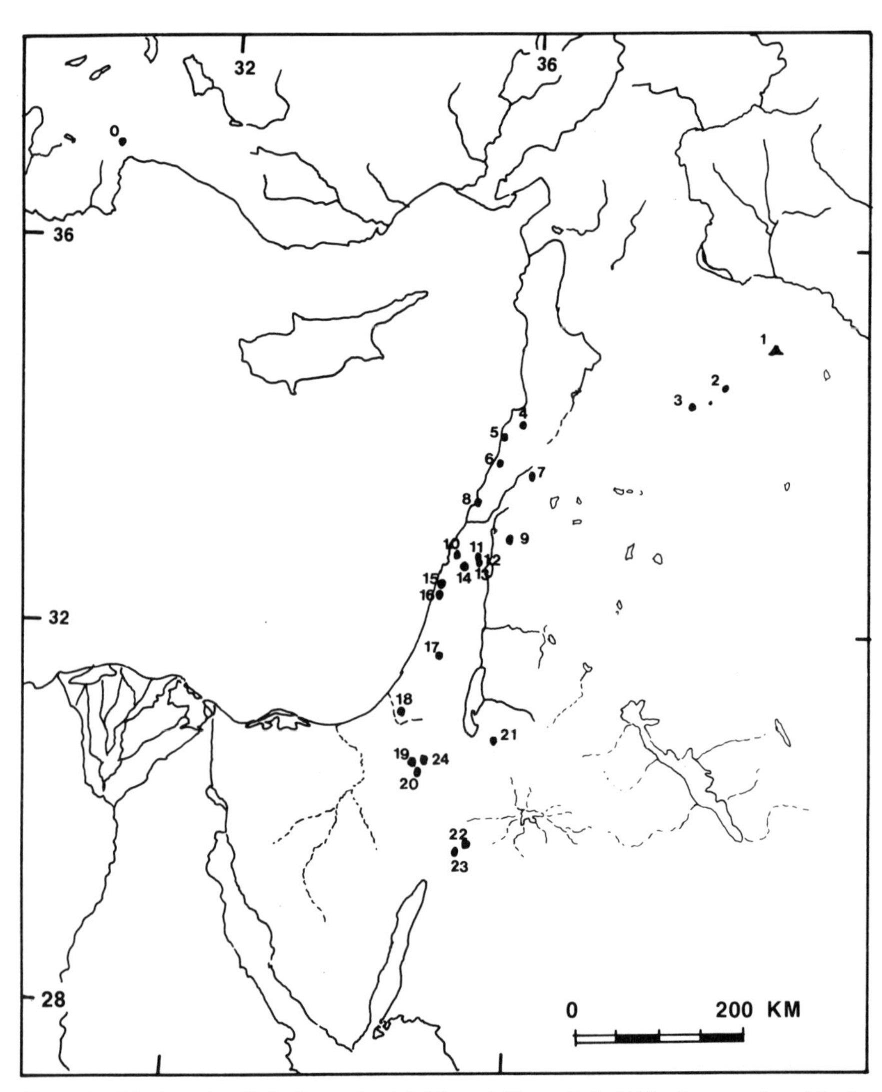

Figure 2. The location of the Levantine Middle and Upper Paleolithic sites mentioned in the text: 0 - Karain; 1 - El Kowm; 2 - Douara; 3 - Jerf 'Ajla; 4 - Keooe; 5 - Nahr Ibrahim; 6 - Ksar Akil; 7 - Yabrud I; 8 - Adlun caves (Bezez and Zumoffen); 9 - Quneitra; 10 - Hayonim; 11 - Amud; 12 - Zuttiyeh; 13 - Emireh; 14 - Qafzeh; 15 - El Wad, Skhul, Tabun; 16 - Kebara; 17 - Shukbah; 18 - Fara B; 19 - Rosh Ein Mor; 20 - Ain Aqev; 21 - Ain Difla; 22 - Tor Sabiha; 23 - Tor Faraj; 24 - Boker Tachtit.

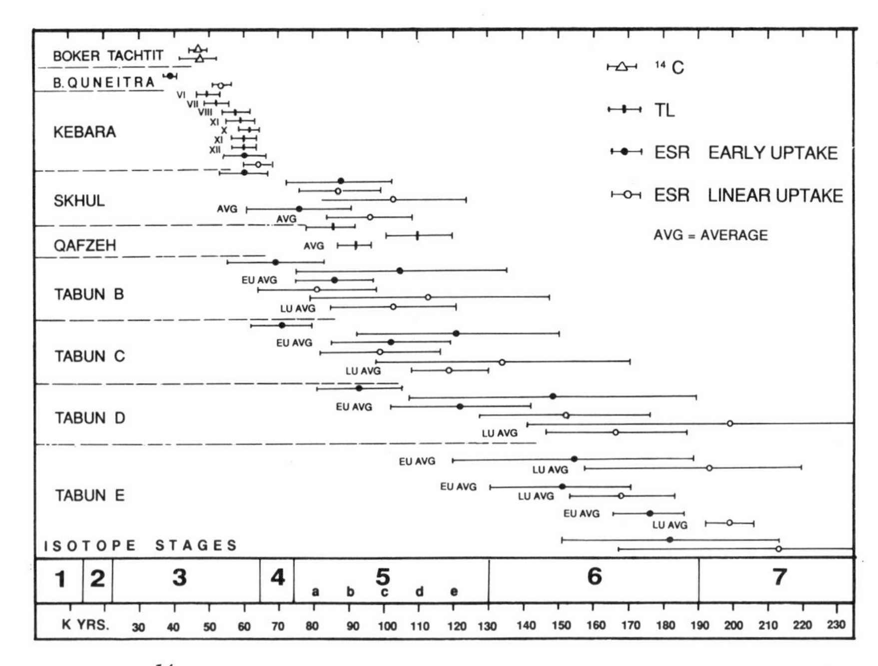

Figure 3. The ^{14}C dates for the early Upper Paleolithic in Boker Tachtit; TL and ESR dates for Mousterian and Acheulo-Yabrudian layers and assemblages.

rary with the Acheulo-Yabrudian. We may toy with the idea that the overall similarity between the Middle Stone Age in South Africa (Singer and Wymer 1982; Thackeray and Kelly 1988) and the "Tabun D type" industry, which is present in the Negev (see below), could have marked the advent of the modern humans in their exodus out of Africa. However, there is no supportive evidence in the current state of our knowledge of the Mousterian in Egypt (Van Peer 1991). Undoubtedly, more research in neighboring regions and additional dates are needed before we can test each of these hypotheses.

Finally, in El-Kowm the Acheulo-Yabrudian is overlain by a new lithic industry, named "Hummalian" (Hours 1983), which occupies the same stratigraphic position as Tabun D above Tabun E. The Hummalian shares with the "Tabun D type" industry the proliferation of blades and points that were produced in El-Kowm, without employing the Levallois technique.

The available radiometric chronology (fig. 3; table 1) indicates that there are difficulties in using U-series together with TL and ESR in dating the sites. The Th/U dates published earlier, in part, do not correlate well with the new ESR and TL dates. However, most Th/U dates indicate that the Acheulo-Yabrudian is earlier than 150 ka. The current situation is reminiscent of the early days of the use of the radiocarbon technique, but when the stratigraphic evidence is placed together with the available dates one may reach the following conclusions (fig. 4):

1. The Upper (or Late) Acheulian is of late Middle Pleistocene age. On the basis of stratified sites (Tabun and Yabrud I), as well as K/Ar dates, it predates the Acheulo-Yabrudian complex or the "Mugharan Tradition."

2. Th/U dates often place the Acheulo-Yabrudian as earlier than 150 ka BP (table 1). It is overlain by dated Mousterian assemblages (Hummal I, Yabrud I, Bezez, Tabun, Zuttiyeh). The age of this entity is now estimated as 250/200-180/150 ka. This estimate is currently supported by TL dates from Yabrud I (Solecki, pers. comm.).

The Mousterian. The definition of the Levantine Mousterian is based on the Tabun sequence and has been subdivided into three phases labelled Tabun D, Tabun C, and Tabun B or Mousterian Phase 3,2,1 (Copeland 1975). The basic technological and morphological characteristics of each phase, some of which were recently restudied (Jelinek 1982a, 1982b; Meignen and Bar-Yosef 1988; Bar-Yosef and Meignen 1992; McBurney 1967; Boutié 1989; Nishiaki, pers. comm.), are as follows:

1. "Tabun D type." Blanks, blades, and elongated points were predominantly removed from Levallois unipolar convergent cores with minimal preparations of the striking platforms. Elongated retouched points, numerous blades, racloirs, and burins are among the common tool types. Rare bifaces occur in Tabun Cave (that could have been a mixture with Tabun layer E). This industry is found in Tabun D, Abu Sif, Sahba, Rosh Ein Mor, Nahal Aqev 3, Jerf 'Ajla, and Douara, layer IV.

2. "Tabun C type." Blanks, often ovaloid, and large flakes were struck from Levallois cores, through radial preparation or by the "lineal method" as defined by Boëda (1988, 1990). Triangular points appear in small numbers. This industry is common in Qafzeh (Boutié 1989), Tabun layer C, Skhul, and Naamé (Fleisch 1970).

3. "Tabun B type." Blanks were mainly removed from unipolar convergent Levallois cores with a minority of radial preparation. Broad-based points, often short, thin flakes, and some blades all made by the same "recurrent method" (Boëda 1988), are common. The best examples for this industry are Kebara units VI-XII, Tabun B, Amud and Bezez B (Meignen and Bar-Yosef 1991).

The suggestion to see "Tabun B type" and "Tabun C type" industries as two facies of the same entity (Ronen 1979) was not established on solid archaeological evidence. The new technological analyses and the radiometric dates make this proposition untenable (e.g., Meignen and Bar-Yosef 1988, 1991).

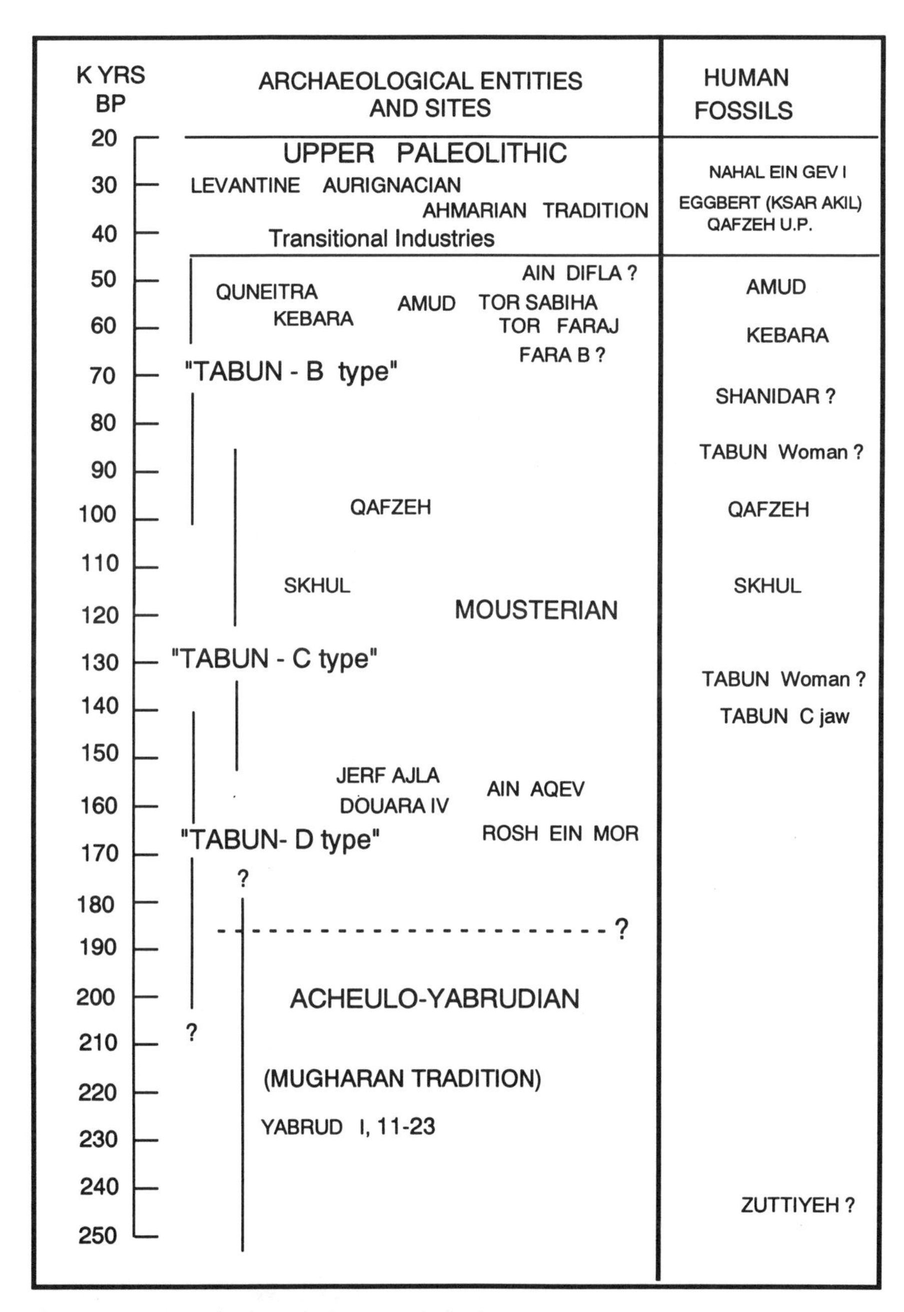

Figure 4. A schematic chronological chart indicating the place of a few various human fossils and industries in relation to the Tabun industrial sequence.

Table 1. Th/U, TL, ESR and ^{14}C dates from the Acheulo-Yabrudian, Mousterian and Early Upper Paleolithic Near Eastern sites.

TH^{230}/U^{234}

Site	Dates	Source
El-Kowm		
Oumm 5 - pre-Yabrudian, travertine	245+16 ka	1
Hummal 2 - Yabrudian, travertine (layer Ib)	156+16 ka	1
Oumm 3 - Yabrudian, travertine	139+16 ka	1
Tell 6 - Yabrudian	99+16 ka	1
Oumm 4 - inter-Yabrudian-Mousterian	76+16 ka	1
Zuttiyeh Cave		
76ZU 1 pre-Acheulo-Yabrudian, travertine	164+21 ka	2
76ZU 4 Acheulo-Yabrudian, travertine	148+ 6 ka	2
76ZU 1 pre-Mousterian, travertine	95+10 ka	2
76ZU 6 pre-Mousterian, travertine	97+13 ka	2
Naamé (Mousterian)		
Strombus level - Enfean II	90+20 ka	3
Strombus level - Enfean II	93+ 5 ka	4
vermet level - Naamean	90+10 ka	3
Nahal Aqev (fossil spring)		
Layer B (average of 3 samples) pre-Mousterian	211+19 ka	4
76NZ6d-4 layer D below Mousterian	85.2+10 ka	4
76NZ1 layer D below Mousterian	74+ 5 ka	4
Ksar Akil Rockshelter		
Layer XXVI B, bone G-88174s	47,000±9,000	11
Layer XXVIB, bone G-88173b	19,000±5,000	11
Layer XXXII, bone G-88177s	51,000±4,000	
Layer XXXII, bone G-88178b	49,000±5,000	

TL

Site	Dates	Source
El-Kowm		
Hummal Well: Three readings for Layer Ib-Yabrudian, pre-Hummalian, Flint	180±25; 145±20; 160±23	12
Average for the Yabrudian	160±22 (Ox85-TLfg 235 Ib)	12
Three readings for Layer 6b (abraded Mousterian flints)	97±8; 105±9; 112±10	12
Average: Abraded Mousterian artifacts	104±9 (Ox85-TLfg235)	12
Kebara Cave (Mousterian)		
Unit VI	48,300±3,500	5
Unit VII	51,900±3,500	5
Unit VIII	57,300±4,000	5
Unit IX	58,400±4,000	5
Unit X	61,600±3,600	5
Unit XI	60,000±3,500	5
Unit XII	59,900±3,500	5
Qafzeh Cave (Mousterian)		
XVII-XXII range of dates	85.400±6.9-109.900±9.9	5
The mean date	92±5 kyr	5

Table 1—*continued*

ESR			
Quneitra (Mousterian), Average	EU 39.2± LU 53.9±1.7 kyr		6
Kebara Cave (Mousterian), Unit X	EU 60±6 kyr LU 64±4 kyr		6
Qafzeh Cave (Mousterian)			
Average XV-XXI	EU 96±13 kyr LU 115±15 kyr		6
Skhul Cave (Mousterian)			
Range of dates	EU 54.6±10.3 to 101.0±19 kyr		6
	LU 77.2±15.7 to 118.0±25.1 kyr		6
Average	EU 81±15 kyr LU 101±12 kyr		6
Tabun (Average dates)			
B Mousterian	EU 86±11 LU 103±18 kyr		6
C Mousterian	EU 102±17 LU 119±11 kyr		6
D Mousterian	EU 122±20 LU 166±20 kyr		6
Ea - Acheulo-Yabrudian	EU 154±34 LU 188±31 kyr		6
Eb - Acheulo-Yabrudian	EU 151±21 LU 168±15 kyr		6
Ec - Acheulo-Yabrudian	EU 176±10 LU 199± 7 kyr		6
Ed - Acheulo-Yabrudian	EU 182±31 LU 213±46 kyr		6
^{14}C			
Kebara Cave (Mousterian			
chB	GrN-2561	41,000+1000	6
Same sample, "rest fraction"	GrN-2551	35,300+500	6
Tabun Cave			
Tabun B Ch	GrN-2534	39,700±800	7
Tabun C Ch	GrN-2729	40,900±1000	7
Tabun Unit I Ch	GrN-7408	>47,000	7
Tabun Unit I Ch	GrN-7409	51,000±4,800/3,800	7
Tabun Unit I Ch	GrN-7410	45,800±2100/1600	7
Tabun Bed 42 very black soil	LJ-2084	38,800±2,400	7
Geula Cave (Mousterian)			
B1 chB	GrN-4121	42,000±1,700	7
Ras el Kelb Cave (Mousterian) chB	GrN-2556	>52,000	7
Ksar Akil Rockshelter (Mousterian)			
Layer XXVI dark clay band	GrN-2579	43,750±1,500	7
Jerf 'Ajla Cave (Mousterian) Ch	NZ-76	42,000±2,000	7
Douara Cave (Mousterian)			
layer E Ch	TK-111	>43,900	8
IIIB Ostrich eggshell	GrN-8638	46,700±2,200 8/1,700	8
IIIB Ostrich eggshell	GrN-8058	>53,800	8
IVB Ch	TK-165	38,900+1,700	8
IVB Ch	TK-166	>43,200	8
IVB Ch	TK-167	>43,200	8
IVB Ch	TK-168	>43,200	8
IVB Ch	GrN-7599	>52,000	8
Fission-track IVB Barite	Kyoto	75,000	8
Racemization IIIB		ca. 85,000	8
IIIB		ca.110,000	8
Rosh Ein Mor (Mousterian)			
Ostrich eggshell	Tx-1119	>37,000	7
Ostrich eggshell	Pta-543	>44,000	7
Ostrich eggshell	Pta-546	>50,000	7
Shanidar Cave (Mousterian)			

Table 1–*continued*

D Ch	GrN-2527	46,800+1,500	10
D Ch	GrN-1495	50,600+3,000	10
Kunji Cave (Mousterian)			
135 cm Ch	SI-247	>40,000	10
145 cm Ch	SI-248	>40,000	10
Earliest Upper Paleolithic			
Boker Tachtit (Transitional Industry)			
Level 1 Ch	GX-3642	>35,000	9
Level 1 Ch	SMU-580	47,280+9,050	9
Level 1 Ch	SMU-259	46,930+2,420	9
Level 1 Ch	SMU-184	>45,490	9
Level 4 Ch Early Ahmarian	SMU-579	35,055+4,100	9
Shanidar Cave (Baradostian)			
C near base, Ch	I-3351	32,300±3,000	10
C-Lower, Ch	GrN-2015	34,540±500	10
C-Lower, Bone fraction	GrN-2016	35,440±600	10
C-Lower, Ch	W-650	32,300±1,000	10
C-Lower, Ch	W-180	>34,000	10
The Ahmarian Tradition			
Negev			
Boker A Ch	SMU-181	>33,600	9
Boker A Ch	SMU-260	>33,420	9
Boker A Ch	SMU-578	37,920±2810	9
Northern Sinai			
Lagama VII Ch	SMU-172	34,170±3670	13
Lagama VII Ch	SMU-185	31,210±278	13
Lagama VII Ch	RT-413A	>19,000	13
Lagama VIII Ostrich eggshell	SMU-119	32,980±2140	13
Lagama IIID Ostrich eggshell	SMU-118	30,050±1240	13
Qadesh Barnea B501 Ostrich eggshell	Pta-2819	33,800±940	14
Qadesh Barnea B601 Ostrich eggshell	Pta-2964	32,470±78	14
Southern Sinai			
Abu Noshra I Ch	B-12125>30,440		15
Abu Noshra I	B-13198	29,580+1610/-1340	15
Abu Noshra I	SMU-2007	35,805±1520	15
Abu Noshra II	SMU-1762	31,536±2275	15
Abu Noshra II	ETH-3076	33,940±790	15

Sources: 1. Hennig and Hours 1982; 2. Schwarcz, Goldberg & Blackwell 1980; 3. Leroi-Gourhan 1980; 4. Schwarcz et al. 1979; 5. Valladas et al. 1987, 1988; 6. Schwarcz et al. 1988; Stringer et al. 1989; Grün and Stringer 1991; 7. Weinstein 1984; 8. Akazawa 1987; 9. Marks 1983; 10. Henry and Servello 1974; 11. van Der Plicht, Van der Wijk & Bartstra 1989; 12. Ancient TL Supplement, date list, 1989; 13. Bar-Yosef and Phillips 1974; 14. Gilead 1983; 15. Phillips 1988.
Legend: EU = Early uptake; LU = Linear uptake; Ch = charcoal; chB = charred bone.
Comments: 1. ^{14}C dates of Mousterian samples later than 38 ka were not included. 2. The dates for Level 6 in Hummal Well are of Mousterian flints derived from the top of the Mousterian sequence in this site. An estimate of 150-110k is for the Hummalian and other Levalloiso-Mousterian. The chronological assignment of the Yabrudian as earlier than 150 ka will corroborate the Th/U dates for Hummal, Oumm, and Zuttiyeh. The date should, therefore, be rejected or its context reexamined in the field.

Radially prepared cores were used and their products are found mainly in the upper contexts of the "Tabun B type" entity (e.g., Kebara VIII-VII, Biqat Quneitra; Goren-Inbar 1990). Thus, there is no clear unidirectional flake/blade trend in the final Mousterian deposits that could have been interpreted as heralding the Upper Paleolithic Transitional Industry. However, the transitional industries in the central Levant are dominated more often by flakes, while the assemblages of Boker Tachtit in the Negev highlands have definitely more blades (Marks 1983; Marks and Kaufman 1983; Ohnuma and Bergman 1990; see below).

The Levantine Mousterian differs markedly from the Mousterian facies in the Zagros and the Taurus mountains, although the industry of Karain Cave (in southwest Turkey) has a more substantial component of radial Levallois preparation (Yalçinkaya 1988) that differs from the Zagros assemblages (Dibble 1984; Dibble and Holdaway 1990). The number of retouched pieces in these assemblages is higher than in the Levant, perhaps due to constraints of raw material.

Mousterian industries that partially resemble those of the Levant can be found in Armenia, where the dominant raw material is obsidian, while the Charentian character is dominant in the entire Caucasus region (Liubin 1989).

The chronology of the Levantine Mousterian has been in recent years a subject for major controversies. The short chronology suggested by Farrand (1971, 1979) was in essence not much different from the earlier one by Howell (1959) in attributing the Acheulo-Yabrudian to the Last Interglacial (Farrand 1971, 1979). The work of Sanlaville (1977, 1981) seemed to support the paleoclimatic sequence as reconstructed by Farrand (1979, 1982). Alternative chronological frameworks which indicated the much older age of Qafzeh (Bar-Yosef and Vandermeersch 1981; Tchernov 1981) were criticized (Jelinek 1982a), but were supported later when TL and ESR readings became available (Valladas et al. 1988; Schwarcz et al. 1988, 1989; Grün, Stringer and Schwarcz 1991; Grün and Stringer 1991; see table 1 and fig. 3). The older age of Tabun D and E, as suggested by me (Bar-Yosef 1989b), gained support from the ESR dates (Grün, Stringer & Schwarcz 1991) but the ages for Tabun C and B as well as for Skhul became older – to ca. 100-120 ka (fig. 2, table 1). While continuously redating the assemblages, especially from a younger to older age, I have kept the basic sedimentological, palynological, and zoological information, wherever available, in accordance with the paleoclimatic interpretation as known today. Thus, dominance of arboreal pollen can be of early interglacial or stadial age when the Mediterranean coastal ranges received the regular winter precipitation (Bar-Yosef 1989a, 1989b, 1992a).

If the available dates are sorted carefully, in spite of the large standard deviations in many ESR readings, and while most of the Th/U dates are currently seen as not valid due to natural contamination (H. P. Schwarcz, pers. comm.), and considered together with the stratigraphic evidence, then it seems that the Levantine Mousterian lasted from about 180/150 ka to about 47/45 ka (figs. 3, 4).

It is presumptuous not to state that certain ambiguities still plague the Levantine chronology. In order to achieve a better resolution one may offer some directions for future research. For example, paucity of microfaunal remains in Tabun Cave, Layer D, makes it impossible to place this assemblage before or after Qafzeh in a relative chronological scale (Tchernov 1981, 1988; Bar-Yosef and Vandermeersch 1981; Bar-Yosef 1989a, 1992a, 1992b). The debate concerning the chronological value of archaic species of rodents would be solved if a faunal-rich layer with a "Tabun D type" industry, underlying a "Tabun C type" industry, were excavated in a geographic neighborhood to Tabun Cave.

I, therefore, suggest referring to the stratigraphic sequence of Tabun as a scale and as a basis for correlation with other sites and industries (fig. 4). In summary, this would place the Late Acheulian in Stages 9 and 8, the Acheulo-Yabrudian as well as "Tabun D type" assemblages in Stages 7 and 6, the "Tabun C type" (Qafzeh, Skhul) will be in the early part of Stage 5 (130/110-90/80 ka), and "Tabun B type" from about 90/80 ka to 47/45 ka. Additional refinement is urgently needed as well as a better dating resolution for the Early Upper Paleolithic.

The Upper Paleolithic. The Levantine Upper Paleolithic started with two versions of a "Transitional Industry," one in the central Levant – Ksar Akil, Lebanon (Ohnuma 1988; Ohnuma and Bergman 1990) – and the other in the southern Levant – Boker Tachtit, Israel (Marks and Kaufman 1983). An earlier examination of part of the Ksar Akil assemblages (stored in the Institute of Archaeology in London) led Copeland (1975) to describe the transition from the Mousterian to the Upper Paleolithic as characterized by the continued use of the unipolar Levallois core reduction strategy. Similar unipolar, convergent blanks, which in the late Mousterian times were shaped into racloirs and points, were modified in the Early Upper Paleolithic (henceforth EUP) into end scrapers, chamfered pieces (an end scraper in which the working edge was obtained by a transverse blow) and burins. The more recent studies by Ohnuma and Bergman (1990) clearly indicated that prismatic, unipolar cores dominate the earliest assemblages of Ksar Akil (XXV-XXI) with diminishing numbers of Levallois cores. This industry directly evolves into the full-blown blady Ahmarian.

The question of whether the technological transition occurred in place motivated Marks and Volkman (1986) to reanalyze the Mousterian assemblages of Ksar Akil, of which the largest collection is stored in the Peabody Museum in Harvard University. Their study demonstrated what is now also documented in the topmost Mousterian assemblage in Kebara (Meignen and Bar-Yosef 1988), namely, that a trend to radial preparation is well expressed in the latest assemblages and, therefore, they could not have been the technological origin of the unipolar and bipolar EUP Transitional Industry. However, it is worth mentioning that there is a stratigraphic gap in Ksar Akil between the Middle and the Upper Paleolithic, of which the earliest assemblages (XXV-XXIV) are quite poor.

The excavations in Boker Tachtit documented, on the basis on numerous refitted cores (Marks 1983; Marks and Kaufman 1983; Volkman and Kaufman 1983), a gradual shift from an assemblage (Level 1) which contains some Levallois cores as well as numerous bipolar blade cores to an assemblage (Level 4) dominated by unipolar blade cores. The earlier tool types include Emireh points, Levallois points, end scrapers, and burins, and can be considered as an Upper Paleolithic assemblage or Transitional Industry. The main differences between this site and Ksar Akil are in the typology of the artifacts. Boker Tachtit is dominated by points and Ksar Akil by chamfered pieces which served as scrapers (Marks and Ferring 1988; Marks 1990).

Boker Tachtit seems to be the best candidate for an autochthonous transition from the Middle to the Upper Paleolithic, especially that it was radiocarbon dated to 47/46 ka BP (table 1). However, the lack of a well-established Mousterian assemblage, stratigraphically immediately under the Boker Tachtit layers, cautions against uncritical acceptance of this conclusion. The Levantine case could have been the result of rapid acculturation similar to the Châtelperronian (Harrold 1989; Mercier et al. 1991). The contention that this EUP evolved from the Negev Mousterian (Marks 1983; Marks and Kaufman 1983; Henry 1986; Lindly and Clark 1987; Clark and Lindly 1989) has to be established on stratified assemblages and additional dates. The suggestion that the "Tabun D type" industry lasted in the Negev and arid southern Jordan during Isotope Stages 5, 4 and early 3 (or at least 4 and early 3) is open to alternative explanation (Bar-Yosef 1989b). The Late Mousterian in Tor Faraj and Tor Sabiha in southern Jordan (Henry 1986) is rather like the Kebara Mousterian ("Tabun B type" industry) and will not support a model for regional continuity in this semiarid/arid area. The assemblage of 'Ain Difla (Lindly and Clark 1987) resembles Tabun D but its age is unknown.

A striking resemblance between the Transitional Industry rich in chamfered pieces in Ksar Akil, Abu Halka, and Abri Antelias in Lebanon and the Early Dabban in Cyrenaica is worth noting (McBurney 1967). The lack of radiometric dates temporarily precludes a conclusion about what direction the bearers of this industry went, except from suggesting that the connection was along today's submerged part of the coastal plain. There is no positive evidence from the region in between that would indicate the presence of even ephemeral encampments.

The EUP entity which follows the initial "Transitional phase" is characterized by a series of blade/bladelet industries, recently named the Ahmarian Tradition (Gilead 1981; Marks 1981), which are radiocarbon-dated to about 35/33 ka through 22/20 ka. What seems to be a continuous sequence is interrupted by a different industry commonly referred to as the "Levantine Aurignacian." The typical Levantine Aurignacian assemblages have been uncovered solely in the central Levant (Lebanon, Anti-Lebanon, Galilee and Mt. Carmel). Its early phase, also known as "Levantine Aurignacian A" (Copeland 1975) or "Early Antelian" (Garrod 1957), is characterized by the dominance of blade/bladelets with carinated and nosed scrapers as well as retouched bladelets and El-Wad points. The later phase ("Levantine Aurignacian B" or "Late Antelian") is characterized by flake production, nosed and carinated scrapers, El-Wad points, and bone and antler industry in cave sites. It was succeeded by flake industries which are often related, due to the proliferation of scrapers and burins, to the Levantine Aurignacian tradition, known as "Levantine Aurignacian C" or "Atlitian" (Gilead 1981, 1989; Marks 1981; Marks and Ferring 1988; Bar-Yosef and Belfer-Cohen 1988).

The Ahmarian blade/bladelet assemblages, which presumably are contemporary, at least with the typical Levantine Aurignacian, dominate the later part of the Upper Paleolithic, especially from 26 ka to 20 ka. As agreed by most workers, it evolved into the variable Epi-Paleolithic complexes (which in Eurasian terminology would be called Late Upper Paleolithic, or Late Stone Age, in Africa).

In summary, the Levantine transition from the Middle to the Upper Paleolithic is dated on the basis of radiocarbon dates. If the published dates from clear-cut Mousterian contexts are taken into account (table 1) they would generally range around 42 ka or earlier than this date. The ^{14}C readings from Boker Tachtit place the earliest Upper Paleolithic around 47/46 k yrs. In both traditional radiocarbon laboratories and those using the Accelerator Mass Spectrometer (AMS), 40/45 ka is generally the current limit of the technique. Therefore, if the Boker Tachtit dates will be supported by additional dates, they would be earlier than any of the published dates for similar transitional industries, such as the Bachokiran (Kozlowski 1988; Kozlowski, Laville & Sirakov 1989), the Bohunician (Svoboda and Simán 1989), the Szeletian (Allsworth-Jones 1986), the Uluzzian (Palma Di Cesnola 1989), or the earliest Aurignacian and Châtelperronian in Spain and France (Straus 1989; Harrold 1989). This transition is as yet undated in the adjacent Anatolian-Iranian or the Caucasus regions.

Both the dates and the traits of the lithic industries related to the

Middle/Upper Paleolithic transition in the Levant raise the issues of interpretation. The assemblages can be seen as either (1) an autochthonous change or (2) the results of acculturation. The case for in situ change was forcefully advocated by Marks (Marks 1983; Marks and Kaufman 1983; Marks and Ferring 1988), and was accepted by others (Bar-Yosef 1989b; Gilead 1991). The acceptance of the alternative hypothesis requires many more radiocarbon dated assemblages as well as stratigraphic evidence for alternate occupations represented by two different industries. In this case one industry, like the Châtelperronian, is the derivative of the Mousterian, such as Boker Tachtit Level 1, and the other is a different blady industry, like the Aurignacian.

Finally, the Southwestern Asian Upper Paleolithic sequence from 35 ka onwards presents cultural changes at a similar rate to other parts of Asia and Africa. It did not provide the richness of artistic expressions, including body decorations, as in temperate Europe. However, in this domain it does not differ much from the South African Late Stone Age or the North African record. In spite of this apparent lack, the rate of change, as observed in the lithics, settlement pattern, hunting strategies, use of vegetal food processing, stone objects, and bone and antler industry, resembles other regions and evinces the same departure from the slow cultural change of the Middle Paleolithic. These aspects are briefly examined in the next section.

MIDDLE AND UPPER PALEOLITHIC HUMAN BEHAVIOR AS REFLECTED IN ARCHAEOLOGICAL ATTRIBUTES

Settlement Pattern. The information concerning the number of sites in both periods is incomplete, but even a cursory survey of the literature indicates that there are fewer EUP sites than Mousterian and LUP sites. This could be a false impression as perhaps many EUP sites are buried under the alluvium in the intermontane valleys. Moreover, the clustering of all Mousterian sites into one entity ignores their temporal range which, by the new chronology, is much longer than suggested in most of the papers published before 1989.

A somewhat clearer picture emerges from the ways sites were used. At least two types of sites can be defined: (1) hunting/gathering ephemeral camps, and (2) some sort of base camps. The first category is often represented by low frequencies of artifacts relative to the excavated volume, and mainly by the large number of curated pieces and lack of primary chipping (absence or almost total absence of cores). Both site types are present in the Middle Paleolithic. Examples for ephemeral sites include Kunji Cave in the Zagros (Speth, pers. comm.), Koudaro 3 in the Caucasus (Liubin 1989), Sefunim in Mt. Carmel (Ronen 1984), lower layers of Qafzeh, rock-shelters in the Judean desert, etc. Base camps or intensively used sites are Tabun, Qafzeh XV-VII, Kebara, El-Wad G-F, Ras el Kelb, and so on. The distribution in the Mediterranean coastal belt was already noticed by Binford (1968), who suggested that human groups were moving across the countryside in an east-west direction following the main wadi courses. Preliminary studies of tooth increments of gazelles (Lieberman, pers. comm.) indicate that even among the cave sites located on the west-facing slopes of the coastal ranges, the season of occupation differs from one locale to the next.

Upper Paleolithic cave occupations within the Mediterranean vegetational belt are often more intensive than those in the semiarid areas such as Yabrud Rockshelter II (Rust 1950). But such a difference is expected when the potential for gathering and hunting is taken into account. Most excavated cave sites are located within the Mediterranean vegetational belt, where over one hundred species of trees, shrubs, and annuals provide edible plant food almost year round. As mentioned above, Ksar Akil Rockshelter was intensively used because it was probably in a more attractive location than other caves or rock-shelters (e.g., Abu Halka, Yabrud II, Hayonim, El-Wad, Kebara, Erq el Ahmar, etc.). While numerous Upper Paleolithic open-air sites have been excavated in the desert regions (Bar-Yosef and Belfer 1977; Gilead 1984b; Marks and Kaufman 1983), their length of occupation, as judged from the numbers of artifacts per volume of sediment, was shorter.

Spatial Organization and Intensity of Occupation. The only change across the transition from the Middle to the Upper Paleolithic is the use of stones for warmth banking in hearths in the latter period. Small round-oval hearths were uncovered in Qafzeh and Kebara. A large hearth, 5 m in diameter, was reported from Douara Cave, layer IV (Akazawa 1987).

The available evidence about spatial distribution of artifacts within the archaeological horizons of the Mousterian is rather scanty. Spatial analysis in Rosh Zin suggested to Stevens and Hietala (1977) a pattern of repeated occupations in which Levallois points, burins, end scrapers, and notches were usually aggregated in the same area over the total thickness of the site (ca. 1 m) within the excavated block (45 m^3). Some patterning appears in the EUP as exposed in Boker Tachtit (Hietala 1983) and in the early Ahmarian at Abu Noshra and Gebel Lagama in Sinai (Phillips and Gladfelter 1991; Gilead 1983). It mainly indicates a clustering of artifacts and debitage in certain places with minimal movement. Epi-Paleolithic sites after 20 ka were intensively occupied and only ephemeral camps reflect the organization of space, often around or beside hearths (Gorring-Morris 1988).

Site size and intensity of occupation (artifacts per cubic meters) are difficult to discuss due to insufficient published information. However, it should be stressed that several sites lend the impression of having been used relatively quite often. For example, in Tabun Cave a total of 8 m accumulated between 180/150-50/60 ka (layers D-B, without the accumulation in the chimney which was about 3 m thick). This sequence, besides slumping and minor gaps, includes a major erosional phase between D and C. In Kebara about 4 m of ashes and fine sediments were deposited from 60/65 to 48 ka. In Ksar Akil an accumulation about 18 m thick represents a time span from ca. 45 ka to 20 ka (based on ^{14}C). Of course, there are many sites in which the intensity of use is less impressive, and a gradual change can be observed as one moves into semiarid/arid regions or high altitudes.

Burials. Human behavior, as reflected in the presence of burials, well-shaped hearths, discrete bone accumulations, tool use, and exploitation of raw material, does not demonstrate any clear differences between the archaeological remains left by Middle Paleolithic modern-looking hominids or the local Mediterranean Neanderthals. It is worth mentioning that we often assume that the hominids were buried in their own camps, whether occupied repeatedly (like Kebara) or ephemerally (like the lower layers at Qafzeh). The clearest burials are Qafzeh 9 and 11, Skhul 4 and 5, Kebara 2, and Amud 1, as well as some of the Shanidar skeletons.

No EUP burials (ca. 45-30 ka), except for "Egbert" in Ksar Akil level 17, where the bones were uncovered under a pile of cobbles brought into the rock-shelter (Bergman and Stringer 1989) were documented. Nahal Ein Gev I (Arensburg 1977), exhibited a flexed disposition of the remains and should be now dated, on the basis of lithic comparisons with Ksar Akil, to about 25-22 ka.

Subsistence. Unfortunately, most of the direct evidence concerning Middle Paleolithic and Upper Paleolithic diets still comes from the analysis of the preserved animal bones. Rare finds such as in Douara and Kebara caves indicate that we miss the plant portion of the prehistoric menu. By comparison to other low- to mid-latitude hunter-gatherers, it is expected that their diet was mainly based on fruits, seeds, leaves, and tubers. In Douara Cave the fruits of the *Celtis* sp. (Akazawa 1987) were found. Legume seeds occur in all the Mousterian layers in Kebara, and indicate the presence of humans in the cave in springtime and the exploitation of carbohydrates at the time of seasonal stress (Speth and Spielmann 1983).

It is quite possible that the difference between Middle and Upper Paleolithic humans was in the ways in which vegetal resources were collected and prepared for food. A fine example, although from the Late Upper Paleolithic (19 ka), is the rich collection of carbonized plant remains from a waterlogged site on the shores of the Sea of Galilee (Kislev, Nadel & Carmi 1992). This assemblage supports the contention of a wide range use of plants encompassing more than 35 species. Sporadic grinding stones in several Upper Paleolithic contexts substantiate this observation (Bar-Yosef 1989a; Wright 1991).

The available faunal reports from southwest Asia concerning hunted and scavenged species were summarized elsewhere in terms of bone frequencies (Bar-Yosef 1989a). Only a few published studies include counts of body parts. The main phenomena reflected in the bone counts are as follows:

1. One-third of the mammalian bones in Acheulo-Yabrudian contexts such as Masloukh Cave, Abri Zumoffen and Tabun E are of large animals (*Bos primigenius*, rhinoceros, and equids).

2. Large mammals are predominant in the two open-air Mousterian sites–Fara II (Gilead 1984a) and Biqat Quneitra (Davis, Rabinovich & Goren-Inbar 1988). It was suggested that such bone collections reflect opportunistic scavenging (Bar-Yosef 1989a).

3. The tendency to consume medium and small mammals, reptiles, and birds continued without apparent change from the Mousterian through the Upper and Epi-Paleolithic. The interpretation that these assemblages represent essentially the results of hunting activities is now supported by detailed analysis of the Kebara collection from both periods (Speth, pers. comm.).

The overall conclusion that hunting techniques developed during Mousterian times is in accordance with similar situation in a Mediterranean environment in coastal Italy (Stiner 1990). In prime situations where scavenging was a time- and energy-saving strategy, the Mousterian people took advantage of the available carcasses.

In sum, although it seems that the above survey minimized the differences between the Middle and the Upper Paleolithic, this would be a misleading impression. The major differences in the Upper Paleolithic include the use of built-up hearths, the exploitation of bones and antlers as raw material for making tools, the use of grinding stones (e.g., Qafzeh, Yabrud II), some artistic expressions, and the higher degree of typological variability among the lithics with well-defined types and specific geographic distribution (e.g., chamfered pieces in Lebanon, Emireh points in the southern Levant). Additional information from the Nile Valley indicates that quarrying flint nodules (Vermeersch et al. 1988) was one of the activities of Upper Paleolithic hunter-gatherers. We may add to the list the exploitation of semiarid areas, which could have been made possible by an improved technology and higher degree of mobility that indicates differences in the territorial expansion of social alliances. I am, therefore, tempted to conclude, despite the flimsy evidence, that Upper Paleolithic human groups had larger territories than their immediate predecessors.

HUMAN FOSSILS AND LONG-RANGE MOVEMENTS

The classification of the Levantine human fossils is in the center of the bioanthropological controversy. These fossils fall into two morphotypes: Western Asia Neanderthals (or Mediterranean Neanderthals), and "Proto-Cro-Magnons" (or modern-looking hominids) (Trinkaus 1984; Vandermeersch 1981, 1989; Stringer 1989). The two different types were present in the region during the Middle Paleolithic. They were either contemporary for part of the sequence, as indicated by the radiometric dates (see below and fig. 3), or were sequential (e.g., Trinkaus 1989; Vandermeersch 1990). The second approach regards the documented morphological and metrical variability as reflecting only the diversity among a single Middle Paleolithic population of Southwestern Asia (Wolpoff 1989; Simmons, Falsetti & Smith 1991; Arensburg 1991). Each school of thought has consequently drawn a different evolutionary scenario (Stringer 1989; Bräuer 1989; Vandermeersch 1989; *contra* Wolpoff 1989; Wolpoff et al., this volume; Smith, Falsetti & Donnelly 1989; Smith, this volume; Jelinek 1982a, this volume; Clark and Lindly 1989). The proponents of the Out-of-Africa model found support and encouragement in the genetic trees based on mitochondrial DNA and protein and enzyme polymorphisms (Cann, Stoneking & Wilson 1987; Cavalli-Sforza et al. 1988; Cann, Rickards & Lum, this volume).

Most scholars do not deny the observable metrical and nonmetrical variability of these hominids and many feel that it surpasses what is documented for the population of the "classical Neanderthals" in Western Europe. The particular set of features of the latter group seem to have been the result of long periods of isolation during the Last Glacial stage. Irrespective of whether these various fossils can be defined as two separate species or subspecies, I see no reason not to regard them as different populations. In a continental overview, I suggest delineation of regions with higher densities of Middle Paleolithic humans ("populations") and areas which were either not inhabited or temporarily exploited by small groups. The two populations, the Southwest Asian and the West European, were at both ends of a geographic axis 3000-3500 km long, from the Atlantic coast to the Near East. The vast continental area in between accommodated at least two to four additional populations, in central and southeast Europe as well as the Mediterranean ones in the Iberian and Italian peninsulas.

In the often hazardous climatic conditions of the Last Glacial, the ecological advantages of the Mediterranean Levant were obvious. The lowlands of this region enjoyed relatively higher temperatures, and stable vegetal and animal food resources which were more predictable, reliable, and easily accessible than in most European environments. The Levant was, therefore, attractive to nearby human groups who lived under conditions of diminishing resources and increasing social stress such as in the Balkans, the Anatolian plateaus, and the Taurus-Zagros ranges. Those who occupied the Caucasus area had their own refugium in the lowlands near the Black Sea and the Caspian Sea.

During ameliorated climatic conditions, the Levantine region could have served as a two-way corridor for movements of humans and animals between Africa and Eurasia (Bar-Yosef 1989a, in press; Trinkaus 1989). However, it seems that since the days of 'Ubeidiya, about 1-1.4 M yrs ago, most of the faunas were of Eurasian origins (Tchernov 1987). Humans, on the other hand, were more successful in using this "corridor."

The movement of *Homo erectus* out of Africa was not a unique event. In the course of colonizing the temperate zone, which is subject to major seasonal fluctuations, failures probably occurred. The record of the European Lower Paleolithic is marked with many gaps which are probably not always due to poor preservation or to the presence of unknown well-buried sites, but may be the result of natural calamities which wiped out many prehistoric groups.

The repeated efforts by *Homo erectus* to colonize the temperate zone of Eurasia indicate that humans were equipped with appropriate food acquisition techniques and social alliances and, therefore, also with certain linguistic abilities that were required for such long-range movements.

A model of continuing development of language abilities during this extended period is described by Deacon (1989), entirely on the basis of neuroanatomy. Deacon argues that "the relative deviation of the prefrontal cortex (and thus language related structures within prefrontal cortex) from the allometric trend exhibited in other primate brains is directly reflected in relative brain/body deviation from the anthropoid primate allometric. The degree of quantitative organization of these language circuits should thus correlate with the degree of brain size disproportion in each species. Evidence for changes in brain/body proportions in these hominid lineages is robustly represented in the fossil record. . . . Three important consequences of this association between gross allometric trends, language circuits reorganization and selection with respect to language functions are that: (1) selection with respect to language began at least 2 million years ago, (2) directional selection continued to produce changes in this system in *habilis*, *erectus* and *sapiens* from this period to approximately the time of the origins of archaic *Homo sapiens,* and (3) during the evolution of distinct lineages of *Homo sapiens*, including the Neanderthals and the period around Neolithic cultural revolution, selection on brain function did not produce any further major neurological changes" (Deacon 1989:395-96). A nonpunctuated evolution of language abilities is conceivable and, therefore, it is worth the effort to model what other cultural and demic factors might account for the changes in the brain. If the technologies, ecological knowledge, language abilities, and social customs of *Homo erectus* were sufficient to enable this hominid to invade the greater part of the Eurasian continent, it is not difficult to imagine that the means were present for cultural diffusion of information and new technologies and that migration patterns enabled some frequency of interdemic contact.

I, therefore, find it rather surprising that some researchers of the Middle Paleolithic of Eurasia cannot envision movements of human groups during a period that lasted about 150,000 years (Isotope Stages 6, 5, 4 and early 3). The view that the prehistoric populations were stable and well established in their region assumes their ability to adapt to most environmental fluctuations. If, indeed, Middle Paleolithic humans had this capacity to endure the climatic vagaries of the northern latitudes why did they not move farther on and colonize the New World? Upper Pleistocene stratigraphic records from Britain, Belgium, The Netherlands, south Germany, Poland, etc., clearly indicate a period of cold desert during Isotope Stage 4, with no human occupation. Where did the people move and how far? Did they succeed in evacuating their territories or did they die out? These questions and how to go about testing them in the archaeological record are rarely discussed. I believe that the impact of the last decades, when "migrations" and "diffusions" as explanatory devices for observable changes in material culture were heavily criticized, should now be over (Anthony 1990; Ammerman 1989). I will return to this subject below. Meanwhile, it is sufficient to suggest that the mixture of physical features of the Levantine human fossils resulted from the region serving as a refugium for several groups during periods of environmental stress.

The chronology of the Middle Paleolithic fossils which contribute to the recent debate is presented in figure 4. I have placed the woman from Tabun in the context of Layer B, where it seems, in view of Garrod's hesitations, to suit better the current interpretation. Fossils in "Tabun C type" industry are those of Skhul, Qafzeh, and Tabun II, and with the "Tabun B type" are Kebara, Tabun I, and Amud. Currently we have no human fossils associated with the "Tabun D type" industry. Uncovering additional human remains will facilitate the testing of the hypothesis that the archaeological assemblages included in this entity, or an earlier one (such as a Late Acheulian of Levallois facies), were produced by an in-coming African or alternatively local population. Even if superficially there seems to be a correlation between the lithic industries and the fossils, the reader should be reminded that there are early manifestations for blade industries (in Stage 6 and 5e) in the Levant and Western Europe (Jelinek 1990; Otte 1990; Otte and Keeley 1990) which are not necessarily associated with modern-looking hominids. In sum, the association between the fossil hominids and the archaeological assemblages indicates that modern-looking hominids such as those from Qafzeh and Skhul produced Mousterian industries which show no hints of developing into the Upper Paleolithic. Therefore, we may conclude that the transition from Middle to Upper Paleolithic was mainly a socio-economic and cultural revolution unrelated to the morphological differences that are observed among the fossils. However, it is possible to hypothesize that this cultural shift occurred within the descendants of populations similar to the Qafzeh-Skhul group. Only additional human relics will enable the testing of such an hypothesis.

Finally, the current available dates and paleontological evidence for the relative chronology of the Levant and South Africa (Grün, Shackleton & Deacon 1990) cannot support or refute the genetic Out-of-Africa model. Realizing this, Stringer has already mentioned the possibility that with the early dates of Qafzeh (and now those of Skhul) "the picture would dramatically change, and the Asian corridor would become a critical area for the *origin* of modern humans, rather for their dispersal" (Stringer 1989:92; italics in original). Viewed from another angle, the Levant could have become part of greater Africa under "pluvial" conditions, either at the beginning or the end of a Glacial cycle and the early part of the ensuing Interglacial. During these periods increased precipitation of both European and monsoonal origins created additional hospitable vegetational belts in North Africa, the Eastern Sahara, the Sinai, and along the edges of the Syro-Arabian desert. The continuous stretches of steppic and Mediterranean vegetation zone facilitated the movements of human groups in both directions (Bar-Yosef 1987).

DISCUSSION

Most authorities agree that the Middle to Upper Paleolithic transition, clearly reflected in the lithic industries and in western Europe by the appearance of body decorations, bone and antler industry, and the long-distance exchange of raw material, is rapid (Harrold 1989; Klein 1989b, this volume; Farizy 1990, this volume; Soffer, this volume; Jelinek, this volume; White 1989; Kozlowski, Laville & Sirakov 1989; Roebroeks, Kolen & Rensink 1988). Other attributes of the transition are the appearance of larger sites and clear intrasite patterns of spatial distribution which were recently interpreted as representing sexual division of labor, social differentiation, the presence of groups of various sizes and kin-based societies (Soffer, this volume). The rate of cultural change during the Upper Paleolithic was faster than in the Middle Paleolithic. Human groups were more mobile and their ranges covered larger territories. A higher degree of social organization, both locally and spatially, is also inferred. A series of inventions and innovations enabled some groups to move into the northern latitudes and colonize the Americas. In the lower latitudes the ability to cross waterways brought about the colonization of Australia (Jones 1989; Bowdler 1992). One may conclude that with the Upper Paleolithic, culture became the extrasomatic means of adaptation for human groups (White 1959). The dates for this major shift demonstrate a temporal and geographic distribution, from early to late, along an east-west axis beginning in the Near East through to Western Europe. Unfortunately, no dates are available for Northeast Africa or Eastern Asia regions (Klein, this volume). Moreover, until now it seems that the emergence of the Upper Paleolithic is not well correlated with a known climatic change.

The archaeological cases for the transition are better recorded in Western Europe and are, therefore, employed as a model for other regions. This can be done *if* we accept the assumption that the Upper Paleolithic emerged from a small core area and then spread through demic diffusion and acculturation. There is now a general agreement that the Châtelperronian of Western Europe is a cultural continuation of the Mousterian of Acheulian Tradition (as asserted by F. Bordes) and its assemblages were produced by the latest Neanderthals. It should be stressed that these humans were capable of producing blades through a complex *chaîne opératoire* (Pelegrin 1990), which means that they were not limited by physical or mental constraints as suggested by Jelinek (this volume). The long-claimed contemporaneity between the Aurignacian and the Châtelperronian was recently supported by the TL dates for Saint-Césaire (Mercier et al. 1991) and the dates for the early Aurignacian in Spain (Straus 1989, 1990; Otte and Keeley 1990). Thus, in this region the phenomenon of an immigrating population seems well documented. Unfortunately, the exact dates of the Early Aurignacian are not well established, because the sites in question were dated by different radiometric techniques and laboratories, and, therefore, do not permit detailed correlations. Furthermore, no human remains were recovered in the earliest Aurignacian levels but only in the somewhat later ones. These classical Cro-Magnons are robust when compared to the gracile Terminal Paleolithic hu-mans of Europe (Frayer 1984; Gambier 1989; Wolpoff 1980) which perhaps would indicate a period of a "bottle neck" at the time of the Late Glacial maximum. However, the discussion of this issue is beyond the scope of the paper. Finally, similar archaeological generalizations were deduced from known Central and Eastern European sequences (Kozlowski 1988; Mellars 1989; Klein 1989b, this volume).

The question that intrigues everyone is, "why" and "where" did it happen? The easiest answer is that it was caused by a biological change, which means accepting the "replacement" hypothesis originally known as "Noah's Ark" (Howells 1976; Stringer and Andrews 1988; Stringer 1989; Klein 1989a, this volume, but see above citation from Deacon 1989; Wolpoff 1989; Wolpoff et al., this volume). The advocates of biological change agree that it did not take place in Europe, perhaps not even in Asia (although see Stringer 1989 for a potential Western Asia locus, and Excoffier 1990 for a Southeast Europe-Western Asia locus). The possibility that the "replacement" was in reality the combined effects of demic diffusion and acculturation was already raised (Bräuer 1989). The evidence from the Near East indicates a rapid change in the lithics and settlement pattern in the Levant around 47/46 ka (Marks 1983). These dates need to be supported by additional data. This leads to the conclusion that *if* the cultural change did not emerge from the Levant then it should be traced to Northeast Africa, East and/or Central Africa. A review of the Middle Paleolithic industries in Egypt and Sudan by Marks (1990, 1992; Van Peer 1991) indicates that this region was probably not the source area and that the Nile Valley was not the route of Upper Paleolithic humans to Asia. No archaeological information is available at present from the Northeast African coastal belt or the alternative route through the Ethiopian-Arabian Peninsula. The relevant records from the vast region of East and Central Africa are extremely poor, although discoveries such as that by Brooks (1991) in Zaire, where a Middle Stone Age industry contained bone harpoons, may suggest a sub-Saharan core area where the cultural change began. Although it is tempting to see the Howieson's Poort in South Africa, now dated to 75-45 ka but considered to have a shorter time range of 75-65/60 ka (but see Parkington 1990; Kaplan 1990 for a different view), as somehow the source of the Eurasian Upper Paleolithic, this industry is overlain by Middle Stone Age assemblages (e.g., Border Cave, Klasies River Mouth).

One way or another, in Eurasia we still face the problem of a core and periphery relationship, meaning locating a core area where the "Middle/ Upper Paleolithic Revolution" took place and establishing that the ensuing cultural changes resulted from demic diffusion or acculturation in the other regions. For further development of this model, I suggest adopting the principles of the "Neolithic Revolution" as a basis for understanding and pursuing research into the origins and spread of the "Middle/Upper Paleolithic Revolution." In order to materialize this approach we have to do what Braidwood did in the 1940s when he sought the core area of the "Neolithic Revolution." In an overall summary he took into account all the available evidence and drew a "Gap Chart," which indicated what was the unknown archaeological time period in the Near East and pointed out where the sites that would fill in the gap should be found. He pursued his conclusions by going himself with numerous associates to the "Hilly Flanks," namely, the foothills of the Zagros and the Taurus, and thus motivated others to follow in his footsteps to other parts of the Near East (Young, Smith & Mortensen 1983).

Today the main elements and processes involved in the "Neolithic or

Agricultural Revolution" are basically known (Moore 1985; van Zeist 1988; Hillman and Harris 1989; Hillman and Davies 1990; Bar-Yosef and Belfer-Cohen 1991, and references therein). In order to employ the lessons learned from the research of this revolution while studying an earlier one, I will summarize them below. The following contentions are, in my view, accurate and based on our most recent archaeological, archaeozoological and archaeobotanical evidence (Bar-Yosef 1991b):

1. The "Neolithic Revolution" took place within a single species, namely, *Homo sapiens sapiens*, with no biological change.

2. The shift to intentional and systematic cultivation of cereals and legumes in the southern Levant occurred within a short time, at the most during several hundred radiocarbon years (ca. 10,300-9,900 BP). The domestication of animals (goats, sheep, cattle, pigs) happened next, as part of the activities of the agricultural communities, within several hundred to 1,000 radiocarbon years (ca. 9,000-8,000 BP).

3. Population growth, expressed in an increase in average site size from 0.2 to 2.0-3.0 hectares, was coincident with the establishment of cereal cultivation, and probably with predictable supplies of weaning foodstuffs. Increased sedentariness and secured supplies caused a drop in the age of menarche of better-fed women and, thus, to population increase.

4. The expansion of Early Neolithic communities was oriented northward and eastward. The introduction of cereal cultivation to the Anatolian basins was rapid (within less than a thousand radiocarbon years). Its introduction to new environments such as Konya Plain created a population explosion and motivated further-westward demic diffusion (Ammerman and Cavalli-Sforza 1984; Ammerman 1989).

5. Neolithic economy spread through the Mediterranean basin by coastal navigation (Cherry 1990), and inland movement through the Danube Valley (Ammerman and Cavalli-Sforza 1984; Renfrew 1987; Sokal, Oden & Wilson 1991). Both demic diffusion and acculturation were the processes responsible for the "Neolithization" of Europe.

6. The eastward expansion of Neolithic subsistence systems reached Pakistan within 1,500 radiocarbon years. It took, however, about 2,000 radiocarbon years for agriculture to penetrate into the Nile valley although it lies within one week's walk from the Jordan Valley.

Observing a rapid technocultural evolution within one region is not unique to the "Neolithic Revolution" or the Middle/Upper Paleolithic transition. Prehistory and history are full of similar examples for local rapid changes. Moreover, hunting-scavenging and gathering groups have been non-equal both on the intra- and inter-societal levels from primordial times because of differences between regional carrying capacities, technologies, and social organizations. The continuous success of one group would probably cause decline of another group. The demographic modelling by Zubrow (1989) indicates that it does not take long for a less successful population, in this case the Neanderthals, to disappear. Even if his suggestion for "kinless" humans is untenable, as no such relationships are known to last for generations among human societies, or indeed apes, it is useful in showing how human groups could become extinct quite quickly. Similar interpretation would apply to the Mesolithic/Neolithic transition in most of the Eurasian lands.

In table 2 I have summarized the basic cultural-technological changes of the "Middle/Upper Paleolithic Revolution" as partially derived from the "Neolithic Revolution." I have omitted the details of the causes which, relative to those of the "Agricultural Revolution," are as yet unknown.

The search for the causes of the "Middle/Upper Paleolithic Revolution" led to the proposition, for example, that the evolution of modern language was necessary for major changes in social organization (Lieberman 1989). This contention is not supported by the fossil evidence since the discovery of the modern hyoid bone in Kebara (Arensburg et al. 1990) as well as by Deacon's (1989) brain studies. One component of the linguistic argument relates to the degree of planning depth (Binford 1989). This can be studied through the decipherment of the Mousterian lithic *chaîne opératoires* that would exhibit the level of scheduling of core reduction strategies among the Eurasian people. The degree of complexity of the Middle Paleolithic *chaîne opératoire* hardly differs from those devised by Upper Paleolithic artisans (Boëda 1988; Geneste 1989; Boëda, Geneste & Meignen 1990; Bar-Yosef and Meignen 1992; Meignen and Bar-Yosef 1991). The differences are in the secondary trimming, the shaping of "tools" from blanks. The number of clearly defined types of retouched pieces in almost any Upper Paleolithic assemblage in Western Asia and Europe is higher than in the Middle Paleolithic. This observation is reflected in the commonly used type-lists (Isaac 1972).

In addition, the Middle to the Upper Paleolithic transition can be explained as resulting from the introduction of new technologies. These include techniques for food acquisition, such as spear throwers and even archery, perhaps basketry, as well as new tools for food preparation such as grinding stones (de Beaune 1989). Possibly new trapping and storing techniques became available, although the evidence is still meager (Soffer 1989). Stable food provisioning in seasons of stress resulted in population increase caused by the better chances of newborns for surviving and reaching adulthood. A slight increase in life expectancy secured the survival of older members of the group thus extending the "living memory" of the group. Through time it meant a better monitoring of the environment and of more distant regions. Long-range social alliances (Gamble 1982) were developed in order to over-

Table 2. A proposed model for the transition from the Middle to the Upper Paleolithic as based on factors derived from a model for the origins of agriculture. Certain innovations were only needed in the northern latitudes.

1. Core Area

High degree of topographic and phytological variability in relation to the absolute geographic dimensions
Predictable, highly accessible, and reliable plant food resources
Dominance of stationary mammalian species over migratory ones

2. The Nature of the Old Pattern

Low degree of mobility especially in lusher areas
Seasonal exploitation of resources of steppic environments such as semi-deserts and high altitudes (indicating the presence of task groups?)
High degree of regionalization
The use of hafted spears in Southwestern Asia
Simple clothing
Collection and processing of selected plant resources
Low level of symbolic behavior (burials, red ocher, rare marine shells)
Low levels of birth rate in temperate zone

3. Change

(causes unknown – ecological/technological/biological/?)

4. The Nature of the New Pattern

Improved subsistence strategies with new technologies/techniques
New hunting devices – spear throwers, earliest archery?
Improved clothing, especially the kind needed in northern latitudes
Improved gathering and transport devices – baskets (?), sledges, etc.
Use of grinding stones for food processing
Increase in the number of exploited raw materials (i.e., antlers and bones, special hard rocks)
Long-distance procurement of raw materials, quarrying activities
Improved systems of long-distance, intergroup communication

4A. Short-term Results

Increased rates of survival of newborns
Prolonged survival for the elders in the group
Better planning depth of subsistence strategies
Changes in the intensity of symbolic behavior reflected in the new expressions of self-awareness, intra- and intersocietal and environmental attitudes, rituals, etc.

4B. Long-term Results

Selective advantages in long-term monitoring of the environments expressed in the prolonged "living memory" of the groups
Formation of long-distance social alliances
Increased rate of technological adaptation to specific regional environments (e.g., the formation of regional cultures identified by their tool kits)
Population increase per square kilometer

5. Expansion

By demic migration and/or technological diffusion

6. Rate of Expansion

Slow and/or rapid

come seasonal or annual economic disasters. They are expressed in the movement of objects and raw material over long distances (Roebroeks, Kolen & Rensink 1988). The identification of groups is also reflected in specialized lithic artifacts (Otte and Keeley 1990) and body decorations (White 1989).

Therefore, one may see dramatic changes, similar to the origin of agriculture, taking place within a single human population. It can be plausibly argued that there is no need to look for a biological threshold for the Upper Paleolithic. Careful studies of the archaeological contexts related to the immediate ancestors of Upper Paleolithic humans will probably reveal what were the technical and organizational preadaptations that made a successful population change its lifeways and leave behind the kind of archaeological residues that we call "Early Upper Paleolithic."

In summary, despite the capabilities of Middle Paleolithic humans, Upper Paleolithic populations within 30 ka did much better, most of the time, in every cultural context. They lasted more successfully in subarctic conditions due to numerous technological innovations (Soffer 1987) and succeeded in colonizing the Americas. The development of preserved expressions of self-awareness and of changing intra- and intersocietal as well as societal/environmental relationships is reflected in body decorations, decorated objects, portable art, and later in rock art. This was the work of Upper Paleolithic human groups, although it is worth mentioning that not all contemporary groups in the Old World expressed themselves in the same way. Art objects are rare outside the European world (Belfer-Cohen 1988), partially due to paucity of excavations and partially due to the possibility that not all human groups reacted to their immediate situation in a similar way and by using similar materials.

CONCLUDING REMARKS

Following a survey of Southwestern Asian fossils and archaeological remains, their chronology, and their contribution to the study of the origins of modern humans and the Upper Paleolithic, I propose a model which may explain various aspects of our Eurasian data sets and will assist us in developing future field projects.

On previous occasions I have mentioned either orally, or briefly in writing (Bar-Yosef 1992a), the potential of employing the model of the origin of agriculture and the dispersal of farming communities as an analogue of the origin of the Eurasian Upper Paleolithic cultural entities. The advantage of this analogy is that we may look for the core area as well as for specific evidence for the presence of technological innovations for the acquisition, preparation, and storage of food. We need to establish, for example, the amount and importance of long-distance procurement of raw materials during the Early Upper Paleolithic, and search for evidence for demographic changes which followed the Middle to Upper Paleolithic transition. Dating the transition to the Upper Paleolithic in various regions will lead us to the source area. We have to find out why there is variation in symbolic behavior between Upper Paleolithic groups across the Old World.

Finally, the question remains open as to whether it was a biological change which caused the cultural-societal revolution, or if it simply happened within the same grade of humans. In this respect the search of the observed biomechanical differences among the human fossils failed to provide convincing answers. Unfortunately, it was assumed that by looking at the available archaeological information, mainly lithics and residues of the meat and marrow part of the menu, we may unveil the behaviors that resulted in the measured skeletal variability. In the lack of other kinds of data, this was the simplest strategy for providing the desired explanation. But if the biological aspects, which today are solely expressed in morphological attributes of fossils, were not the cause for the observable cultural change, then the fossils should be removed from this debate, and a new approach to the study of biomechanical differences is needed.

I strongly believe that the archaeological assemblages which are widely geographically distributed and are often in abundance, despite their elusive nature, are amenable for testing hypotheses. Even when models are not supported by the evidence, their failure may serve as guidelines for future research.

As a closing note, when the issues of the origin of modern humans are examined in detail, one must feel grateful to all the symposia organizers who invited many of us during the last five years to participate in stimulating meetings. I feel that many of the problems involved in resolving the unanswered questions are now much clearer. Perhaps the time has come for a coordinated, worldwide approach to various funding agencies to enable us to organize additional large-scale field operations, especially in crucial geographic locations such as East Africa, Southwestern Asia, and Southern and Central Asia.

ACKNOWLEDGMENTS

I am grateful to all my colleagues in the Kebara project with whom I have discussed many of the aspects summarized in this paper during our joint field endeavor from 1982 to 1990. I thank David Pilbeam and Terrence Deacon (Peabody Museum, Harvard University) who made many critical comments and editorial suggestions on an earlier version. Martha Tappen and Sally Shearman skillfully helped me with the inclusion of numerous references and proofread the final manuscript. Needless to say, I alone am responsible for the opinions expressed in this text.

REFERENCES

Akazawa, T. 1987. The ecology of the Middle Paleolithic Occupation at Douara Cave, Syria. *University Museum, University of Tokyo Bulletin* 29:155-66.

Allsworth-Jones, P. 1986. *The Szeletian and the Transition from Middle to Upper Palaeolithic in Central Europe*. Oxford: Clarendon Press.

Ammerman, A. J. 1989. On the Neolithic transition in Europe: A comment on Zveilbel and Zveilbel 1988. *Antiquity* 63:162-65.

Ammerman, A. J., and L. L. Cavalli-Sforza. 1984. *The Neolithic Transition and the Genetics of Populations in Europe*. Princeton: Princeton University Press.

Anderson-Gerfaud, P. 1990. Aspects of behaviour in the Middle Palaeolithic: Functional analysis of stone tools from Southwest France, 389-418. In *The Emergence of Modern Humans: An Archaeological Perspective*, ed. P. Mellars. Edinburgh: Edinburgh University Press.

Anthony, D. W. 1990. Migration in archaeology: The baby and the bathwater. *American Anthropologist* 192:895-914.

Arensburg, B. 1977. New Upper Palaeolithic human remains from Israel. *Eretz Israel* 13:208-15.

Arensburg, B. 1991. From *sapiens* to *neandertals*: Rethinking the Middle East. *American Journal of Physical Anthropology*. Supplement 12:44.

Arensburg, B., L. Schepartz, A. M. Tillier, B. Vandermeersch, and Y. Rak. 1990. A reappraisal of the anatomical basis for speech in Middle Palaeolithic hominids. *American Journal of Physical Anthropology* 83:137-46.

Bar-Yosef, D. E. 1989. Late Paleolithic and Neolithic marine shells in the southern Levant as cultural markers, 169-74. In *Shell Bead Conference*, ed. C. F. Hayes. Rochester, NY: Rochester Museum and Science Center.

Bar-Yosef, O. 1975. Archaeological occurrences in the Middle Pleistocene of Israel, 571-604. In *After the Australopithecines*, ed. K. W. Butzer and G. L. Isaac. Chicago: University of Chicago Press.

Bar-Yosef, O. 1987. Pleistocene connexions between Africa and Southwest Asia: An archaeological perspective. *The African Archaeological Review* 5:29-38.

Bar-Yosef, O. 1988. The date of Southwest Asian Neanderthals, 31-38. In *L'Homme de Néandertal*, ed. M. Otte. Liège: Etudes at Recherches Archéologiques de l'Université Liège.

Bar-Yosef, O. 1989a. Geochronology of the Levantine Middle Palaeolithic, 589-610. In *The Human Revolution: Behavioural and Biological Perspectives on the Origins of Modern Humans*, ed. P. Mellars and C. Stringer. Edinburgh: Edinburgh University Press.

Bar-Yosef, O. 1989b. Upper Pleistocene cultural stratigraphy in Southwest Asia, 154-80. In *The Emergence of Modern Humans: Biocultural Adaptations in the Later Pleistocene*, ed. E. Trinkaus. Cambridge: Cambridge University Press.

Bar-Yosef, O. 1991a. Stone tools and social context in Levantine prehistory, 371-95. In *Paradigmatic Biases in Mediterranean Hunter-Gatherers Research*, ed. G. A. Clark. Philadelphia: University of Pennsylvania Press.

Bar-Yosef, O. 1991b. The early Neolithic of the Levant: Recent advances. *The Review of Archaeology* 12:1-18.

Bar-Yosef, O. 1992a. Middle Palaeolithic chronology and the transition to the Upper Palaeolithic in southwest Asia, 261-72. In *Continuity or Replacement: Controversies in* Homo sapiens *Evolution*, ed. G. Bräuer and F. H. Smith. Rotterdam: A. A. Balkema.

Bar-Yosef, O. 1992b. Middle Palaeolithic human adaptations in the Mediterranean Levant. In *The Evolution and Dispersal of Modern Humans in Asia*, ed. T. Akazawa, K. Aoki, and T. Kimura. Tokyo: Hokusen-sha.

Bar-Yosef, O. In press. The role of Western Asia in modern human origins. *Philosophical Transactions of the Royal Society*, London.

Bar-Yosef, O., and A. Belfer. 1977. The Lagaman Industry, 23-41. In *Prehistoric Investigations in Gebel Maghara, Northern Sinai*, ed. O. Bar-Yosef and J. L. Phillips. Jerusalem: Institute of Archaeology, Hebrew University.

Bar-Yosef, O., and A. Belfer-Cohen. 1988. The early Upper Palaeolithic in Levantine Caves, 23-41. In *The Early Upper Paleolithic*, ed. J. H. Hoffecker and C. A. Wolf. Oxford: British Archaeological Reports International Series 437.

Bar-Yosef, O., and A. Belfer-Cohen. 1991. From sedentary hunter-gatherers to territorial farmers in the Levant, 181-202. In *Between Bands and States*, ed. S. A. Gregg. Carbondale, IL: Center for Archaeological Investigations.

Bar-Yosef, O., and L. Meignen. 1992. Insights into Levantine Middle Paleolithic cultural variability, 163-82. In *The Middle Paleolithic: Adaptation, Behavior and Variability*, ed. H. L. Dibble and P. Mellars. Philadelphia: The University Museum, University of Pennsylvania.

Bar-Yosef, O., and J. L. Phillips. 1974. Prehistoric sites in Nahal Lavan, Western Negev. *Paléorient* 2:477-82

Bar-Yosef, O., and B. Vandermeersch. 1981. Notes concerning the possible age of the Mousterian Layers at Qafzeh Cave, 555-69. In *Préhistoire du Levant*, ed. J. Cauvin and P. Sanlaville. Paris: Centre National de la Recherche Scientifique.

Beaune, S. A. de. 1989. Essai d'une classification typologique des galets et plaquettes utilises au Paléolithique. *Gallia Préhistoire* 31:27-64.

Belfer-Cohen, A. 1988. The appearance of symbolic expression in the Upper Pleistocene of the Levant as compared to Western Europe, 25-29. In *L'Homme de Néandertal.* Vol. 5, *La Pens'ee*, ed. M. Otte. Liège: Etudes et Recherches Archéologiques de l'Université de Liège.

Bergman, C. A., and C. B. Stringer. 1989. Fifty years after: Egbert, an early Upper Paleolithic juvenile from Ksar Akil, Lebanon. *Paléorient* 15:99-112.

Beyries, S. 1988. Functional variability of lithic sets in the Middle Paleolithic, 213-23. In *Upper Pleistocene Prehistory of Western Eurasia*, ed. H. L. Dibble and A. Montet-White. Philadelphia: The University Museum, University of Pennsylvania.

Binford, L. R. 1989. Isolating the transition to cultural adaptations: An organizational approach, 18-41. In *The Emergence of Modern Humans*, ed. E. Trinkaus. Cambridge: Cambridge University Press.

Binford, L. R. 1991. Review of *The Human Revolution*, edited by P. Mellars and C. B. Stringer. *Journal of Field Archaeology* 18:111-15.

Binford, S. R. 1968. Early Upper Pleistocene adaptations in the Levant. *American Anthropologist* 70:707-17.

Boëda, E. 1988. Approche Technologique du Concept Levallois et Evaluation de son Champ d'Application: Etude de trois gisements Saliens et Weichseliens de la France Septentionale. Ph.D. diss., Université de Paris.

Boëda, E. 1990. De la surface au volume: Analyse de conceptions de débitage Levallois et laminaire, 63-68. In *Paléolithique moyen récent et Paléolithique supérieur ancien en Europe*, ed. C. Farizy. Nemours: Centre National de la Recherche Scientifique.

Boëda, E., J. M. Geneste, and L. Meignen. 1990. Identification de chaines operatoires lithiques du Paleolithique ancien et moyen. *Paléo* 2:43-80.

Bordes, F. 1955. Le Paléolithique inférieur et moyen de Jabrud (Syrie) et la question du Pré-Aurignacien. *L'Anthropologie* 59:486-587.

Bordes, F. 1961. Typologie du Paléolithique Ancien et Moyen. Bordeaux: *Publications de L'Institut de Préhistoire L'Universite de Bordeaux Memoire* 1:1-86.

Bordes, F. 1977. Que sont le Pré-Aurignacien et le Iabroudien? 49-55. In *Moshé Stekelis, Memorial Volume*, ed. B. Arensburg and O. Bar-Yosef. Jerusalem: Israel Exploration Society.

Boutié, P. 1989. Etude technologique de l'industrie Moustérienne de la grotte de Qafzeh près de Nazareth, Israël, 213. In *Investigations in South Levantine Prehistory: Préhistoire du Sud-Levant*, ed. O. Bar-Yosef and B. Vandermeersch. Oxford: British Archaeological Reports International Series 497.

Bowdler, S. 1992. *Homo sapiens* in Southeast Asia and the antipodes: Archaeological versus biological interpretations. In *The Evolution and Dispersal of Modern Humans in Asia*, ed. T. Akazawa, K. Aoki, and T. Kimura. Tokyo: Hokusen-sha.

Bräuer, G. 1989. The evolution of modern humans: Recent evidence from Southwest Asia, 123-54. In *The Human Revolution: Behavioural and Biological Perspectives on the Origins of Modern Humans*, ed. P. Mellars and C. Stringer. Edinburgh: Edinburgh University Press.

Brooks, A. S. 1991. New Human Remains from Ishango, Zaire. Paper given at the Annual Meeting of the American Association of Physical Anthropologists, Milwaukee.

Cann, R., M. Stoneking, and A. C. Wilson. 1987. Mitochondrial DNA and human evolution. *Nature* 325:31-36.

Caton-Thompson, G. 1952. *Kharga Oasis in Prehistory*. London: Athlone Press.

Cavalli-Sforza, L. L., A. Piazza, P. Menozzi, and J. Mountain. 1988. Reconstruction of human evolution: Bringing together genetic, archaeological, and linguistic data. *Proceedings of the National Academy of Sciences, USA* 85:6002-6.

Chase, P. G. 1989. How different was Middle Palaeolithic subsistence? A zooarchaeological perspective on the Middle to Upper Palaeolithic transition, 321-37. In *The Human Revolution: Behavioural and Biological Perspectives on the Origins of Modern Humans*, ed. P. Mellars and C. Stringer. Edinburgh: Edinburgh University Press.

Cherry, J. F. 1990. The first colonization of the Mediterranean Islands: A review of recent research. *Journal of Mediterranean Archaeology* 3:145-221.

Clark, G. A., and J. M. Lindly. 1989. The case for continuity: Observations on the biocultural transition in Europe and Western Asia, 626-76. In *The Human Revolution: Behavioural and Biological Perspectives on the Origins of Modern Humans*, ed. P. Mellars and C. Stringer. Edinburgh: University of Edinburgh Press.

Copeland, L. 1975. *The Middle and Upper Palaeolithic of Lebanon and Syria in the Light of Recent Research*. Dallas: Southern Methodist University Press.

Copeland, L., and F. Hours. 1981. La fin de l'Acheuléen et l'avèntment du Paléolique moyen en Syrie, 225-38. In *La Préhistoire du Levant*, ed. J. Cauvin and P. Sanlaville. Paris: Centre National de la Recherche Scientifique.

Copeland, L., and F. Hours. 1983. Le Yabroudien d'El Kowm (Syrie) et sa place dans le Paléolithique du Levant. *Paléorient* 9:39-54.

Davis, S. M. J., R. Rabinovich, and N. Goren-Inbar. 1988. Population increase in Western Asia: The animal remains from Biq'at Quneitra. *Paléorient* 14:95-106.

Deacon, T. D. 1989. The neural circuity underlying primate calls and human language. *Human Evolution* 4:367-401.

Dibble, H. L. 1984. Interpreting typological variation of Middle Paleolithic scrapers: Function, style, or sequence of reduction? *Journal of Field Archaeology* 11:431-36.

Dibble, H. L., and S. J. Holdaway. 1990. Le Paléolithique moyen de L'Abri Sous Roche de Warwasi et ses relations avec le Moustérien du Zagros et du Levant. *L'Anthropologie* 94:619-42.

Excoffier, L. 1990. Evolution of human mitochondrial DNA: Evidence for departure from a pure neutral model of population at equilibrium. *Journal of Molecular Evolution* 30:125-39.

Farizy, C. 1990. The transition from Middle to Upper Palaeolithic at Arcy-sur-Cure (Yonne, France): Technological, economic and social aspects, 303-26. In *The Emergence of Modern Humans: An Archaeological Perspective*, ed. P. Mellars. Edinburgh: Edinburgh

University Press.
Farrand, W. R. 1971. Late Quaternary Paleoclimates of the Eastern Mediterranean Area. In *The Late Cenozoic Glacial Ages*, ed. K. K. Turekian. New Haven: Yale University Press.
Farrand, W. R. 1979. Chronology and paleoenvironment of Levantine prehistoric sites as seen from sediment studies. *Journal of Archaeological Science* 6:369-92.
Farrand, W. R. 1982. Environmental conditions during the Lower/Middle Paleolithic transition in the Near East and the Balkans, 105-8. In *The Transition from Lower to Middle Paleolithic and the Origin of Modern Man*, ed. A. Ronen. Oxford: British Archaeological Reports International Series 151.
Fleisch, H. 1970. Les habitats du Paléolithic Moyen à Naame (Liban). *Bulletin di Musée de Beyrouth* 23:25-93.
Frayer, D. W. 1984. Biological and cultural change in the European Late Pleistocene and Early Holocene, 211-50. In *The Origins of Modern Humans: A World Survey of the Fossil Evidence*, ed. F. H. Smith and F. Spencer. New York: Alan R. Liss.
Gambier, D. 1989. Fossil hominids from the early Upper Palaeolithic (Aurignacian) of France, 194-211. In *The Human Revolution: Behavioural and Biological Perspectives on the Origins of Modern Humans*, ed. P. Mellars and C. Stringer. Edinburgh: Edinburgh University Press.
Gamble, C. S. 1982. Interaction and alliance in Palaeolithic society. *Man* 17:92-107.
Garrod, D. A. E. 1956. Achéuléo-Jabroudien et "Pré-Aurignacien" de la Grotte du Taboun (Mont Carmel); Etude stratigraphique et chronologique. *Quaternaria* 3:61-70.
Garrod, D. A. E. 1957. Notes sur le Paleolithique superieur du Moyen-Orient. *Bulletin de la Société Préhistorique Française* 54:439-46.
Garrod, D. A. E., and D. M. Bate. 1937. *The Stone Age of Mount Carmel*. Oxford: Clarendon Press.
Geneste, J.-M. 1989. Economie des resources lithiques dans le Moustérien du Sud-Ouest de la France, 75-97. In *L'Homme de Néandertal*. Vol. 6, *La Subsistance*, ed. M. Otte. Liège: Etudes et Recherches Archéologiques de l'Université de Liège.
Gilead, D. 1973. Cleavers in Early Paleolithic industries in Israel. *Paléorient* 1:73-88.
Gilead, I. 1981. Upper Palaeolithic tool assemblages from the Negev and Sinai, 331-42. In *Préhistoire du Levant*, ed. J. Cauvin and P. Sanlaville. Paris: Centre National de la Recherche Scientifique.
Gilead, I. 1983. Upper Palaeolithic occurrences in Sinai and the transition to the Epi-Palaeolithic in the Southern Levant. *Paléorient* 9:39-54.
Gilead, I. 1984a. Farah II: A Middle Palaeolithic open air site in the northern Negev, Israel. *Proceedings of the Prehistoric Society* 50:71-97.
Gilead, I. 1984b. Is the term "Epipalaeolithic" relevant to Levantine prehistory? *Current Anthropology* 25:227-29.
Gilead, I. 1989. The Upper Palaeolithic in the Southern Levant: Periodization and terminology. In *Investigations in South Levantine Prehistory*, ed. O. Bar-Yosef and B. Vandermeersch. Oxford: British Archaeological Reports International Series 497.
Gilead, I. 1991. The Upper Paleolithic period in the Levant. *Journal of World Prehistory* 5:105-54.
Goren-Inbar, N. 1990. *Quneitra: A Mousterian Site on the Golan Heights*. Jerusalem: QEDEM, Monographs of the Institute of Archaeology, Hebrew University.
Goren-Inbar, N. 1992. The Acheulian site of Gesher Benot Ya'aqov–An Asian or an African entity. In *The Evolution and Dispersal of Modern Humans in Asia*, ed. T. Akazawa, K. Aoki, and T. Kimura. Tokyo: Hokusen-sha.
Gorring-Morris, A. N. 1988. Trends in the spatial organization of terminal Pleistocene hunter-gatherer occupations as viewed from the Negev and Sinai. *Paléorient* 14:231-44.
Grün, R., N. J. Shackleton, and H. J. Deacon. 1990. Electron-spin-resonance dating of tooth enamel from Klasies River Mouth Cave. *Current Anthropology* 31:427-32.
Grün, R., and C. B. Stringer. 1991. Electron spin resonance dating and the evolution of mod-

ern humans. *Archaeometry* 33:153-99.
Grün, R., C. B. Stringer, and H. P. Schwarcz. 1991. ESR dating of teeth from Garrod's Tabun Cave collection. *Journal of Human Evolution* 20:231-48.
Harrold, F. B. 1989. Mousterian, Châtelperronian and Early Aurignacian in Western Europe: Continuity or discontinuity? 677-713. In *The Human Revolution: Behavioural and Biological Perspectives on the Origins of Modern Humans*, ed. P. Mellars and C. Stringer. Edinburgh: Edinburgh University Press.
Hennig, G. J., and F. Hours. 1982. Dates pour le passage entre l'Acheuléen et le Paléolithique moyen à El Kowm (Syrie). *Paléorient* 8:81-83.
Henry, D. O. 1986. The prehistory and palaeoenvironments of Jordan: An overview. *Paléorient* 12:5-26.
Henry, D. O., and F. Servello. 1974. Compendium of C^{14} determinations derived from Near Eastern prehistoric sites. *Paléorient* 2:19-44.
Hietala, H. 1983. Boker Tachtit: Intralevel and interlevel spatial analysis. In *Prehistory and Paleoenvironments in the Central Negev, Israel*. Vol. 3. *The Advat/Agav Area*, ed. A. E. Marks. Dallas: Southern Methodist University Press.
Hillman, G. C., and M. E. Davies. 1990. Domestication rates in wild-type wheats and barley under primitive cultivation. *Biological Journal of the Linnaean Society* 39:39-78.
Hillman, G. C., and D. R. Harris. 1989. *Foraging and Farming, The Evolution of Plant Exploitation*. London: Unwin Hyman.
Hours, F. 1983. Quatre colloques récents sur la Préhistoire du Levant. *Paléorient* 9:110-11.
Howell, F. C. 1959. Upper Pleistocene cultural stratigraphy and early man in the Levant. *Proceedings of the American Philosophical Society* 103:1-65.
Howells, W. W. 1976. Explaining modern man: Evolutionists *versus* migrationists. *Journal of Human Evolution* 5:477-95.
Isaac, G. L. 1972. Chronology and tempo of culture change during the Pleistocene, 381-430. In *Calibration in Hominoid Evolution: Recent Advances in Isotopic and Other Dating Methods Applicable to the Origin of Man*, ed. W. W. Bishop and J. Miller. Edinburgh: Scottish Academic Press.
Jelinek, A. J. 1982a. The Middle Palaeolithic in the southern Levant with comments on the appearance of modern *Homo sapiens*, 57-104. In *The Transition from the Lower to the Middle Paleolithic and the Origin of Modern Man*, ed. A. Ronen. Oxford: British Archaeological Reports International Series 151.
Jelinek, A. J. 1982b. The Tabun Cave and the Paleolithic man in the Levant. *Science* 216:1369-75.
Jelinek, A. J. 1990. The Amudian in the context of the Mugharan tradition at the Tabun Cave (Mount Carmel), Israel, 81-90. In *The Emergence of Modern Humans: An Archaeological Perspective*, ed. P. Mellars. Edinburgh: Edinburgh University Press.
Jones, R. 1989. East of Wallace's line: Issues and problems in the colonisation of the Australian continent, 743-82. In *The Human Revolution: Behavioural and Biological Perspectives on the Origins of Modern Humans*, ed. R. Mellars and C. Stringer. Edinburgh: Edinburgh University Press.
Kaplan, J. 1990. The Umlatuzana rock shelter sequence: 100,000 years of Stone Age prehistory, 1-94. In Cape Town: *Natal Journal of Humanities*.
Kislev, M. E., D. Nadel, and I. Carmi. 1992. Epipalaeolithic (19,000 BP) cereal and fruit diet at Ohalo II, Sea of Galilee, Israel. *Review of Palaeobotany and Palynology* 73:161-66.
Klein, R. G. 1989a. Biological and behavioural perspectives on modern human origins in Southern Africa, 529-46. In *The Human Revolution: Behavioural and Biological Perspectives on the Origins of Modern Humans*, ed. P. Mellars and C. Stringer. Edinburgh: Edinburgh University Press.
Klein, R. G. 1989b. *The Human Career*. Chicago: University of Chicago Press.
Kozlowski, J. K. 1988. The transition from the Middle to the early Upper Paleolithic in central Europe and the Balkans, 193-235. In Oxford: British Archaeological Reports International Series 437.

Kozlowski, J. K., H. Laville, and N. Sirakov. 1989. Une nouvelle séquence géologique et archéologique dans les Balkans: La Grotte Temnata à Karloukovo (Bulgarie nord). *L'Anthropologie* 93:159-72.
Leroi-Gourhan, A. 1980. Les analyses pollinques au moyen Orient. *Paléorient* 6:79-91.
Lieberman, P. 1989. The origins of some aspects of human language and cognition, 391-414. In *The Human Revolution: Behavioural and Biological Perspectives on the Origins of Modern Humans*, ed. P. Mellars and C. Stringer. Edinburgh: Edinburgh University Press.
Lindly, J., and G. Clark. 1987. A preliminary lithic analysis of the Mousterian site of 'Ain Difla (WHS Site 634) in the Wadi Ali, West-Central Jordan. *Proceedings of the Prehistoric Society* 53:279-92.
Lindly, J., and G. Clark. 1990. Symbolism and modern human origins. *Current Anthropology* 31:233-62.
Liubin, V. P. 1989. *Palaeolithic of Caucasus*. Leningrad: Academy of Sciences of USSR.
Marks, A. E. 1981. The Middle Paleolithic of the Negev, 287-98. In *Préhistoire du Levant*, ed. J. Cauvin and P. Sanlaville. Paris: Centre National de la Recherche Scientifique.
Marks, A. E. 1983. The Middle to Upper Paleolithic transition in the Levant. *Advances in World Archaeology* 2:51-98.
Marks, A. E. 1990. The Upper and Middle Palaeolithic of the Near East and the Nile Valley: The problem of cultural transformations, 56-80. In *The Emergence of Modern Humans: An Archaeological Perspective*, ed. P. Mellars. Edinburgh: Edinburgh University Press.
Marks, A. E. 1992. Upper Pleistocene archaeology and the origins of modern man: A view from the Levant and adjacent Areas. In *The Evolution and Dispersal of Modern Humans in Asia*, ed. T. Akazawa, K. Aoki, and T. Kimura. Tokyo: Hokusen-sha.
Marks, A. E., and C. R. Ferring. 1988. The Early Upper Paleolithic of the Levant, 43-72. In *The Early Upper Paleolithic: Evidence from Europe and Near East*, ed. J. F. Hoffecker and C. A. Wolf. Oxford: British Archaeological Reports International Series 437.
Marks, A. E., and D. Kaufman. 1983. Boker Tachtit: The artifacts, 69-126. In *Prehistory and Paleoenvironments in the Central Negev, Israel*, ed. A. Marks. Dallas: Southern Methodist University Press.
Marks, A. E., and P. Volkman. 1986. The Mousterian of Ksar Akil. *Paléorient* 12:5-20.
McBurney, C. B. M. 1967. *The Haua Fteah (Cyrenaica) and the Stone Age of the South-East Mediterranean*. Cambridge: Cambridge University Press.
Meignen, L., and O. Bar-Yosef. 1988. Kebara et le Paléolithique moyen du Mont Carmel. *Paléorient* 14:123-30.
Meignen, L., and O. Bar-Yosef. 1991. Les industries Mousterienne de Kebara. In *Une Squelette mousteriénne á Kebara, Mt. Carmel Israel*, ed. O. Bar-Yosef and B. Vandermeersch. Paris: Centre National de la Recherche Scientifique.
Mellars, P. 1989. Major issues in the emergence of modern humans. *Current Anthropology* 30:349-85.
Mercier, N., H. Valladas, J.-L. Joron, J. L. Reyss., F. Lévêque, and B. Vandermeersch. 1991. Thermoluminescence dating of the Late Neanderthal remains from Saint-Césaire. *Nature* 351:737-39.
Moore, A. 1985. The development of Neolithic societies in the Near East, 1-69. In *Advances in World Archaeology*, ed. F. Wendorf and A. E. Close. New York: Academic Press.
Ohnuma, K. 1988. *Ksar Akil, Lebanon: A Technological Study of the Earlier Upper Palaeolithic Levels at Ksar Akil*. Vol. III. *Levels XXV-XIV*. Oxford: British Archaeological Reports International Series 426.
Ohnuma, K., and C. A. Bergman. 1990. A technological analysis of the Upper Palaeolithic levels (XXV-VI) of Ksar Akil, Lebanon, 91-138. In *The Emergence of Modern Humans: An Archaeological Perspective*, ed. P. Mellars. Edinburgh: Edinburgh University Press.
Otte, M. 1990. From the Middle to the Upper Palaeolithic: The nature of the transition, 438-56. In *The Emergence of Modern Humans: An Archaeological Perspective*, ed. P. Mellars. Edinburgh: Edinburgh University Press.

Otte, M., and L. H. Keeley. 1990. The impact of regionalisation on the Palaeolithic studies. *Current Anthropology* 31:577-82.

Palma Di Cesnola, A. 1989. L'Uluzzuen: Faciès Italien du Leptolithique Archaïque. *L'Anthropologie* 93:783-812.

Parkington, J. 1990. A critique of the consensus view on the age of Howieson's Poort assemblages in South Africa, 34-55. In *The Emergence of Modern Humans: An Archaeological Perspective*, ed. P. Mellars. Edinburgh: Edinburgh University Press.

Pelegrin, J. 1990. Observations technologiques sur quelques séries du Châtelperronien et du MTA B du sud-ouest de la France, une hypothèse d'evolution. In *Paléolithique moyen récent et Paléolithique supérieur ancien en Europe,* ed. C. Farizy. Nemours: Mémoires du Musée de Préhistoire Ile de France 3.

Phillips, J. L. 1988. The Upper paleolithic of the Wadi Feiran, Southern Sinai. *Paléorient* 14:183-200.

Phillips, J. L., and B. G. Gladfelter. 1991. The refitting of lithics and site formation processes in the Upper Paleolithic of Southern Sinai. Paper presented at the Annual Meeting of the Society for American Archaeology, New Orleans.

Plisson, H. 1988. Technologie et tracéologie des outils lithiques moustériens en Union Sovietique: les travaux de V. E. Schelinski, 12-168. In *L'Homme Néandertal*, ed. M. Otte. Liège: Etudes et Recherches Archéologiques de l'Université de Liège.

Renfrew, C. 1987. *Archaeology and Language: The Puzzle of Indo-European Origins*. Cambridge: Cambridge University Press.

Roebroeks, W., J. Kolen, and E. Rensink. 1988. Planning depth, anticipation and the organization of Middle Palaeolithic technology: The 'Archaic Natives' meet Eve's descendants. *Helinium* 28:17-34.

Ronen, A. 1979. Paleolithic industries, 296-307. In *The Quaternary of Israel*, ed. A. Horowitz. New York: Academic Press.

Ronen, A. 1984. *Sefunim Prehistoric Sites, Mount Carmel, Israel*. Oxford: British Archaeological Reports International Series 230.

Rust, A. 1950. *Die Höhlenfunde von Jabrud (Syrien)*. Neumunster: K. Wacholtz.

Sanlaville, P. 1977. *Etude Géomorphologique de la Region Littorale du Liban*. 2 Vols. Beyrouth: Université Libanaise.

Sanlaville, P. 1981. Stratigraphie et chronologie du Quartinaire marin du Levant, 21-32. In *Préhistoire du Levant*, ed. J. Cauvin and P. Sanlaville. Paris: Centre National de la Recherche Scientifique.

Schwarcz, H. P., B. Blackwell, P. Goldberg, and A. E. Marks. 1979. Uranium series dating of travertine from archaeological sites, Nahal Zin, Israel. *Nature* 277:558-60.

Schwarcz, H. P., W. M. Buhay, R. Grün, H. Valladas, E. Tchernov, O. Bar-Yosef, and B. Vandermeersch. 1989. ESR dating of the Neanderthal site, Kebara Cave, Israel. *Journal of Archaeological Science* 16:653-59.

Schwarcz, H. P., P. Goldberg, and B. Blackwell. 1980. Uranium series dating of archaeological sites in Israel. *Israel Journal of Earth-Sciences* 29:157-65.

Schwarcz, H. P., R. Grün, B. Vandermeersch, O. Bar-Yosef, H. Valladas, and E. Tchernov. 1988. ESR dates for the hominid burial site of Qafzeh in Israel. *Journal of Human Evolution* 17:733-37.

Shea, J. J. 1988. Spear points from the Middle Paleolithic of the Levant. *Journal of Field Archaeology* 15:441-50.

Shea, J. J. 1989. A functional study of the lithic industries associated with hominid fossils in the Kebara and Qafzeh caves, Israel, 611-25. In *The Human Revolution: Behavioural and Biological Perspectives on the Origins of Modern Humans*, ed. P. Mellars and C. Stringer. Edinburgh: Edinburgh University Press.

Simmons, T., A. B. Falsetti, and F. H. Smith. 1991. Frontal bone morphometrics of southwest Asian Pleistocene hominids. *Journal of Human Evolution* 20:249-69.

Singer, R., and J. J. Wymer. 1982. *The Middle Stone Age at Klasies River Mouth in South Africa*. Chicago: University of Chicago Press.

Skinner, J. 1970. El-Masloukh, a Yabroudian site in Lebanon. *Bulletin du Musée de Beyrouth* 23:143-72.
Smith, F. H., A. B. Falsetti, and S. M. Donnelly. 1989. Modern human origins. *Yearbook of Physical Anthropology* 32:35-68.
Smith, P. E. L. 1988. *Palaeolithic Archaeology in Iran*. Philadelphia: The American Institute of Iranian Studies, The University of Pennsylvania Museum.
Soffer, O. 1989. The Middle to Upper Palaeolithic transition on the Russian plain, 714-42. In *The Human Revolution: Behavioural and Biological Perspectives on the Origins of Modern Humans*, ed. P. Mellars and C. Stringer. Edinburgh: Edinburgh University Press.
Soffer, O., ed. 1987. *The Pleistocene Old World: Regional Perspectives*. New York: Plenum Press.
Sokal, R. R., N. L. Oden, and C. Wilson. 1991. Genetic evidence for the spread of agriculture in Europe by demic diffusion. *Nature* 351:143-45.
Speth, J. D., and K. A. Spielmann. 1983. Energy source, protein metabolism, and hunter-gatherer subsistence strategies. *Journal of Anthropological Archaeology* 2:1-31.
Stekelis, M. 1960. The Paleolithic deposits of Jist Banat Yaqub. *Bulletin of the Research Council of Israel* 9G:61-90.
Stevens, D. S., and H. J. Hietala. 1977. Spatial analysis: Multiple procedures in pattern recognition. *American Antiquity* 42:539-59.
Stiner, M. 1990. The use of mortality patterns in archaeological studies of hominid predatory adaptations. *Journal of Anthropological Archaeology* 9:305-51.
Straus, L. G. 1989. Age of the modern Europeans. *Nature* 342:476-77.
Straus, L. G. 1990. The Early Upper Palaeolithic of Southwest Europe: Cro-Magnon adaptations in the Iberian peripheries, 40 000-20 000 BP, 276-302. In *The Emergence of Modern Humans: An Archaeological Perspective*, ed. P. Mellars. Edinburgh: Edinburgh University Press.
Stringer, C. B. 1989. Documenting the origin of modern humans, 67-96. In *The Emergence of Modern Humans: Biocultural Adaptations in the Late Pleistocene*, ed. E. Trinkaus. Cambridge: Cambridge University Press.
Stringer, C. B., and P. Andrews. 1988. Genetic and fossil evidence for the origin of modern humans. *Science* 239:1263-68.
Stringer, C. B., R. Grün, H. P. Schwarcz, and P. Goldberg. 1989. ESR dates for the hominid burial site of Es Skhul in Israel. *Nature* 338:756-58.
Svoboda, J., and K. Simán. 1989. The Middle-Upper Paleolithic transition in Southeastern Central Europe (Czechoslovakia and Hungary). *Journal of World Prehistory* 3:283-322.
Tchernov, E. 1981. The impact of the post-glacial on the fauna of southwest Asia. In *Contributions to the Environmental History of Southwest Asia.* Wiesbaden: Beihefte zum Tubinger Atlas des Vorderen Orients.
Tchernov, E. 1987. The age of the 'Ubeidiya Formation, an early Pleistocene hominid site in the Jordan Valley, Israel. *Israel Journal of Earth Sciences* 36:3-30.
Tchernov, E. 1988. Biochronology of the Middle Paleolithic and dispersal events of hominids in the Levant, 153-68. In *L'Homme de Néandertal*, ed. M. Otte. Liège: Etudes et Recherches Archéologiques de l'Université de Liège 34.
Thackeray, A. I., and A. J. Kelly. 1988. A technological and typological analysis of Middle Stone Age assemblages antecedent to the Howieson's Poort at Klasies River Main Site. *South African Journal of Science* 43:15-26.
Trinkaus, E. 1984. Western Asia, 251-93. In *The Origins of Modern Humans*, ed. F. H. Smith and F. Spencer. New York: Alan R. Liss.
Trinkaus, E. 1989. The Upper Pleistocene transition, 42-66. In *The Emergence of Modern Humans: Biocultural Adaptations in the Later Pleistocene*, ed. E. Trinkaus. Cambridge: Cambridge University Press.
Valladas, H., J.-L. Joron, G. Valladas, B. Arensburg, O. Bar-Yosef, A. Belfer-Cohen, P. Goldberg, H. Laville, L. Meignen, Y. Rak, E. Tchernov, A.-M. Tillier, and B. Vandermeersch. 1987. Thermoluminescence dates for the Neanderthal burial site at Kebara

in Israel. *Nature* 330:159-60.

Valladas, H., J. L. Reyss, J. Joron, G. Valladas, O. Bar-Yosef, and B. Vandermeersch. 1988. Thermoluminescence dating of Mousterian 'Proto-Cro-Magnon' remains from Israel and the origin of modern man. *Nature* 331:614-16.

van der Plicht, J., A. Van der Wijk, and G.-J. Bartstra. 1989. Uranium and thorium in fossil bones: Activity ratios and dating. *Applied Geochemistry* 4:339-42.

Van Peer, P. 1991. Interassemblage variability and Levallois styles: The case of the Northern African Middle Palaeolithic. *Journal of Anthropological Archaeology* 10:107-51.

van Zeist, W. 1988. Some aspects of Early Neolithic plant husbandry in the Near East. *Anatolica* 15:49-68.

Vandermeersch, B. 1981. *Les Hommes Fossiles de Qafzeh (Israel)*. Paris: Centre National de la Recherche Scientifique.

Vandermeersch, B. 1989. L'extinction des Néandertaliens, 11-21. In *L'Homme de Néandertal*, ed. M. Otte. Liège: Etudes et Recherches Archéologiques de l'Université de Liège 34.

Vandermeersch, B. 1990. Réflexions d'un anthropologie à propos de la transition Mousterian Paléolithique Supérieur, 25-28. In *Paléolithique moyen récent et Paléolithique superieur ancien en Europe*, ed. C. Farizy. Nemours: Centre National de la Recherche Scientifique.

Vermeersch, P. M., E. Paulissen, M. Otte, G. Gijselings, and D. Drappier. 1988. Middle Paleolithic in the Egyptian Nile Valley. *Paléorient* 14:245-52.

Volkman, P. W., and D. Kaufman. 1983. A reassessment of the Emireh point as a possible type fossil for the technological shift from the Middle to the Upper Paleolithic in the Levant, 35-51. In *The Mousterian Legacy, Human Biocultural Change in the Upper Pleistocene*, ed. E. Trinkaus. Oxford: British Archaeological Reports International Series 164.

Weinstein, J. M. 1984. Radiocarbon dating in the Southern Levant. *Radiocarbon* 26:297-366.

Wendorf, F., and R. Schild. 1980. *Prehistory of the Eastern Sahara*. New York: Academic Press.

White, L. A. 1959. *The Evolution of Culture*. New York: McGraw-Hill.

White, R. 1989. Production complexity and standardization in early Aurignacian bead and pendant manufacture: Evolutionary implications, 366-90. In *The Human Revolution: Behavioural and Biological Perspectives on the Origins of Modern Humans*, ed. P. Mellars and C. Stringer. Edinburgh: Edinburgh University Press.

Wolpoff, M. H. 1980. *Paleoanthropology*. New York: Alfred A. Knopf.

Wolpoff, M. H. 1989. Multiregional evolution: The fossil alternative to Eden, 62-108. In *The Human Revolution: Behavioural and Biological Perspectives on the Origins of Modern Humans*, P. Mellars and C. Stringer. Edinburgh: Edinburgh University Press.

Wright, K. 1991. The origins and development of ground stone assemblages in late Pleistocene Southwest Asia. *Paléorient* 17:19-45.

Yalçinkaya, I. 1981. Le Paléolithic Inférior de Turquie, 207-18. In *Préhistoire du Levant*, ed. J. Cauvin and P. Sanlaville. Paris: Centre National de la Recherche Scientifique.

Yalçinkaya, I. 1988. Resultats recents des Fouilles a Karain en Anatolie. *L'Homme de Neandertal* 8:257-71.

Young, T. C., P. E. L. Smith, and P. Mortensen, eds. 1983. *The Hilly Flanks and Beyond: Essays on the Prehistory of Southwestern Asia*. Chicago: University of Chicago Press.

Zubrow, E. 1989. The demographic modelling of Neanderthal extinction, 212-31. In *The Human Revolution: Behavioural and Biological Perspectives on the Origins of Modern Humans*, ed. P. Mellars and C. Stringer. Edinburgh: Edinburgh University Press.

Chapter 3

Hominids, Energy, Environment, and Behavior in the Late Pleistocene

ARTHUR J. JELINEK

From the first recognition of Western European Neandertals as physically distinct from present-day *Homo sapiens*, the significance of these differences for the relationships between the two kinds of hominids has been a topic of continuing debate among students of human evolution. The nature and substance of these arguments have been described in detail elsewhere (e.g., Klein 1989; Trinkaus and Howells 1979) and will not be treated extensively here. Some of the basic questions around which the discussions have centered concern possible differences in intellectual and physical abilities, and the genetic relationships of the two forms. Despite the length of time these questions have been considered, there is still relatively little general agreement on the interpretation of this Late Pleistocene fossil and cultural evidence. The discussion below is an attempt to integrate some of the cultural, biological, and paleoenvironmental evidence in an interpretive framework that emphasizes the kinds of potentially meaningful relationships among these variables that may take us closer to answers to the basic questions of the relationships between Neandertals and early examples of hominids like ourselves.

It is important to consider, through this discussion, how deeply the concepts of human biological and cultural development in the literature have been shaped by the relative abundance of evidence from Western Europe and, to a lesser extent, Southwest Asia. This accident of good preservation of human fossils and artifacts, in the presence of the emerging scientific curi-

ARTHUR J. JELINEK • Department of Anthropology, University of Arizona, Tucson, AZ 85721
Origins of Anatomically Modern Humans, edited by Matthew H. Nitecki and Doris V. Nitecki. Plenum Press, New York, 1994.

osity associated with the development of industrial civilization in the nineteenth and early twentieth century, led to broad generalizations about cultural and biological development based almost entirely on the Western European evidence. These ideas continue to influence our global interpretations, albeit in sometimes subtle ways. We can consider what our present general scheme of human development would be like had industrial civilization first developed in north China, Java, or southern Africa, with only a few curious (and obviously aberrant!) human fossils and stone industries known from an undeveloped Western Europe. Nonetheless, the fact remains that Western Europe and the Levant *have* produced the bulk of our present evidence (and questions) bearing on Late Pleistocene human development. So, despite the perhaps atypical nature of this record when compared to that from the rest of the Old World, this is the evidence that is now most likely to allow some restricted degree of generalization and interpretation. In fact, it may be that the peculiar nature of the extreme conditions under which some of this evidence accumulated will allow insights that would not be as evident in other regions of the world. After all, only when we can compare and understand a variety of contrasting regional evidence in some depth will we be able to begin to generalize about global developments in this period.

THE CHRONOLOGICAL AND ENVIRONMENTAL BACKGROUND

The most useful scale to employ in setting the time limits of the materials under discussion here is the deep-sea core oxygen-isotope chronology for major episodes of ice accumulation (Shackleton 1987; Martinson et al. 1987). This scale has been widely confirmed by independent tests and reflects a reliable succession of major episodes of ice accumulation and melting, which presumably relate to global temperature conditions that can be calibrated in time when lag-time in sea water depletion and replenishment of ^{16}O are considered. This evidence shows a relatively warm period from about 125 kyr (thousand years ago) to 80 kyr (stage 5), punctuated by two pronounced cold intervals at about 110 kyr (stage 5b) and 90 kyr (stage 5d). It is, however, unlikely that global temperatures approached present conditions between late stage 5e (ca. 112 kyr) and late stage 2 at about 12 kyr. Stage 4, at about 75-60 kyr, represents a major cold interval, and, while stage 3 (ca. 60-24 kyr) is seen as a relatively warmer period, particularly from about 60-45 kyr, ice accumulations remained at a very high level through the entire episode. Confirming climatic evidence, reflecting terrestrial conditions in several regions, has been produced in the form of floral changes as seen in long pollen records, of which that from Grand Pile is most appropriate to this discussion (Woillard and Mook 1982).

The problem of the absolute dating of individual archaeological discoveries and contexts within this sequence of climatic change is particularly frustrating. The widespread deep interest in this period has led to an intense desire for an absolute time frame. Although many dates have been made available for this period by radiocarbon, and the still largely experimental thermoluminescence, electron-spin-resonance, and uranium decay series techniques, all of these results should be treated with considerable caution. Most of the relevant evidence lies beyond the effective range of present radiocarbon techniques, and the control of variables that affect the results of the other techniques has yet to be convincingly demonstrated. The result has been the release of numerous age determinations that are at odds with paleoclimatic and stratigraphic evidence (Jelinek 1992). The fact that some of these dates have been widely promulgated in the secondary literature before they have been adequately confirmed has only added to the confusion. We simply do not yet know the absolute age (or even the relative age) of most of the hominid fossils, even though we would very much like to. In the present discussion the absolute chronological placement of materials, except for the most recent evidence within the range of reliable radiocarbon dating, is generally restricted to the most probable context on the basis of stratigraphic evidence and the paleoenvironmental record.

CURRENT ARCHAEOLOGICAL EVIDENCE AND THE GENERAL PROBLEM

There is now sufficient stratigraphic, paleontological, and cultural evidence from Western Europe and the Levant to allow us to reconstruct some basic relationships between the hominids and Paleolithic industries of the early Late Pleistocene within regional paleoenvironmental contexts. Much of this evidence has been compiled and discussed recently in the contributions to a series of major symposia and conferences on these topics (Dibble and Montet-White 1988; Otte 1988; Mellars and Stringer 1989; Trinkaus 1989b; Mellars 1990; Farizy 1990c). The extent of this literature, published in the brief period of a few years, is in itself a reflection of the recent intense professional interest in the problems of this aspect of prehistory. The picture that emerges from our current knowledge can be summarized as follows.

WESTERN EUROPE – THE BIOLOGICAL EVIDENCE

Here the most extreme examples of the Neandertals are found, with virtually all characteristic physical features well developed. These include the cranial features of a markedly projecting face, large anterior teeth, large nasal aperture, large orbits, marked supraorbital torus, low frontal region, occipital bun and suprainiac depression, mandibular retromolar gap and sloping symphysis, and heavily developed nuchal muscular attachments. The peculiar postcranial features include generally heavy muscularity, especially in the cervical region, shoulder girdle, limb bones, and hands, also reflected in a marked robusticity of the diaphyses of the limb bones with thick cortical bone, heavy muscle attachments and large epiphyses, a relatively enormous chest cavity, and an elongation of the pubic bone in the pelvis. In addition, low crural and brachial indices reflect short distal long bones relative to the femur and humerus.

In a landmark paper written almost 40 years ago, Howell (1952) pointed out the probability that many of the extreme features that characterize the Western European Neandertals may have resulted from a period of near genetic isolation of a small population of hominids under heavy selective stress in the periglacial environment that accompanied the early last glacial advance. The implications of this interpretation for the probably unique position of Western European Neandertals in a global view of human development are still not widely appreciated.

In a recent review of interpretations of Neandertal physical characteristics, Trinkaus (1989a) has considered the possible reflections of cold-adaptation in skeletal morphology. The only evidence that he finds convincing relates to the low crural and brachial indices, which parallel those of modern arctic-adapted peoples (Trinkaus 1989a:61). In his comparison of other characteristic Neandertal skeletal morphology with that of hominids more similar to present *Homo sapiens*, he sees evidence of a greater use of a power grip, as opposed to more precise manipulation, in the hand structure; a more powerful, but more restricted, use of the arm and shoulder; and evidence of use of the lower limbs for markedly more prolonged locomotion, that may have involved more irregular movement (over irregular terrain?) and less consistent directional striding than characterizes anatomically modern *Homo sapiens* (however, see Wolpoff 1989:123-24 for alternative interpretations). In the facial skeleton, he interprets the very large and projecting nasal aperture as a possible adaptation to minimize moisture loss and *expel excess heat* from energetic activity in the cold dry periglacial environment (Trinkaus 1989a:56-57). In agreement with most other current workers, he sees the massively projecting face and large anterior dentition as a response to heavy loading for paramasticatory (nondietary) tasks. Such habitual use appears to be confirmed in a number of older individuals by heavy wear on the anterior dentition. The heavy nuchal musculature is interpreted as linked to this activity and to compensating for the projecting face at the front of the skull.

Beyond the distinctive morphological features that differentiate the

Neandertals from *Homo sapiens* similar to present-day populations, skeletal evidence has revealed several other aspects of Neandertal activities and life. Trinkaus (1989a) cites more frequent evidence of trauma in the bones of Neandertals and evidence for lower extreme ages for Neandertals (30s to mid-40s as opposed to 50s). Considering the apparent shorter maximal life span, the greater evidence of trauma suggests a markedly more hazardous life for the Neandertals. The presence of frequent periods of severe nutritional stress up to maturity (and presumably subsequently) for the Neandertals, comparable to recent human extremes, is shown by a recent study of dental enamel hypoplasia (Ogilvie, Curran & Trinkaus 1989). Altogether, this evidence suggests a significantly more stressful existence for Neandertals than for structurally more modern populations.

While it is generally accepted that the hominids who produced the early Aurignacian industries of Western Europe which stratigraphically follow the industries associated with Neandertals were similar to modern populations of *Homo sapiens*, this assumption is based to a considerable degree on an acceptance of the traditional association of the skeletons from the Cro-Magnon shelter with (Late) Aurignacian artifacts from that site. It is important to note that there is still little firm evidence for the morphological characteristics of the earliest Western European successors of the Neandertals. The situation is summarized by Gambier (1989), who concludes that, while the few fragmentary remains that can be most reliably associated with the early Aurignacian are essentially modern in form, they exhibit greater robusticity than is characteristic of present populations and occasional archaic traits. While this evidence appears to favor a replacement (as opposed to a continuity) model for the Paleolithic hominids in the region, the poverty of specimens that can be firmly placed in this period, and their somewhat ambiguous morphology, severely limits reliable conclusions in this regard.

WESTERN EUROPE – THE CULTURAL EVIDENCE

The hominid remains in Western Europe that possess the distinctive Neandertal features are primarily associated with Mousterian (Middle Paleolithic) lithic industries, which first appear in the region prior to stage 5e and continue into early stage 3. These industries are dominated by two major kinds of tools, scrapers (*racloirs*) and denticulates, which seem to be to some extent mutually exclusive in abundance (e.g., Jelinek 1988:202). Other aspects of the technology and typology include varying frequencies of Levallois flake manufacture, biface (hand ax) manufacture, and the production of relatively low frequencies of other flake tool categories, such as points and blade tools. Patterns in the association of relative frequencies of the tool forms, and the presence or absence of biface production, form the basis for distinguishing between the four major recognized industrial variants separated by Bordes (1953, 1961).

The significance of the differences in these tool frequencies has been the focus of considerable discussion (e.g., Bordes 1961, 1981; Binford and Binford 1966; Mellars 1970; Rolland 1981; Dibble 1987; Jelinek 1988; Rolland and Dibble 1990) that has suggested underlying causes ranging from contemporary ethnic (i.e., stylistic) differences (Bordes), to functional preferences (Binford and Binford), to evolutionary change (also implying stylistic change) (Mellars), to material access and intensity of utilization (Rolland, Dibble, Jelinek). The weight of the current evidence appears to favor the presence of a single Mousterian industry, whose particular expression in archaeological sites was differentially influenced by the above factors. It now seems likely that combinations of all of these factors may have contributed to the differences in tool frequencies to varying degrees at different times and places, and it seems unlikely that any one of them was responsible for all of the observed differences. Little evidence of deliberately shaped bone (other than chipped pieces) is found in the Mousterian, although unmodified bone pieces used for percussion flaking of stone are not uncommon.

The Mousterian industries contrast with the succeeding early Upper Paleolithic Aurignacian industries, which are generally characterized by relatively large numbers of tools manufactured on prismatic blades, such as burins and end scrapers, generally few flake tools of Mousterian type, and the presence of symmetrical bone tools shaped by carving and abrading, frequently in the form of points and pointed pieces. These utilitarian assemblages are associated with clear evidence of personal ornament in the form of shaped and perforated beads and pendants of bone, ivory, and animal teeth (White 1989). These last kinds of artifacts are virtually absent in Mousterian contexts and have been taken as evidence of a qualitatively different social environment in the Upper Paleolithic, as has the existence of portable and mural representational art, which also has not been found in the Mousterian (Chase and Dibble 1987). Another difference in the Upper Paleolithic, with similar social/conceptual implications, is what has been described as a greater level of standardization in artifact manufacture than characterizes the Middle Paleolithic, and the ways in which this seems linked to successions of distinctive tool forms that may reflect socially significant stylistic differences (or restricted socially acceptable patterns) in artifact manufacture through time and across space. Finally, there appear to be significant differences in the spatial organization and use of occupation surfaces in the Upper Paleolithic, involving the segregation of artifact categories, surfaces cleared of refuse, and more clearly structured hearths, although as Rigaud (1989:150) has pointed out, this kind of evidence is not yet clearly described for the earliest Aurignacian.

An exception to an exclusive association of Neandertal fossils with Middle Paleolithic Mousterian industries is the find at Saint-Césaire of an apparent burial of a Neandertal associated with a Châtelperronian industry (Lévêque and Vandermeersch 1981). This discovery confirmed a postulated earlier association of Neandertal and Châtelperronian at the Grotte du Renne, based on isolated hominid teeth (Leroi-Gourhan 1958). Although Châtelperronian assemblages frequently contain a strong representation of characteristic Mousterian scrapers and denticulates, this industry is generally classified as Upper Paleolithic (Farizy 1990a, 1990b; Harrold 1989:678-79) on the typological evidence of significant frequencies of blade tools, as well as the not infrequent presence of shaped bone artifacts. It occurs stratigraphically only above the Mousterian. While the generic relationships between the Mousterian and the subsequent Châtelperronian have been treated as equivocal (Farizy 1990a, 1990b; Harrold 1989), recent excavations at the cave of the Grande Roche de la Plématerie (see Lévêque and Miskovsky 1983) appear to provide a convincing demonstration of a gradual transition through an impressive sequence of deposits there. Whether this transition is best explained as an in situ evolution or as a gradual acculturation through contact with early Aurignacian Upper Paleolithic peoples, whose contemporaneity with the Châtelperronian seems demonstrated by interstratified deposits of the two industries at Roc de Combe, Le Piage, and El Pendo, may be better understood when the site is fully published. It may also be of some interest here that the Châtelperronian occupations at Grotte du Renne appear to show spatial patterning more characteristic of the Upper rather than the Middle Paleolithic (Farizy 1990a).

SOUTHWEST ASIA – THE BIOCULTURAL EVIDENCE

Evidence from this region contrasts to a significant degree with that from Western Europe. The industrial sequence has recently been described by Bar-Yosef (1989, this volume) and will not be treated in detail here. The hominid evidence, aside from the Shanidar Neandertals, has virtually all been derived from a tiny geographical area (< 2500 km^2) within the borders of northern Israel in the southern Levant. Here the six sites of Zuttiyeh, Tabun, Skhul, Qafzeh, Amud, and Kebara have produced human remains associated, or probably associated (Zuttiyeh), with Middle Paleolithic industries. The most problematic of the Levantine fossil hominids is the Zuttiyeh partial cranium, recovered during a poorly controlled excavation from deposits that probably contained an early Middle Paleolithic Mugharan industry (Gisis and Bar-Yosef 1974; Jelinek 1981). It is generally accepted as significantly earlier than the other hominids and has been described as Archaic *Homo sapiens* rather than Neandertal (Vandermeersch 1989:160-62). The morphology of the other hominids ranges from robust Neandertal (Tabun, Amud, Kebara), to less robust archaic (Skhul), to fully modern (Qafzeh). All of these forms are clearly, or probably (Tabun), associated with a similar Levantine Mousterian industry (Tabun type B/C) that is characterized by the frequent production of broad Levallois flakes and generally short, wide Levallois points. This suggests that the hominids from these five sites shared a single basic technological system. In all of the described stratigraphic sequences throughout the Levant that contain this industry, and show evidence of Middle Paleolithic industrial change, it is chronologically the latest. The questions concerning age estimates for the Levantine sequence have been recently treated elsewhere (Jelinek 1992). The Tabun B/C industry may have been present immediately following the last high sea stand at Ras-el-Kelb in coastal Lebanon (stage 5a?), and its association with a warm fauna at Qafzeh may also point to an age as early as stage 5a. On the other hand, local environmental circumstances, such as differential uplift on the Lebanese coast, and differential relict faunal survival in a buffered habitat near Qafzeh, could produce misleading evidence suggesting an association with this stage. The terminal date for this industry is equally questionable; all that can be said at present is that it seems to lie beyond the effective range of radiocarbon dating (at least as old as early stage 3). In sum, the evidence suggests a general contemporaneity between the hominids from the five sites in Israel and the Western European Neandertals, but a more specific time relationship cannot yet be established with confidence.

The evidence for a Middle to Upper Paleolithic transition in this region is best described by Marks (1983) from the site of Boker Tachtit, and is not associated with any hominid remains. The industrial change appears to fall in midstage 3, perhaps around 45 kyr, and is signaled primarily by a change from the bidirectional flaking techniques that characterize the Negev Mousterian (essentially Tabun type "D") to the unidirectional production of prismatic blades.

The only other site in Southwest Asia to produce significant hominid remains relevant to this discussion is the Shanidar Cave in the Zagros foothills of northeastern Iraq (Solecki 1963). While the industry has never been fully described, it is clearly similar to the Zagros Mousterian described from other sites (Dibble 1984), and differs significantly from the industries of the Levant in the intense reduction of tools down to very small discarded pieces. The hominids from Shanidar have been described by Trinkaus (1983), and constitute the largest group of Neandertal fossils from a single site yet published in detail. They exist, however, in an isolated context, with little firm evidence of chronological placement or links with a well-defined cultural or biological succession in the region. They serve as examples of what can and did happen to Neandertals that periodically made use of the large cave in a relatively rugged terrain, under conditions that were probably colder and more rigorous than the present.

The Southwest Asian Neandertals have at one time or another been remarked upon as differing to some extent from the Western European group in their less pronounced development of the distinctive features that characterize the latter (Trinkaus 1981). Whether this is an indication of lower selective pressures or a wider gene pool remains an unresolved issue.

WHY WERE THERE NEANDERTALS?

What this question is really asking is whether there is a reasonable explanation for the association of morphological features that characterize Neandertals. When Trinkaus posed this same question in a recent symposium contribution, his answer involved the posing of another question, "What were they doing with their teeth that neither non-Neandertal archaic *H. sapiens* nor anatomically modern humans habitually did?" (Trinkaus 1988:24). He thus saw only the facial structure, and primarily the use of the anterior dentition, as distinctively Neandertal. He did not remark on the enormous nasal aperture in this regard other than to state that its *projection* was a reflection of ancestral features (Trinkaus 1988:17) and did not remark on the large orbits at all, perhaps assuming that their presence was an allometric response to the enlargement of other facial features. The size (both height and width) of the nasal aperture of Neandertals falls near or beyond the upper limits of earlier and later hominids and may well represent a distinctive Neandertal feature, albeit to some extent related structurally to the loading on the projecting dentition. While the size difference of the orbits of Neandertals with other populations is less marked, they, too, lie well above most other earlier and later specimens for which data are available.

In his discussion, Trinkaus dismissed most of the postcranial evidence of robusticity as already somewhat reduced from more archaic levels and, therefore, not a distinctive direction in Neandertal development. It is particularly difficult to understand the basis for this judgment as regards femoral robusticity, since the sample of *Homo erectus* specimens suitable for calculation of an index of robusticity, which compares length with midshaft dimensions, is almost nil. Given the recent suggestion that the thickness of cortical bone on the femora of early *Homo erectus* may be more related to nutrition than stress (Kennedy 1983, 1984), the interpretation of this aspect of robusticity reduction may be open to some question. Trinkaus did not consider the relatively short distal limb proportions as relevant to this question either, since he saw them as an aspect of "biogeographical patterning and thermal adaptation characteristic of all human groups." How this removes them from relevance in considering the combination of characters that typify Neandertal morphology is not apparent. Thus, I am not sure that it is a good idea to dismiss such features as nasal aperture and orbital size, or the index of femoral robusticity, or the low crural and brachial indices as unimportant or irrelevant to an understanding of "Why were there Neandertals?" All of these features may well have some role in an explanation of the presence of the Neandertals and their behavior.

WHAT WAS THE WORLD OF THE NEANDERTALS OF WESTERN EUROPE LIKE?

There is little question that the Mousterian industries and their Neandertal producers represent a successful adaptation to the cold-temperate-to-periglacial conditions of Western Europe through some 70 to 80 kyr of the early Late Pleistocene. This is a time interval equivalent to approximately twice the length that their successors have occupied this region. They were, however, not the first hominids to survive in this region under rigorous periglacial conditions. Such sites as Combe Grenal, the Abri Vaufrey, and La Chaise all show good evidence of a human presence under very cold conditions equivalent to stage 6. The relatively short interval (ca. 10 kyr) that separated stage 6 from 5d gave only a brief respite from colder conditions than those at present, and it is selection through this longer history of gradually successful hominid adaptation to these conditions that ultimately produced the biological creature that was fully Neandertal.

There seems little doubt that hominids were successfully killing and butchering medium-sized herd animals by the end of stage 6; the dominant herbivore at Combe Grenal in the Riss III layers is reindeer (Bordes and Prat 1965). An even greater dominance of reindeer is present in the stage 3 deposits of Bed 8 at La Quina, where over 90% of the identifiable bone, from a dense layer of this material, is assignable to *Rangifer*. So, if the Neandertals of La Quina were so successful at hunting reindeer, why weren't they more like the Magdalenians than like the Acheulians? Was it because they lacked an ability to plan in depth (Binford 1989:19)? The recent discovery of a clear excavated pit covered with 90 kg of large rocks in the upper deposits at La Quina, and the presence of other enigmatic pits at Combe Grenal (Bordes 1972), La Ferrassie, and Pech de l'Azé, all suggest that the Neandertals who inhabited these sites might have had some system of storage that would indicate that they were thinking ahead in terms of time and space. In fact, it is highly unlikely that they could have survived in the environment in which they are found, and with the equipment that we can assign to them, without some significant degree of anticipation of where and when food and shelter might most likely be available. Was it random accident that brought them back repeatedly over many years to the same spots, away from kill sites, to leave their food remains and discarded tools? Probably not. Why, then, if they were capable of a significant degree of planning, were they not more like the Upper Paleolithic Perigordians, or Solutreans?

One thing that the biological evidence indicates is that life was not easy for the Neandertals. They show a higher incidence of physical trauma than later people, and an extreme level of what appears to be periodic nutritional stress. Even the most aged individuals appear not to have survived much beyond 40 years, and the great majority seem to have died at least a decade earlier. So, the successful adaptation of the Neandertals appears to have included a shorter and more difficult life than that of later Upper Paleolithic hunters, but not so short and difficult as to prevent them from surviving as a biological population.

That some system of social support was present seems evident in the prolonged existence of such disabled individuals as the "old man" of La Chapelle-aux-Saints (probably mid-30s maximum; Trinkaus and Thompson 1987:127). If life was so brutish, why were such apparently nonproductive individuals kept alive? altruism and humanitarian instincts? Among modern hunter-gatherers under stress, the literature is replete with examples of aged or disabled nonproductive individuals being abandoned, or voluntarily drifting away. The Neandertal evidence, then, may suggest that some of these physically disabled individuals were not seen as nonproductive, but that they were able to contribute some significant service, perhaps in essential maintenance tasks for which other individuals had little time, or in attending children too small for the more intense activities of prolonged food procurement trips, all of which might have enhanced the chances of survival of the group. Altruism need not have played a significant role in the continuing support of partially functional individuals in groups so small, and with so many needs that in lean periods every functioning hand might mean the difference between survival and extinction.

The skeletal evidence may allow some additional inferences (perhaps "speculations" is a more proper term at this point) regarding the nature of the physical activities involved in food procurement and survival by the Neandertals. It is evident that the Neandertals, including men, women, and children, were endowed with a greater muscle mass than their later Upper Paleolithic counterparts. The extent to which they, at least seasonally, also transported significant quantities of body fat (a relatively efficient insulation, as well as a source of reserve calories) is at present unknown, although perhaps clues might eventually be found in some muscle development. *If* the robusticity of the femur is a reflection of both load-bearing and prolonged activity in transporting this relatively great body mass, the consumption of a significant amount of energy, in the form of food calories, is implied for each of these individuals. This energy would have been derived through metabolism, fed by calorie-rich food, presumably in the form of large quantities of animal fat, marrow, brains, etc., and oxygen. Now how much greater than for more gracile hominids were the demands for oxygen intake to burn sufficient caloric fuel to simply maintain the Neandertal muscle system, let alone produce sufficient heat in the muscles to maintain a viable core temperature in a periglacial environment with a lower level of effective cultural buffering than was available to Upper Paleolithic hominids? Might not these factors explain several of the anatomical peculiarities of the Neandertals? It would have been a considerable advantage under cold conditions, in the absence of adequate artificial or natural insulation, to maintain prolonged physical activity (walking or trotting) to employ the massive muscle system in heat production (Shephard 1978:49, 79; Åstrand and Rodahl 1977:541, 547). Might not the relatively large nasal aperture have functioned more efficiently as an oxygen-intake/CO_2-exhaust valve for such a system, with its relatively enormous lung capacity, than would a smaller nose, or more heat-wasteful mouth-breathing? Perhaps what this leads to is whether, given their body structure and energy needs, Neandertals could have survived with smaller noses.

In any event, it seems obvious that, even without a more prolonged schedule of activity, the Western European Neandertals would have required significantly more calories per capita simply to keep those great masses of muscle and bone in operation than would more gracile hominids (Klein 1989:330). If Trinkaus is correct about the regular, more prolonged diurnal physical schedule (not unknown for occasional periods of activity among recent arctic hunters; Shephard 1978:79), then how many additional calories would be required to keep each individual going, how much useful animal biomass within individual accessible daily range was available to the (apparently limited) hunting technology of the Neandertals, and what are the implications of these figures for group size and the amount of time that could be spent in campsites as opposed to in movement? Might the smaller size of Neandertal hunting groups and the shorter average time spent in camps help to explain the apparent differences in degree of preparation of their habitation surfaces that distinguish them from those of later hunters? If the sites that survive in natural shelters were primarily winter habitations, occupied during periods of greatest stress, might we not expect to find more evidence of surface preparation and segregation of activities in the open sites occupied during less stressful seasons, if we can discover such sites undamaged by cryoturbation? Or were these people so frequently on the move that most of their campsites will show relatively random distributions of material?

Also pertinent to the implications of the suggested prolonged period of diurnal activity is the question of whether it is not possible that the greater size of the orbits could relate to a selection for more acute vision in periods of reduced light early and late in the day. Does the allometric response to the projection of the lower face require such large orbits, or does their expansion exceed expected effects in this regard? Did the supraorbital torus develop as an overhanging protection for these enlarged orbits?

Finally, in this regard, if the osteological interpretations that imply more prolonged activity and significantly higher metabolic rates are correct, might this more intensive use of the biological system not have resulted in relatively premature aging and mortality with respect to the more gracile hominids? Is this a partial explanation for the current postulated differences in maximum age reported by Trinkaus and Thompson (1987)?

Much of our ability to infer the extent to which the Neandertals were forced to rely on extreme biological adaptation to survive in periglacial Western Europe depends on our ability to answer questions about the effectiveness of the technologically produced insulation from the cold environment that was available to them. That they controlled fire to some significant extent is not in doubt; it is unlikely that their predecessors in midlatitude Europe could have survived without some control of fire, even under the most favorable conditions. The presence, in the deepest postinterglacial layers at Combe Grenal, of localized ash deposits and concentrated patches of fire-produced oxidation on the bedrock offers one example of strong direct support for the control of fire from early in the last glacial cycle. While most Neandertal hearths were less effectively structured than those of the Upper Paleolithic (Rigaud 1989:150), the presence of a deliberate hearth at the Grotte du Bison at Arcy-sur-Cure that consisted of "an ashey grey oval area (20 x 30 cm in diameter) containing bone fragments and burnt stones enclosed quite regularly by a circle of blackened and heated blocks" (Farizy 1990b:305), attests to the fact that, on occasion, they could be well-defined and effective heating structures. The most pressing question, for which there is at present no strong supporting evidence, is the effectiveness of clothing and artifical shelter available to the Neandertals. The apparent structure in Lazaret (de Lumley 1969) in the late penultimate glacial period (stage 6) appears to demonstrate that this kind of technology was known in Western Europe by that time. The single posthole at Combe Grenal and the low wall at Pech de l'Azé (Bordes 1972) suggest the supplementary use of artificial shelters at the mouths of caves and rock shelters by Neandertals. The presence of many open-air sites from this period, most of which have been badly disturbed by cryoturbation, also indicates that some form of artificial shelter was available, although how extensive or effective it was, there is no way to say.

Our knowledge of the extent to which effective clothing was available to the Neandertals depends most directly on interpretations of use-wear on their stone tools (e.g., Beyries 1988; Anderson-Gerfaud 1990). While these studies indicate that hide-scraping was represented among the activities detected, it does not appear to have been a predominant activity; woodworking is most commonly in evidence (but see cautions by Vaughn [1985] in this regard). From the animals represented in the faunal remains in Neandertal sites during colder periods, it appears that some of the most efficient fur-bearing animal hides for insulation, i.e., *Rangifer* and *Alopex* (Scholander et al. 1950), were available in some abundance. However, in the absence of evidence of tools other than scrapers and a few denticulates that appear to have been used on hides, or the presence of other tools suitable for hide-working, it is difficult to envision any very complex use of these materials by softening and sewing into effective garments. Despite the occasional suggestion that denticulates might have been employed as sinew frayers, no clear evidence of this employment has yet been reported in the use-wear studies. So that, while hides do appear to have been used by Neandertals, the effectiveness of their employment for personal insulation and structural components may have been limited.

The effectiveness of the hunting technology of the Neandertals was a major determinant of the amount of animal biomass that was available to satisfy individual caloric needs in any particular environment. Although there is now some strong evidence to support the presence of hafting on some Mousterian scraping and/or cutting tools, there is as yet no strong evidence from the Western European Middle Paleolithic for the use of chipped stone points on projectiles or on thrusting spears (Anderson-Gerfaud 1990). The abundance of use-wear related to woodworking, including apparent shaft-scraping, does suggest the probable use of some kind of sharpened spear, as does the well-known Lehringen discovery from an earlier time period (Movius 1950). The Neandertals' habitual use of such a weapon may have been more restricted to thrusting rather than throwing if Trinkaus's (1989a:48-50) interpretation of the functional morphology of the arm and hand is correct. It appears, from their morphology and technology, that Neandertal hunting strategy may have relied more on strength than speed and complex weapons. Because of this, we might expect that Klein's (1979) pioneering interpretations of restricted hunting effectiveness for the Middle Stone Age in South Africa will eventually be found to hold true for the Western European Mousterian as well.

It is difficult to conceive of the kinds of evidence that might give us insights into the structure of Neandertal hunting parties. There are only a few known kill and primary butchering sites, and little opportunity to isolate the individual episodes in these that might reflect on group size. Perhaps, when more data are available from these contexts, the presence of such evidence as evulsed deciduous teeth might suggest the presence of full family groups during these activities, but even this interpretation includes assumptions that would be difficult or impossible to confirm. At present, the bulk of the evidence seems to favor very small groups spread thinly over the more favorable regions. The restricted circulation of lithic raw materials during this period in Southwest France (e.g., Geneste 1988), and the virtual absence of exotic objects, does not seem to support the occupation of extended territorial ranges or the presence of extensive social contacts.

The picture of the Western European Neandertals outlined here, then, is one of a population near the limits of hominid ability to adapt biologically to cold conditions, with only minimal sheltering technology and aided by a limited hunting and processing technology. In this, perhaps worst, case reconstruction, each healthy individual normally would have spent prolonged intervals of vigorous activity retrieving sufficient calories to support the continuing physical work necessary to maintain a survival core temperature. These individuals would have had relatively few opportunities for prolonged social contact and interaction aside from that with the few relatively short-lived other members of their subsistence group. This impoverished social environment could well have been a more crucial factor in limiting the formulation of abstract ideas and syntheses than any differential ability for linguistic expression.

While the most extreme environmental pressures and human responses may have been present for only a portion of each year, and may have fluctuated in intensity from year to year, these factors could have constituted a repeated formidable barrier or constriction for the development of larger social units and more complex behavior patterns among the Neandertals.

WHAT HAPPENED TO THE WESTERN EUROPEAN NEANDERTALS?

Given the fragmentary nature and questionable contexts for the physical evidence of the morphology of the bearers of the early Aurignacian technology that replaced the Mousterian industries of the Western European Neandertals, it seems futile to try to address the question of replacement, continuity, or hybridization for the biological population. The definitive evidence necessary to resolve this issue is simply not yet at hand.

Nor is there yet definitive evidence of the place of initial development of the early Aurignacian industry. While radiocarbon evidence has recently been presented to show the presence of Aurignacian in Spain significantly earlier than in southwestern France (Cabrera Valdes and Bischoff 1989; Bischoff et al. 1989), the dates were derived by methods not yet applied to the French evidence and thus may reflect a greater potential for age resolution with the new techniques rather than a significant age difference. Comparable dating methods at present favor an earlier central European appearance for cultures similar to the Western European Aurignacian. In particular, the appearance of well-made bone points and blade-based industries classified as Aurignacian by shortly after 40 kyr in eastern and central Europe (Kozlowski 1990), and the earlier appearance in that region of Aurignacian-like cultures such as the Bachokiran (Kozlowski 1982), lend support to a possible later spread of this industry to the west.

If this industry was carried by a more gracile *Homo sapiens* than the Neandertals, how would this creature have been able to survive as a hominid, with that vulnerable physiognomy in the periglacial habitat of such regions as Western France? How did they, as frequently robust but essentially anatomically modern *Homo sapiens*, survive the rigors of Central Europe contemporary with the Western Neandertals? The answer is, of course, most probably with more effective artificial insulation against the cold. And what evidence do we possess that might indicate the presence of such effective insulation?

The major new item in the technology, since Neandertals on occasion demonstrated their ability to make prismatic blades and the full range of retouched blade-tools, is the symmetrical carved bone point, frequently described in the literature as a *sagaie*, or spear-point. This weighted functional nomenclature has been strongly influenced by the idealized "noble hunter" image of the traditional (though erroneous) Cro-Magnon bearer of the early Aurignacian. It is, in fact, likely to be viewed as a minor heresy and threat to the established order to even raise a question of whether the function of these symbols of the clever triumph of our bold ancestors might not have been different from this now (completely) accepted traditional view. But then it is perhaps not all that long ago that questions were raised about the function of that apparent symbol of social status – the chiefly distinction, the *bâton de commandement* – now generally functionally relegated to the relatively lowly status of a shaft wrench. At the risk of an accusation of heresy, it is proposed here that the majority of these implements were probably used in the preparation of at least a primitive version of tailored hide clothing that made it possible for the gracile hominids to, like the Eskimo, surround ". . . themselves with a little piece of the same tropical microclimate upon which we also depend" (Scholander 1955:23). While Klein (1989:370) has proposed this in a general sense, he has not gone so far as to question specifically the function of those artifacts traditionally classified as spear-points, unless this is what he means by "other pointed bone objects." It is suggested here that the burden of proof of the hunting function of the split- and bevel-based points of the Aurignacian and succeeding cultures should be placed on the adherents of that view, although, of course, examination for evidence of the proposed leather-working function is equally appropriate. In fact, these tools might well have been used for multiple tasks, but we need more objective tests to confirm their most probable function.

How might the development of a significant level of personal insulation have made it possible for hominids with this system to replace the Neandertals, or for Neandertals with this system to lose their extreme adaptative features (as may have happened at least in east Central Europe)? The answer is basically one of more effective conservation of heat, which reduced the necessary level of caloric intake and metabolic activity to maintain a normal core temperature. Thus, there would be an initial gain per individual in the amount of time freed from the direct procurement of caloric energy. The nonsubsistence time could have been effectively employed in the manufacture and maintenance of the more elaborate technology of the Upper Paleolithic, with much of this activity likely to take place in a social setting that ultimately may have led to more effective communication and synthesis of ideas and, eventually, abstraction. The reduced individual caloric need, the more effective technology, and the reduced strain on the individual biological systems would result in some greater carrying capacity within a given habitat for more hominids with a greater generational range of experience, and the developing social communication mentioned above. The opportunity for at least some individuals to spend more time in maintenance tasks may be reflected in the better-cleaned and organized occupation surfaces that differentiate the Upper from Middle Paleolithic sites. This could well have signaled an initial important segregation of labor, with women wielding the bone points in essential clothing manufacture and repair, while men's tasks were more centered on procurement. We may see an earlier reflection of a different division of tasks in the Mousterian, in the heavily worn teeth of elderly and infirm individuals and some children, who may have aided the group by softening hides or employing their teeth in some other essential task.

While the above reconstruction is frankly hypothetical, and is restricted in its relevance for *general* human development, it may raise some points of interest for an explanation of the succession of events that accompanied the disappearance of Middle Paleolithic industries and Neandertal morphology in that curious and unique isolate in human development that was early Late Pleistocene Western Europe.

A BRIEF PERSPECTIVE FROM THE LEVANT

The situation in the very small region of northern Israel that has yielded the early Late Pleistocene hominid evidence in the Levant stands in striking contrast to Western Europe in many ways, as described earlier in this chapter. Here an almost complete morphological range of hominids is present, and there appears to be a single general succession of industries that holds true in a wider sense in all of the deeply stratified sites of the Levant.

Pleistocene climatic change in this region was occasionally dramatic, though not as extreme as in Western Europe. Still, there is a reasonable representation of many of the characteristic features of the European Neandertals present here in the skeletons from Tabun, Kebara, and Amud, although none of them can be firmly linked to the most rigorous cold periods in the region. There is little obvious typological difference in the nature of the industries associated with all of the early Late Pleistocene hominids, except Zuttiyeh, and thus there appears to be little hope of resolving the relative chronology of these forms with that kind of evidence. However, at Tabun, the site with the deepest and longest sequence of deposits, it has been possible to study several aspects of change in the production of the lithic industries through a prolonged time period. No site with as complete a sequence has yet been excavated in this part of the world.

One aspect of the lithic analysis from Tabun has produced evidence of a regular time trend; this concerns the relative thickness compared to width of all complete flakes with a maximum diameter > 2.5 cm. The variance of the mean of this ratio shows a gradual increase through time (Jelinek 1982). What this increase in variance reflects is a gradual increase in the relative quantity of thinner flakes, and a gradual expansion of the range of relative thinness in the direction of thinner flakes. The directionality of the trend is regular and statistically significant beyond serious question, but what does it mean? An initial proposal that the variance values (widely misunderstood as the *mean* values) might be shown to have chronological significance beyond Tabun has been generally rejected (e.g., Clark and Lindly 1989:646-47; Binford 1989:41).

The values for the several statistical parameters of the width/thickness ratio for stratigraphic contexts in the recent and earlier Tabun collections, and for the other sites from which hominid remains have been recovered (except Zuttiyeh and Amud) are shown in table 1, where they are listed by descending order of the variance.

In this table, the ordering of the site sample contexts that contain hominid remains shows a downward trend in the placement of those fossils from the most anatomically modern to the most Neandertaloid. It is suggested that this correlation is not without significance for the variations in human morphology represented. It may be recalled that Trinkaus has most recently discussed the difference between the hand of the Neandertals and that of morphologically modern individuals as showing a contrast between an emphasis on a power grip and a precision grip. Is it not reasonable, then, to see the ability to produce more proportionately thin flakes as a reflection of

Table 1. Statistics on Width/Thickness of Complete Flakes for Levant Samples

Context	Mean	Median	Variance	N
Qafzeh XVI-XXIV	6.39	5.71	9.08	599
Qafzeh VII-XV	5.97	5.40	8.92	2188
Qafzeh I + L*	6.74	6.17	8.76	1234
Skhul B*	6.50	6.00	8.67	532
Kebara F*	6.04	5.50	7.64	1770
Tabun Chimney + B*	6.20	5.67	7.10	338
Tabun I ('B' + 'C')	4.64	4.25	5.15	1346
Tabun II ('C')	4.55	4.18	4.06	331
Tabun III-VIII ('C' + 'D')	4.36	4.00	3.97	1236
Tabun IX ('D')	4.26	4.00	3.13	745
Tabun X ('D'/'Ea')	3.95	3.63	2.90	446
Tabun XI ('Ea' + 'Eb')	3.45	3.13	2.67	2342
Tabun XII ('Eb'?)	3.55	3.25	2.56	959
Tabun XIII ('Ec' + 'Ed')	3.52	3.25	2.39	1897
Tabun XIV ('G'?)	3.21	2.92	1.95	761
Tabun G*	3.22	3.00	1.55	153

* Samples from earlier excavations in which flake recovery may not have been comparable to that in later excavations and was generally biased in favor of thinner flakes. The Skhul B sample incorporates material from the dumps of the old excavation which may partially compensate for this effect.

a greater achievement of a precision grip? What better evidence are we likely to find in the archaeological record to confirm these anatomical differences? The stratigraphic sequence at Tabun shows a regular progression through time toward thinner flakes. Given the relationships just described, is it unreasonable to see this as a gradual evolutionary change in the manipulative abilities of the hominids that made these flakes? This long record shows no evidence of interruption by earlier hominids making thin flakes prior to the level in which the Neandertal burial was recovered (Unit I or II most probably). The evidence from the collections from the earlier excavations at the site suggests that the trend continued above this level. Since all of these hominids and sites have been found within an area with a radius of less than 25 km, with access to similar lithic resources, is it likely that the deep time-trend at Tabun was independent of what the hominids in the other sites in that very small area were doing? There is no way to deny the stratigraphically confirmed time-trend at Tabun, and the variations in flake morphology that produce it; it exists as verifiable fact. It is suggested here that, in the absence of *conclusive contrary evidence*, this trend strongly supports a local development of later more gracile hominids from earlier more robust forms in this restricted area of the southern Levant.

Now one final question can be posed here; might this kind of information provide some insights into the Western European evidence? A sequence

Table 2. Statistics on Width/Thickness for Complete Flakes

Context (Industry)	Mean	Median	Variance	N
		Combe Grenal		
Layer 11 (Denticulate)	4.30	4.15	2.46	312
Layer 13 (Denticulate)	4.08	3.55	4.46	119
Layer 14 (Denticulate)	3.83	3.50	2.42	163
Layer 17 (Quina)	4.08	3.67	2.90	293
Layer 23 (Quina)	4.91	4.38	5.50	326
Layer 27 (Ferrassie)	5.18	4.80	5.77	278
Layer 35 (Ferrassie)	4.94	4.40	4.99	311
		Würm I/Würm II		
Layer 38 (Denticulate)	5.09	4.56	6.30	474
Layer 50 (Typical)	4.92	4.66	3.50	303

of nine samples from layers in which more than 100 complete flakes with maximum diameter > 2.4 cm were present in the collections from the excavations of François Bordes through the long Mousterian sequence at Combe Grenal was examined for the width/thickness ratio. The sample statistics for this ratio are presented, in stratigraphic order, in table 2.

It is immediately apparent that there is no unidirectional trend in table 2. It is important, however, to note that while the total time span represented here is probably comparable to a large portion of the record at Tabun, the time units represented by the layers at Combe Grenal are probably much briefer than the major stratigraphic units at Tabun in table 1. The means and medians of the ratio do suggest the presence of two groups of flakes with generally similar ratios: Layers 11 to 17, and 23 to 50. The variance also supports this grouping, except for Layer 50. Although these differences might, at first glance, be taken as evidence for a development of manipulative abilities opposite to that shown in the trends from the Levant, the lack of gradual change and the relatively short time interval implied for the abrupt transition between Layers 23 and 17 appear to rule out a bioevolutionary transition. The absolute size of the flakes remains relatively constant through the sequence and shows no relationship to the width/thickness ratios. There is no clear typological (industrial) or technological (i.e., more Levallois Ferrassie vs. less Levallois Quina) correlation with these differences. Nor is there any obvious relationship to faunal or other paleoclimatic data (Chase 1986). The impression derived during the examination of the collections is that there is little significant difference in the raw materials employed through the sequence. The most that we can say from this evidence, without better absolute chronological controls, is that it appears that the flakes left at Combe Grenal show some change in flaking techniques that may reflect less use of a precision grip in the later occupations, but the abruptness of the transition suggests a technologically based rather than a bioevolutionarily based cause. It can be seen from these data, then, that the directional technological development that can be demonstrated at Tabun cannot be demonstrated at Combe Grenal, the best comparable lithic sequence in Western Europe. In the latter site the only suggestion of directionality is opposite to that at Tabun, and, in general, prolonged intervals of technological stability are implied for Combe Grenal.

CONCLUSIONS

This chapter brings together anatomical, environmental, and archaeological information to explore some possible interrelationships of these kinds of evidence that may lead us toward an additional understanding of the developments that took place during the earlier Late Pleistocene in Western Europe. Many of the interpretations drawn from these data are speculative, but most are based on a core of material and contextual evidence and follow from generally defensible probable relationships. There is little disagreement among students of this period regarding significant differences between the Neandertals and their Mousterian culture and most of the succeeding Upper Paleolithic cultures and hominids.

There is now some reason to see the Châtelperronian industries, with their essentially Upper Paleolithic technology, as a final generic episode of the Neandertal-produced Mousterian. And there is evidence that this final Mousterian was contemporary with an Upper Paleolithic Aurignacian industry similar to one that existed prior to this period in central Europe, and perhaps in Spain. We do not yet have strong evidence of the physical morphology of these first Aurignacians in Western Europe, nor do we know whether the Upper Paleolithic aspects of the Châtelperronian were developed independently or were a result of observation of, or even exchange with, groups with an Aurignacian technology. It does not now appear likely that the Aurignacian arose independently from the Châtelperronian or any other industry in the region, as has occasionally been proposed, although it is not impossible that this technological complex was adapted by a succession of local groups between Central and Western Europe rather than having been carried by migrating populations of herd-following hunters. The lack of strong differences in hunting patterns between the Middle and early Upper Paleolithic (Chase 1989:333) may support this interpretation.

The technological changes evident in the early Upper Paleolithic imply, most significantly, an increased sophistication in the preparation of artifical insulation in the form of tailored clothing and tents. The acquisition of this kind of technology by (essentially tropical) hominids in a cold-temperate to periglacial setting would have reduced the need for the extreme biological response to cold that seems to be especially well reflected in the morphology of the Western European Neandertals. This kind of extreme biological response implies significantly more physical activity to retain viable core temperature levels through the intake of food calories and oxygen in relatively prodigious quantities to manufacture heat in a veritable metabolic furnace of muscle mass. This kind of prolonged diurnal activity pattern, in the setting of periglacial Europe, with a simple hunting/processing technology, implies small social units and minimal social contact and stimulation. In contrast the development of more effective thermal insulation would have reduced the person/time devoted to caloric retrieval, encouraged periods of social exchange and more segregation of tasks during the manufacture and maintenance of the more complex technology, and allowed larger social units to survive in this setting.

The most obvious preserved feature of this new technology is the abundance of pointed bone tools, many of which are frequently classified as tips for weapons. It is proposed here that a re-examination of many of these objects may well demonstrate that, for these early Upper Paleolithic groups, the awl was mightier than the spear as a vital element for survival, and perhaps as a reflection of the beginning, in this region, of the mutually supportive integrated segregation of tasks and activities that characterize modern arctic-adapted hunting groups of *Homo sapiens sapiens*. Perhaps it would not be inappropriate to end this hypothetical scenario with the observation that, for a successful biocultural transition from the Middle Paleolithic in Western Europe, it was awls or nothing!

ACKNOWLEDGMENTS

I am most grateful to the many individuals who have responded constructively to several preliminary presentations of some of these ideas at the University of Arizona and at Indiana University. I am particularly grateful to Stephen L. Zegura, who has given me access to useful information, and to A. T. Steegman, Jr., for his guidance to appropriate literature on cold-adaptation. Harold L. Dibble assisted me with flake measurements in France and Israel those many years ago, and coded and programmed these data for the computer analysis. I am deeply grateful to Mme Denise de Sonneville-Bordes and her family for permission to use information from François Bordes' collections. The research at Tabun mentioned in this chapter was supported by the Smithsonian Foreign Currency Program, the Ford Foundation, the National Science Foundation, and the Universities of Arizona and

Michigan. The work at La Quina has been supported by the National Geographic Society, the National Science Foundation, the Wenner-Gren Foundation for Anthropological Research, and the University of Arizona.

REFERENCES

Anderson-Gerfaud, P. 1990. Aspects of behaviour in the Middle Palaeolithic: Functional analysis of stone tools from Southwest France, 389-418. In *The Emergence of Modern Humans: An Archaeological Perspective*, ed. P. Mellars. Edinburgh: Edinburgh University Press.

Åstrand, P.-O., and K. Rodahl. 1977. *Textbook of Work Physiology*. New York: McGraw-Hill.

Bar-Yosef, O. 1989. Upper Pleistocene cultural stratigraphy in southwest Asia, 154-80. In *The Emergence of Modern Humans: Biocultural Adaptations in the Later Pleistocene*, ed. E. Trinkaus. Cambridge: Cambridge University Press.

Beyries, S. 1988. Functional variability of lithic sets in the Middle Paleolithic, 213-23. In *Upper Pleistocene Prehistory of Western Eurasia*, ed. H. L. Dibble and A. Montet-White. Philadelphia: The University Museum, University of Pennsylvania.

Binford, L. R. 1989. Isolating the transition to cultural adaptations: An organizational approach, 18-41. In *The Emergence of Modern Humans: Biocultural Adaptations in the Later Pleistocene*, ed. E. Trinkaus. Cambridge: Cambridge University Press.

Binford, L. R., and S. R. Binford. 1966. A preliminary analysis of functional variability in the Mousterian of Levallois Facies. *American Anthropologist* 68:238-95.

Bischoff, J. L., N. Soler, J. Maroto, and R. Julia. 1989. Abrupt Mousterian/Aurignacian boundary at c. 40 ka BP: Accelerator ^{14}C dates from L'Arbreda Cave (Catalunya, Spain). *Journal of Archaeological Science* 16:563-76.

Bordes, F. 1953. Essai de classification des industries Moustériennes. *Bulletin de la Société Préhistorique Française* 50:457-66.

Bordes, F. 1961. Mousterian cultures in France. *Science* 134:803-10.

Bordes, F. 1972. *A Tale of Two Caves*. New York: Harper and Row.

Bordes, F. 1981. Vingt-cing ans après: Le complexe Moustérien révisité. *Bulletin de la Société Préhistorique Française* 78:77-87.

Bordes, F., and F. Prat. 1965. Observations sur les faunes du Riss et du Würm I en Dordogne. *L'Anthropologie* 69:31-45.

Cabrera Valdes, V., and J. L. Bischoff. 1989. Accelerator ^{14}C dates for early Upper Paleolithic (Basal Aurignacian) at el Castillo Cave (Spain). *Journal of Archaeological Science* 16:577-84.

Chase, P. G. 1986. The hunters of Combe Grenal: Approaches to Middle Paleolithic subsistence in Europe. Oxford: British Archaeological Reports International Series S286.

Chase, P. G. 1989. How different was Middle Palaeolithic subsistence? A zooarchaeological perspective on the Middle to Upper Palaeolithic transition, 321-37. In *The Human Revolution*, ed. P. Mellars and C. Stringer. Edinburgh: Edinburgh University.

Chase, P. G., and H. L. Dibble. 1987. Middle Palaeolithic symbolism: A review of current evidence and interpretations. *Journal of Anthropological Archaeology* 6:263-96.

Clark, G. A., and J. M. Lindly. 1989. The case for continuity: Observations on the biocultural transition in Europe and Western Asia, 626-76. In *The Human Revolution*, ed. P. Mellars and C. Stringer. Edinburgh: Edinburgh University Press.

Dibble, H. L. 1984. The Mousterian industry from Bisitun Cave (Iran). *Paléorient* 10:23-34.

Dibble, H. L. 1987. The interpretation of Middle Paleolithic scraper morphology. *American Antiquity* 52:109-17.

Dibble, H. L., and A. Montet-White, eds. 1988. *Upper Pleistocene Prehistory of Western Eurasia*. Philadelphia: The University Museum, University of Pennsylvania.

Farizy, C. 1990a. Du Moustérien au Châtelperronien à Arcy-sur-Cure: un état de la question, 281-89. In *Paléolithique moyen récent et Paléolithique supérieur ancien en Europe*, ed. C. Farizy. Nemours: Mémoires du Musée de Préhistoire d'Ile de France 3.

Farizy, C. 1990b. The transition from Middle to Upper Palaeolithic at Arcy-sur-Cure (Yonne, France): Technological, economic and social aspects, 303-26. In *The Emergence of Modern Humans: An Archaeological Perspective*, ed. P. Mellars. Edinburgh: Edinburgh University Press.

Farizy, C., ed. 1990c. *Paléolithique moyen récent et Paléolithique supérieur ancien en Europe*. Nemours: Mémoires du Musée de Préhistoire d'Ile de France 3.

Gambier, D. 1989. Fossil Hominids from the early Upper Palaeolithic (Aurignacian) of France, 194-211. In *The Human Revolution*, ed. P. Mellars and C. Stringer. Edinburgh: Edinburgh University Press.

Geneste, J.-M. 1988. Systèmes d'approvisionnement en matières premières au Paléolithique moyen et au Paléolithique supérieur en Aquitaine, 61-70. In *L'Homme de Néandertal*. Vol. 8, *La Mutation*, ed. M. Otte. Liège: Etudes et Recherches Archéologiques de l'Université de Liège 35.

Gisis, I., and O. Bar-Yosef. 1974. New excavations in Zuttiyeh Cave, Wadi Amud, Israel. *Paléorient* 2:175-80.

Harrold, F. B. 1989. Mousterian, Châtelperronian and Early Aurignacian in Western Europe: Continuity or discontinuity? 677-713. In *The Human Revolution*, ed. P. Mellars and C. Stringer. Edinburgh: Edinburgh University Press.

Howell, F. C. 1952. Pleistocene glacial ecology and the evolution of "Classic Neandertal" man. *Southwestern Journal of Anthropology* 8:377-410.

Jelinek, A. J. 1981. The Middle Paleolithic in the Southern Levant from the perspective of the Tabun Cave, 265-80. In *Préhistoire du Levant*, ed. J. Cauvin and P. Sanlaville. Paris: Centre National de la Recherche Scientifique.

Jelinek, A. J. 1982. The Tabun Cave and the Paleolithic man in the Levant. *Science* 216:1369-75.

Jelinek, A. J. 1988. Technology, typology, and culture in the Middle Paleolithic, 199-212. In *Upper Pleistocene Prehistory of Western Eurasia*, ed. H. L. Dibble and A. Montet-White. Philadelphia: The University Museum, University of Pennsylvania.

Jelinek, A. J. 1992. Problems in the chronology of the Middle Paleolithic and the first appearance of Early Modern *Homo sapiens* in Southwest Asia, 153-75. In *The Evolution and Dispersal of Modern Humans in Asia*, ed. T. Akazawa, K. Aoki, and T. Kimura. Tokyo: Hokusen-sha.

Kennedy, G. E. 1983. Some aspects of femoral morphology in *Homo erectus*. *Journal of Human Evolution* 12:587-616.

Kennedy, G. E. 1984. Bone thickness in *Homo erectus*. *Journal of Human Evolution* 14:699-708.

Klein, R. G. 1979. Stone age exploitation of animals in Southern Africa. *American Scientist* 67:151-60.

Klein, R. G. 1989. *The Human Career*. Chicago: University of Chicago Press.

Kozlowski, J. 1990. A multiaspectual approach to the origins of the Upper Palaeolithic in Europe, 419-37. In *The Emergence of Modern Humans: An Archaeological Perspective*, ed. P. Mellars. Edinburgh: Edinburgh University Press.

Kozlowski, J., ed. 1982. *Excavation in the Bacho Kiro Cave (Bulgaria): Final Report*. Warsaw: Państowowe Wydawnictowo Naukowe.

Leroi-Gourhan, A. 1958. Etude des vestiges humains fossiles provenant des grottes d'Arcy-sur-Cure. *Annales de Paléontologie* 44:87-148.

Lévêque, F., and J.-C. Miskovsky. 1983. Le Castelperronien dans son environnement géologique. *L'Anthropologie* 87:369-91.

Lévêque, F., and B. Vandermeersch. 1981. Le Néandertalien de Saint-Césaire. *La Recherche* 119:242-44.

Lumley, H. de. 1969. Une Cabane Acheuléene dans la Grotte du Lazaret. *Mémoires de la Société Préhistorique Française* 7:1-234.

Marks, A. E. 1983. The Middle to Upper Paleolithic Transition in the Levant. *Advances in World Archaeology* 2:51-98.
Martinson, D. G., N. G. Pisias, J. D. Hays, J. Imbrie, T. C. Moore, Jr., and N. J. Shackleton. 1987. Age dating and the orbital theory of the Ice Ages: Development of a high-resolution 0 to 300,000-year chronostratigraphy. *Quaternary Research* 27:1-29.
Mellars, P. 1970. The chronology of Mousterian industries in the Périgord Region of South-West France. *Proceedings of the Prehistoric Society* 35:134-71.
Mellars, P., ed. 1990. *The Emergence of Modern Humans: An Archaeological Perspective.* Edinburgh: Edinburgh University Press.
Mellars, P., and C. Stringer, eds. 1989. *The Human Revolution: Behavioural and Biological Perspectives on the Origins of Modern Humans.* Edinburgh: Edinburgh University Press.
Movius, H. L. 1950. A wooden spear of Third Interglacial Age from Lower Saxony. *Southwestern Journal of Anthropology* 6:139-42.
Ogilvie, M. D., B. K. Curran, and E. Trinkaus. 1989. Incidence and patterning of dental enamel hypoplasia among the Neandertals. *American Journal of Physical Anthropology (n.s.)* 79:25-41.
Otte, M., ed. 1988. *L'Homme de Neandertal: Actes du colloque international de Liège (4-7 décembre 1986).* Liège: Etudes et Recherches Archéologiques de l'Université de Liège, 28-35 (Vols. 1-8).
Rigaud, J.-Ph. 1989. From the Middle to the Upper Paleolithic: Transition or convergence? 142-53. In *The Emergence of Modern Humans*, ed. E. Trinkaus. Cambridge: Cambridge University Press.
Rolland, N. 1981. The interpretation of Middle Paleolithic variability. *Man* 16:15-42.
Rolland, N., and H. L. Dibble. 1990. A new synthesis of Middle Paleolithic variability. *American Antiquity* 55:480-99.
Scholander, P. F. 1955. Evolution of climatic adaptation in homeotherms. *Evolution* 9:15-26.
Scholander, P. F., V. Walters, R. Hock, and L. Irving. 1950. Body insulation of some arctic and tropical mammals and birds. *Biological Bulletin* 99:225-36.
Shackleton, N. J. 1987. Oxygen isotopes, ice volume, and sea level. *Quaternary Science Reviews* 6:183-90.
Shephard, R. J. 1978. *Human Physiological Work Capacity.* International Biological Programme 15. Cambridge: Cambridge University Press.
Solecki, R. S. 1963. Prehistory in the Shanidar Valley, Northern Iraq. *Science* 139:179-93.
Trinkaus, E. 1981. Neandertal limb proportions and cold adaptation, 187-224. In *Aspects of Human Evolution*, ed. C. Stringer. London: Taylor and Francis.
Trinkaus, E. 1983. *The Shanidar Neandertals.* New York: Academic Press.
Trinkaus, E. 1988. The evolutionary origins of the Neandertals or, why were there Neandertals? 11-29. In *L'Homme de Néandertal.* Vol. 3, *L'Anatomie*, ed. M. Otte. Liège: Etudes et Recherches Archéologiques de l'Université de Liège 30.
Trinkaus, E. 1989a. The Upper Pleistocene transition, 42-66. In *The Emergence of Modern Humans: Biocultural Adaptations in the Later Pleistocene*, ed. E. Trinkaus. Cambridge: Cambridge University Press.
Trinkaus, E., ed. 1989b. *The Emergence of Modern Humans: Biocultural Adaptations in the Later Pleistocene.* Cambridge: Cambridge University Press.
Trinkaus, E., and W. W. Howells. 1979. The Neanderthals. *Scientific American* 241:118-33.
Trinkaus, E., and D. D. Thompson. 1987. Femoral diaphyseal histomorphometric age determinations for the Shanidar 3, 4, 5, and 6 Neandertals and Neandertal longevity. *American Journal of Physical Anthropology* 72:123-29.
Vandermeersch, B. 1989. The evolution of modern humans: Recent evidence from Southwest Asia, 155-64. In *The Human Revolution*, ed. P. Mellars and C. Stringer. Edinburgh: Edinburgh University Press.
Vaughn, P. 1985. *Use-Wear Analysis of Flaked Stone Tools.* Tucson: University of Arizona.
White, R. 1989. Production complexity and standardization in Early Aurignacian bead and

pendant manufacture: Evolutionary implications, 366-90. In *The Human Revolution*, ed. P. Mellars and C. Stringer. Edinburgh: Edinburgh University Press.

Woillard, G. M., and W. G. Mook. 1982. Carbon-14 dates at Grande Pile: Correlation of land and sea chronologies. *Science* 215:159-61.

Wolpoff, M. H. 1989. The place of the Neandertals in human evolution, 97-141. In *The Emergence of Modern Humans: Biocultural Adaptations in the Later Pleistocene*, ed. E. Trinkaus. Cambridge: Cambridge University Press.

Chapter **4**

Behavioral and Cultural Changes at the Middle to Upper Paleolithic Transition in Western Europe

CATHERINE FARIZY

For a long time, archaeologists studying the Old World Paleolithic believed that Lower and Middle Paleolithic were providing the same kind of global evidence. Only the Upper Paleolithic was considered to be strikingly different. However, the last decade of research and discoveries, have made it clear that the gap between Middle and Upper Paleolithic was not unique, but that a huge step exists also between Lower and Middle Paleolithic. It is no longer possible to consider Lower and Middle Paleolithic as a simple unit, because there is a tremendous amount of difference between these stages. We cannot explain what the Middle Paleolithic/Early Upper Paleolithic transition may have been without knowing what the Middle Paleolithic is and how it is different from the Lower Paleolithic.

Most Lower Paleolithic sites, including the recently discovered open-air sites from northern Europe (northern France, southern United Kingdom) or the Garonne Valley (France), the Tagus Valley (Spain), and central and southern Italy, show that there are many sites where bifaces are unexpectedly numerous and generally not found with any debris left behind by their manufacture. This could mean that they had been transported before they were used. I think Binford may be right when he says "we have a technologically aided, biologically based, panspecific form of adaptation" (Binford 1989:29). Bifaces were probably a biological extension of hominid hands. This prolon-

---

CATHERINE FARIZY • Laboratoire D'Ethnologie Préhistorique, Université de Paris, Paris, France 75014.
Origins of Anatomically Modern Humans, edited by Matthew H. Nitecki and Doris V. Nitecki. Plenum Press, New York, 1994.

gation of hands is well known to have involved a long time and would explain their presence all over the world at a certain prehuman stage. While many open-air Acheulian sites are very rich in bifaces, the rock-shelters usually provide poorly elaborated tools made on blanks coming from a very low (in terms of labor investment) technological knapping sequence. Presence or absence of bifaces seems to be strongly correlated with the raw material available on, or very close to, the site. One key explanation of bifaces may be in the faunal remains. However, the remains associated with Acheulian tools are very little known: some northern sites do not contain any, and recently excavated sites, such as Cagny-la-Garenne or Boxgrove, are not yet studied. Most of the Mediterranean Lower Paleolithic sites containing large mammals show mostly scavenging and occupations of short duration. Considering the long time span of the Lower Paleolithic, these sites are very rare – still quite few – and even fewer, if any, of them show evidence of fire. In northern and southern parts of Europe the boundaries between Lower and Middle Paleolithic seem to be nonsynchronous. However, the Middle Paleolithic exhibits distinctive new features between isotopic stages 8 and 7. Not only do the lithic assemblages show great behavioral differences, but the faunal remains and the presence of fire show many more differences than before. The Levallois knapping technic is adopted wherever the raw material allows it; through time, when good raw material is not present on the site, it is brought from the distance of 5 to over 50 kilometers, and the flint is used for purposeful tools such as sidescrapers or unretouched Levallois blanks. The sites become more and more numerous and through time show evidence of hunting large mammals.

Recently excavated open-air occupations in southern France show clear evidence of regional patterns (Farizy and Jaubert 1991). These sites are in the eastern part of the famous Perigord region where the geography is different: the morphology of the landscapes shows no comparable features with those of the Dordogne Valley. There are no rock-shelters, but only poorly preserved karst topography, and good flint is not found. Most of the sites are dated from the beginning of early Würm. The large sites appear to be rather numerous, the indeposit is always a colluvium, and the layers are thick with only discarded lithic and faunal remains. The evidence of fire is the presence of burnt bones, and the faunal remains are always numerous, and dominated by large mammals. The choice of the sites was important: local groups moved through the same location for many generations. The lithic industry is distinctive, consisting of locally derived raw material of poor quality, and the technological investment is poor. These are purposeful specific sites where the lithic assemblages seem to be strongly economic ones, even if the basic Mousterian sidescraper is always present. In addition to those geographic patterns, there could be a chronological pattern (Mellars 1989).

On the one hand, Middle Paleolithic should not be considered the end of Lower Paleolithic. Middle Paleolithic seems to be a specific stage, related not only to a specific technology in flint or other raw materials, but also to an improved relation with the environmental utilization. A new adaptation to the environment seems to emerge with an enlarged control of food resources; many Middle Paleolithic sites show hunting, elaborate butchering, and meat processing which do not change before late Upper Paleolithic. At the same time, several early Weichselian sites provide blade technology whose cores and blanks are not basically different from the early Upper Paleolithic (Châtelperronian and Aurignacian).

On the other hand, it is quite clear that Middle Paleolithic people, especially Mousterians from Western Europe, approximately at the beginning of the last glaciation (from the end of Eemian up to 40 000 BP) appear to have evolved to a top level, as far as technology and typology of flint tools are concerned. These people abandoned blade technology, which seemed of no special use for them, and they regionally developed sidescrapers whose shapes could be interpreted locally as signs (Wobst 1977), that is, ways of communication, how one local group shows its differences from far neighbors.

A crucial chronological problem with the Middle Paleolithic can be called the "time uncertainty." Despite the progress in the last decade due to thermoluminescence dating, we still cannot compare the ages of two sites at a period when we would need a better overlap than 10,000 years, and we usually have no idea of the duration of most of the occupations on which our assumptions are based. However, it is clear that we have an evolution of behavior through time, from the appearance of the first burials to the early Upper Paleolithic. This is not shown by any evolution of the technotypology, which only seems to become more and more Mousterian, but can be shown through the numerous specific sites and through the emergence of less undifferentiated use of space. Spatial distribution of categories of remains analyzed at Champlost, for instance, show the space is not uniformly occupied: heads are totally missing at some places while they are found at others; the complete knapping sequence can be found at some places and not at others; and tools seem to be more strongly correlated with faunal long-bone shafts than with other items, thus showing meat and marrow consumption.

In late Mousterian levels at Arcy-sur-Cure, more broken bones were found than in earlier levels, where many shafts were not broken. In these recent occupation levels Leroi-Gourhan found curiously shaped fossils, possibly introduced into the occupation sites because of their peculiar form.

What can be said about Middle Paleolithic behavior: Were people able to anticipate events a long time before the events happened? Were they able to be successful in hunting by choosing the best time, place and hunters? To what degree was the technology maintained? Many scholars associate

Neandertal groups to early African hominids and deny them any of these possibilities. As Binford (1989:22) reported, "it is primarily the scheduling variations in both space and time in the accessibility of resources, coupled with the incongruent patterns of availability for needed suites of resources, that condition the degree of planning depth, tactical depth and curation," therefore, the global pattern can be seen as an ecological one.

The peculiar spatial pattern of Middle Paleolithic sites may be strongly correlated with their age: that is, longtime accumulations of mixed activities with poor preservation due to bias from weathering or any other agent. If many factors – each the consequence of a specific activity – produced an accumulation in one place, and if the same location was seasonally occupied by the same group, the structural pattern of the discarded remains is inextricable rather than complex. The seeming randomness of Middle Paleolithic accumulations, however, may be explained only by postdepositional biases. The duration of occupation is a factor we cannot control, although I think these superimposed accumulations provide one of the best keys to Middle Paleolithic patterning: the activities were not differentially patterned because of their undifferentiated use of space.

Recent excavations show that, from the beginning of Middle Paleolithic, Mousterian groups were following a previously established itinerary, moving through the same area and the same sites. They also show that men were then able to live by hunting rather than by scavenging as in Lower Paleolithic times. Rock-shelters and caves were inhabited by men and carnivores, and their middens are often mixed. The numerous bones have been studied from other than an archaeological point of view and we know little about meat processing in rock-shelters and caves. On the contrary, recently excavated open-air sites reveal specific kill and butchering sites where large mammals were dismembered and eaten. The taphonomic studies of these remains show that adult or young adult animals were hunted more than juvenile or old animals, and the butchering process was done the same way throughout the archaeological deposits.

The studies of Middle Paleolithic accumulations have not revealed any dwelling structures; only evidence of behavior and habits are repeatedly found in the depth of time. At some places, knapping was done and a few tools were discovered, at others are tools and faunal remains. We cannot really separate the different sites. We can only distinguish between sites with few and specific tools largely correlated with killing or butchering sites, and those where more elaborate tools are found together with fireplaces, knapping debris, and long bones broken for marrow consumption.

Globally, Middle Paleolithic allows for standardization. The tools are not evolving toward the Upper Paleolithic. The monotony of lithic assemblages is reinforced by the overwhelming presence of only one tool: the sidescraper from western Europe, and the bifacial scraper from central and eastern Europe.

The Middle to Upper Paleolithic transition is of particular interest because of its apparent coincidence in western Europe with the end of Neandertal people and the emergence of modern man. How this replacement occurred is as yet unknown, and so are the exact links between lithic industries and biological features.

In the Middle Paleolithic we are clearly faced with the evolution leading to a behavior that cannot be explained without a certain amount of cultural adaptation and social patterning. Large mammals cannot be hunted like small animals are, nor they can be eaten without sharing because of the large amount of meat available, as archeozoological analysis has shown. The open-air sites show that they were occupied seasonally once or twice a year, and also that the Mousterians came to their site with the raw material they would not find there, but which they knew they would need there. This means that they planned their movement all year round. A site such as Mauran could have been a meeting site in which several family groups shared the place and its resources.

Therefore, we are dealing with the "modern" way of life of Mousterian peoples: they hunted, they shared part of their food, they taught their juveniles how to knap correctly, and perhaps how to make an up-to-date sidescraper. They are regionally patterned: the Mousterian lithic industry from eastern France is different from those from northern Burgundy and Brittany. There seems to have been a rather long time span (over 50,000 years) when Mousterians were well adapted to their environment, even though some children show evidence of stress, and, therefore, presumably of bad nutrition.

It is possible that, at least in some cultural rather than economic Denticulate Mousterian sites of western Europe, the lack of sidescrapers means the loss of cultural identity. The arrival of new behavioral patterns, possibly influenced by distant new traditions, may have provoked tremendous change all over Europe. The Mousterians who passed through this stage may have adapted to a new way of life, and this would explain the presence of Denticulate Mousterian before most of the Châtelperronian sites. However, as far as I know, the Middle Paleolithic Mousterian did not evolve to early Upper Paleolithic through lithic assemblages, as was claimed earlier (Bordes 1953). If it evolved towards early Upper Paleolithic, it was through latent features that are difficult for us to see in the Denticulate Mousterian assemblages, and through a dramatic loss of integrity.

While we are now able to distinguish behavioral differences between the Lower and Middle Paleolithic stages, it becomes more and more difficult to place exactly the rupture between Middle and Upper Paleolithic, and to know of what it consists. Behavioral and cultural changes are not as abrupt as expected; some features continued on through the transition while others stopped before it. We do not know much about the boundaries, but we are confident of them in the Early Weichselian, Middle Paleolithic, there are many open-air and cave sites rich in faunal remains and lithic assemblages.

One example of the different kind of transition we may be faced with, and which, in my opinion, reveals the dramatic behavioral change that occurred 35,000 years ago at the Châtelperronian site of Grotte du Renne at Arcy-sur-Cure (northern Burgundy, France; Farizy 1990a and 1990b; Otte 1989). Clearly, Châtelperronian from Arcy-sur-Cure is unique in that it is strikingly different from the Mousterian, even if a small part of the lithic assemblage is the same. The difference is much more obvious than at Saint-Césaire (Lévêque and Vandermeersch 1981), for instance, which lacks living floor structure, whose industry may be archaic, and which lacks ornament and bone tools, and has only a few Upper Paleolithic items.

The Châtelperronians occupied a living floor of more than 80 m^2 at Renne's terraced entrance. The living area was neither a cave nor an open-air site, but rather a roofless shelter whose rear wall faced due south, thus warming up at the slightest ray of sunlight. Three Châtelperronian levels form a sequence very rich in lithic and bone fragments (more than 50,000 lithic pieces have been counted). Because of continuous occupation, the Châtelperronian living floors at Arcy form complex structures, not yet interpreted, and might represent an occupation time of hundreds of years. Our analysis is concerned mainly with the spatial organization that appeared during Leroi-Gourhan's excavation, but only the obvious habitation structures have been studied. Despite successive readjustments in the living areas, some intact structures have been studied, particularly those of the oldest Châtelperronian living floors whose plan was circular, surrounded by limestone plaques forming benches. Leroi-Gourhan (1983) suggested that mammoth tusks and a wooden frame held the structure together, and that the huts covered about 12 m^2.

Therefore, the first, and main, striking characteristic of Châtelperronian is how men conceived their living space in a new way: the throwing-away of large bones, to clean the floor. For the first time the fireplace appears to be more than just the area where the fire is made, as structures appear around the fireplace. The use of ocher is abundant: many ocher markers have been found, and the sediment is colored red by ocher.

Herbivore remains are the same as in the Late Mousterian levels, and the ways they were dismembered and broken for marrow consumption were the same. Carnivores were also present, but they were not hunted nor eaten by men. Only their claws and teeth are present, and all carnivore phalanges from early Upper Paleolithic levels provide cut-marks. These cut-marks, and the distribution of the claws on the floors, suggest the presence of furs on the living floors and, therefore, their utilization by the group. In addition, an important bone tool assemblage of more than 120 bone and ivory tools has been discovered. Some of these are slightly worked bones which have been selected for their natural robust and flat shape, such as ribs used as picks or for skin working. The most numerous implements are bone awls, entirely or partially worked. The manufacturing techniques are remarkably well developed, and most of the extracting or shaping techniques seem to have been used at Arcy (Baffier and Julien 1990). Bird bones of different diameters have been cut transversely. Some bird bones have been incised, and some Aurignacian-like sagaies may also be present. All these elaborate bone tools and artifacts could have been manufactured elsewhere, perhaps by Aurignacian neighbors, and brought to the cave. The entire working sequence is found in the Châtelperronian levels, though it should not be.

Animals are no longer killed for subsistence only – subsistence not only implying to eat, but also complex social behavior. The cave is no more just a dumped accumulation of middens. The huts are cleaned, and there is now a distinct outside as opposed to an inside; the fire is constructed, and domestic activities are around the hearth, which seems for the first time to be the heart of group life.

In addition to these new preoccupations, Châtelperronian people at Arcy commonly made ornament pieces (Taborin 1990), such as perforated and grooved carnivore canines, and pendants of perforated teeth, grooved fossils, and ivory.

In conclusion, spatial distribution demonstrates many similarities between the various Mousterian occupations at Arcy and in any open-air sites, regardless of the kind of site. In all of them we find the same, seemingly random distribution of remains. There is no evidence that the living surface was even rearranged. This may have a clear implication to the understanding of Neandertal behavior. The lack of cleanliness of their living space appears not to have bothered them, though at the end of the Mousterian time (at Arcy-sur-Cure) some levels yielded several curiously shaped fossils.

It must be stressed that the behavioral changes cannot be directly related to a different environmentally induced life style. The animals hunted were the same, and most of the lithic technologies employed by the Châtelperronians were already known to the Mousterians. Bone tools and nonutilitarian objects such as pendants and ocher may be a new cultural expression of items existing but which cannot be found by an archaeologist: wooden tools or corporal painting. What is new is their huge number: one or two of these objects would, by themselves, be insignificant. The changes seem rather to relate to a different quality of life, in which the immediate surroundings of the human groups – the habitation zone – was perceived in a totally different way.

ACKNOWLEDGMENTS

I warmly thank Dr. Matthew H. Nitecki for inviting me to the Symposium, pressing me to write this paper, and for kindly rewriting my English version of it.

REFERENCES

Baffier, D., and M. Julien. 1990. L'outillage en os des niveaux Châtelperroniens d'Arcy-sur-Cure (Yonne), 329-34. In *Paléolithique moyen récent et Paléolithique supérieur ancien en Europe*, ed. C. Farizy. Nemours: Mémoires Musée Préhistoire Ile de France 3.

Binford, L. 1989. Isolating the transition to cultural adaptations: An organizational approach, 18-41. In *The Emergence of Modern Humans: Biocultural Adaptations in the Later Pleistocene,* ed. E. Trinkaus. Cambridge: Cambridge University Press.

Bordes, F. 1953. Essai de classification des industries Moustériennes. *Bulletin de la Société Préhistorique Française* 50:457-66.

Farizy, C. 1990a. The transition from Middle to Upper Palaeolithic at Arcy-sur-Cure (Yonne, France): Technological, economic and social aspects, 303-26. In *The Emergence of Modern Humans: An Archaeological Perspective,* ed. P. Mellars. Edinburgh: Edinburgh University Press.

Farizy, C. 1990b, ed. *Paléolithique moyen récent et Paléolithique supérieur ancien en Europe.* Nemours: Mémoires Musée Préhistoire Ile de France 3.

Farizy, C., and J. Jaubert. 1991. Middle Palaeolithic specialised sites: The case of Western France-French Pyrenees. Paper presented at the 56th Annual Meeting of the Society for American Archaeology. New Orleans, 24-28 April 1991.

Leroi-Gourhan, A. 1983. *Le Fil du Temps.* Paris: Fayard.

Lévêque, F., and B. Vandermeersch. 1981. Le Néandertalien de Saint-Césaire. *La Recherche* 119:242-44.

Mellars, P. A. 1989. Chronologie du Moustérien du Sud-Ouest de la France: Actualisation du débat. *L'Anthropologie* 93:53-72.

Otte, M., ed. 1988. *L'Homme de Néandertal.* Vols. 1-8. Liège: Etudes et Recherches Archéologiques de l'Université de Liège 28-35.

Taborin, Y. 1990. Les prémices de la parure, 335-44. In *Paléolithique moyen récent et Paléolithique supérieur ancien en Europe*, ed. C. Farizy. Nemours: Mémoires Musée Préhistoire Ile de France 3.

Wobst, M. H. 1977. Stylistic behavior and information exchange, 317-42. In *For the Director: Research Essays in Honor of James B. Griffin*, ed. C. Cleland. Ann Arbor: University of Michigan Museum of Anthropology, Anthropological Papers 61.

Chapter 5

Ancestral Lifeways in Eurasia – The Middle and Upper Paleolithic Records

OLGA SOFFER

The origin of modern humans is a hot topic today and, as can be seen from the number of major symposia and volumes devoted to the topic (e.g., Smith and Spencer 1984; Mellars and Stringer 1989; Mellars 1990; Trinkaus 1989), has been such for the last few years when it successfully displaced questions of human origins as the premier paleoanthropological issue. This shift in concern brought with it not only more recent chronological contexts, but also a geographic relocation from Africa to Europe and Asia. What is at issue today is the meaning of the Late Pleistocene record which documents a change both in hominid types and their cultures. In Eurasia this time span – a period between some 100,000 and 30,000 years ago saw anatomically modern humans replace the Neanderthals, and a transition from the Middle to the Upper Paleolithic. The significance and synchrony of these changes have been a subject of many recent debates which have focused on two issues: first, what exactly happened at this transition, and second, which hominid types were responsible for the changes noted in the archaeological record. As a perusal of any of the volumes cited above indicates, no consensus has been reached on these questions – a dilemma stemming not only from the ambiguity of the relevant records but also from excessive focus on the question of what happened. Much less attention has been paid to the potentially more enlightening questions about why the changes occurred. This chapter examines a number of pertinent data sets – biological, ecological, ethological, and archaeological – to argue first, that the transition involved very specific and major sociocultural transformations, second, that these changes did not

OLGA SOFFER • Department of Anthropology, University of Illinois, Urbana, IL 61801.
Origins of Anatomically Modern Humans, edited by Matthew H. Nitecki and Doris V. Nitecki. Plenum Press, New York, 1994.

and major sociocultural transformations, second, that these changes did not happen at any particular Rubicon, and third, that the transformations outlined below ultimately permitted successful and continuous colonization of open northern landscapes of Eurasia and beyond.

THE EURASIAN RECORD

Biological Data. A number of significant biological differences have been noted between the Neanderthals and early anatomically modern humans (EAMHs). First, Neanderthals are considered to have been much more robust and strong, and adapted to significantly greater physical exertion (Trinkaus 1983a, 1983b, 1987, 1989; Trinkaus and Smith 1985). Evidence for this is observed not only in adults but also in children, suggesting that the latter were under considerable physical stress from early childhood on (Trinkaus 1991). Second, their survival only to ages of late thirties to early forties indicates that Neanderthal longevity was considerably less than that of anatomically modern humans, with both sexes living to postreproductive age (Nemeskeri 1989; Skinner 1981; Trinkaus 1986, 1987, 1989 with references). Paleoanthropological literature also documents changes in crania and in dentition. The consensus of opinion is that the reduction in tooth size, especially of anterior dentition, and accompanying changes in masticatory and cranial configurations in EAMHs are attributable to reduction in dental loading, which in turn is seen as a reflection of a decrease in the paramasticatory use of dentition (Frayer 1984; Smith 1983; Trinkaus 1987, 1989; Trinkaus and Smith 1985). A great debate still exists concerning Neanderthal speaking abilities, with differences of opinion reducible to disputes about capacities which totally ignore equally important contextual questions about the use of this capacity – in other words, about performance. Finally, there is no consensus whether these changes reflect population replacement, regional evolution, or a mix of gene flow and population movement (cf. Stoneking and Cann 1989; Stringer 1989; Wolpoff 1979; Smith, Simek & Harrill 1989).

Interpopulational comparisons have noted the greater gracility of Near Eastern Neanderthals in lower latitudes when compared to European populations in higher ones – and accounted for most of the differences by invoking Bergman's and Allen's rules about body proportions and climate (Trinkaus 1987, 1989). The more robust Neanderthals in northern latitudes are seen as reflecting morphological adaptations to life in more stressed environments. Intrapopulational comparisons between the sexes have been few in number but have brought to light even more interesting variability. Frayer (1986), working with data from Central Europe, has argued for differential gracilization rates in EAMHs – with females gracilizing much earlier than males.

While the universality of Frayer's observations has recently been questioned by the observations of Ben-Itzhak, Smith, and Bloom (1988) of greater dimorphism in cortical thicknesses of humeri in AMHs than in the Neanderthals, this study is flawed by the use of mixed European and Near Eastern fossil remains, which obscures significant latitudinal differences in the behavior of Late Pleistocene hominids (Schoeninger 1982). Frayer's observations on asynchronic gracilization rates suggest that, whatever behavioral changes occurred at the time, their impact was far more dramatic on the females whose lifeways no longer required the same amounts of physical exertion as did those of Neanderthal females. Since females are more vulnerable than males to reproductive penalties of stress, this reduction in stress would have clearly benefitted them, their young, and their reproductive partners as well.

While in the past little attention has been paid to morphological evidence for stress in Late Pleistocene hominids (Jacobs 1985a, 1985b), present studies note a significantly higher incidence of dental hypoplasia among Neanderthals than observed among Holocene groups (Molnar and Molnar 1985; Ogilvie, Curran & Trinkaus 1989), as well as a decrease in dental attrition in immature Upper Paleolithic individuals when compared to the Neanderthals (Skinner 1981). Since dental hypoplasia develops in juveniles between the ages of 2 and 5, these data suggest considerably greater stress on Neanderthal children between weaning and adolescence than on their anatomically modern equivalents. The most inclusive study of stress to date, based on skeletal remains of Neanderthal and EAMHs in southwestern France, similarly shows a dramatic decrease in such evidence as Harris lines and dental hypoplasia between the Neanderthals and EAMHs (Brennan 1991). These findings then also indicate a reduction in stress through time, one which would necessarily have had more impact on the female and the more vulnerable immature segment of the population.

The age structure of the hominid remains themselves also underscores the benefits of whatever it was that happened at the transition for the subadults.

Table 1 compares the proportion of juveniles to adults among Pleistocene hominids of northern Eurasia and shows a significant decrease in juvenile mortality that can be noted across the transition (see Soffer 1992 for discussion of data selection). While 43% of the Neanderthals are under 12 years of age (defined as juveniles or subadults), less than 30% of EAMHs are subadults. Since the sizes of the sampled populations are roughly comparable, disparities in sample sizes cannot be invoked as explanations. Furthermore, both samples were primarily derived from comparable cave contexts. Finally, differential preservation favoring adults also does not account for the observed differences, because, if anything, it would produce a steady increase in the proportion of juveniles through time – a situation which is clearly as ex-

Table 1. Pleistocene Hominid Remains from Northern Eurasia (after Soffer 1992)

	Middle Pleistocene	Neanderthals[a]		EAMHs	
		MNI	MaxNI	MNI	MaxNI
Sample size:					
Total number found	85.5[b]	215	298	166	201
number aged	85.5	202	238	155	180
Juveniles (< 12 years):					
no.	22	87	96	46	51
%	25.7	43.1	40.3	29.7	28.3
Adults (> 13 years):					
no.	63.5	115	142	109	129
%	74.3	56.9	59.7	70.3	71.7

Middle Pleistocene Hominids vs Neanderthals:

1. MNI: X^2 value = 7.84; d.f. = 1; $0.001 < p < 0.01$; G statistic = 8.12
   X^2 with continuity correction = 7.11; d.f. = 1; $0.001 < p < 0.01$
2. MaxNI: X^2 value = 5.94; d.f. = 1; $0.01 < p < 0.02$; G statistic = 6.16
   X^2 with continuity correction = 5.32; d.f. = 1; $0.02 < p < 0.05$

Middle Pleistocene Hominids vs EAMHs:

1. MNI: X^2 value = 0.46; d.f. = 1; $0.50 < p < 0.70$; G statistic = 0.46
   X^2 with continuity correction = 0.28; d.f. = 1; $0.50 < p < 0.70$
2. MaxNI: X^2 value = 0.22; d.f. = 1; $0.50 < p < 0.70$; G statistic = 0.22
   X^2 with continuity correction = 0.11; d.f. = 1; $0.70 < p < 0.80$

Legend: a = Krapina MNI counts (Wolpoff 1988, pers. comm.); b = estimated MNI values for Chengyang of 1-2 individuals tabulated as 1.5; MNI = minimum number of individuals (for selection and calculations, see Soffer 1992); MaxNI = maximum number of individuals (for selection and calculations, see Soffer 1992); EAMH = early anatomically modern humans.

pected when we compare Middle Pleistocene hominids to the Neanderthals and the reverse for the Neanderthals vs EAMHs.

The data set used in this table is admittedly problematic. First, it is a very ambiguous indicator of past mortality, because there is no way of reliably factoring out possible biasing cultural practices such as preferential interments of just some segments of the population (children among the Neanderthals, for example). Secondly, the Krapina sample included in this table is also problematic because no consensus of opinion exists on the number of individuals represented by these highly fragmented remains.

Table 2 compares Neanderthal remains to EAMHs at four levels of Krapina exclusion or inclusion (for detailed discussion, see Soffer 1992). The observed X^2 values for all relevant permutations consistently show a statistically significant greater proportion of juveniles among Neanderthal remains.

Table 2. Late Pleistocene Hominid Remains from Northern Eurasia (after Soffer 1992)

	HSN#1		HSN#2		HSN#3		HSN#4	
	MNI	MaxNI	MNI	MaxNI	MNI	MaxNI	MNI	MaxNI
Sample size:								
Total number found	150	233	215	298	178	261	170-175	253-258
number aged	137	173	202	238	165	201	152	188
Juveniles (< 12 years)								
number	59	68	87	96	69	78	64	73
%	43.1	39.3	43.1	40.3	41.8	38.8	42.1	38.8
Adults (> 13 years)								
number	78	105	115	142	96	123	88	115
%	56.9	43.1	56.9	59.7	58.2	61.2	57.9	61.2

X^2 values: d.f. = 1

1. EAMH vs HSN # 1 - Neanderthal sample without Krapina remains.
   A. MNI: $X^2 = 5.66$, $0.01 < p < 0.02$, G statistic = 5.66, X^2 with continuity correction = 5.09, $0.02 < p < 0.05$
   B. MaxNI: $X^2 = 4.75$, $0.02 < p < 0.05$, G statistic = 4.76, X^2 with continuity correction = 4.27, $0.02 < p < 0.05$
2. EAMH vs HSN # 2 - Neanderthal sample with Wolpoff's (1988, pers. comm.) estimates for Krapina.
   A. MNI: $X^2 = 6.73$, $0.001 < p < 0.01$, G statistic = 6.80, X^2 with continuity correction = 6.17, $0.01 < p < 0.02$
   B. MaxNI: $X^2 = 6.48$, $0.01 < p < 0.02$, G statistic = 6.55, X^2 with continuity correction = 5.96, $0.01 < p < 0.02$
3. EAMH vs HSN # 3 - Neanderthal sample with White's (1988) estimates for Krapina.
   A. MNI: $X^2 = 5.12$, $0.02 < p < 0.05$, G statistic = 5.14, X^2 with continuity correction = 4.60, $0.02 < p < 0.05$
   B. MaxNI: $X^2 = 4.65$, $0.02 < p < 0.05$, G statistic = 4.68, X^2 with continuity correction = 4.19, $0.02 < p < 0.05$
4. EAMH vs HSN # 4 - Neanderthal sample with Trinkaus's (1988) estimates for Krapina.
   A. MNI: $X^2 = 5.16$, $0.02 < p < 0.05$, G statistic = 5.17, X^2 with continuity correction = 4.63, $0.02 < p < 0.05$
   B. MaxNI: $X^2 = 4.54$, $0.02 < p < 0.05$, G statistic = 4.08, X^2 with continuity correction = 4.07, $0.02 < p < 0.05$

Legend: HSN = *Homo sapiens neanderthalensis*; EAMH = early anatomically modern humans; MNI = minimum number of individuals; MaxNI = maximum number of individuals. For calculations see Soffer 1992.

These conclusions are in accord with Nemeskeri (1989), who also noted a decline in subadult mortality between the Neanderthal and AMHs – one clearly visible when these populations were divided by latitude. His data led him to conclude not only generally longer life spans for Neanderthal females in the Near East (an indirect indicator of lesser stress when compared to their European counterparts), but also a significant interregional difference in mortality across European observation, which correlates well with different intensities across the continent of environmental stress (see Musil 1985).

Given the problems with the data presented here, I neither claim their demographic accuracy nor that they unambiguously indicate a major change in the pattern of juvenile mortality. Combined with other indicators, however, they do suggest a reduction of stress on juveniles across the Middle to Upper Paleolithic transition. I also suspect, but cannot demonstrate, because of the ambiguity in sexing fragmentary skeletal remains, that the same decrease in stress and in mortality should be apparent in the EAMHs females inhabiting northern latitudes as well. The diverse evidence for greater stress on Neanderthal juveniles and females is significant, because it would have made these populations less viable through greater mortality than EAMHs. Such a disparity would have given a reproductive edge to EAMHs – an edge which recent demographic modeling predicts could have, in cases of sympatry, brought on Neanderthal extinction in as short a period as 30 generations or 1,000 years (Zubrow 1989).

Archaeological Data. The archaeological evidence from the transition is equally instructive about the differences between the respective lifeways. A cursory look at the distribution of Middle and Early Upper Paleolithic sites across Eurasia suggests regional differences in the intensity of occupation. Gamble (1986) has convincingly argued that the Southwestern and Mediterranean provinces witnessed a continuous human presence throughout the Middle and Late Pleistocene while the northern areas were not permanently occupied before the Upper Paleolithic.

Similar disparities can also be monitored in the settlement record of Eastern Europe. They are most clearly seen on the East European Plain, a vast flat expanse of land rimmed by north-south running mountain chains which restrict vertical biotic differentiation to its western, southeastern, and eastern margins. Paleoenvironmental reconstructions indicate that climatic conditions during the Last Interglacial and subsequent interstadials were differentially distributed across Europe, and that the Plain had significantly more precipitation and pronounced warming which resulted in its forestation (Gerasimov and Velichko 1982; Musil 1985; Frenzel, Pécsi & Velichko 1992). At stadial times, the Plain changed to an open landscape covered by a periglacial steppe – one with a dramatic reduction in latitudinal biotic differentiation and diversification and an increase in the spatial and temporal unpredictability of the biotic resources. The only exceptions to this were found in the Dnestr-Prut region in the west, in the Crimea in the southeast, and, of course, in the nearby Caucasus. The proximity of the plains, foothills, and mountain ranges created a number of ecotones with more complex, diverse, and productive biotic communities during both stadial and interstadial times. Musil (1985) has demonstrated that the Crimea and the Caucasus regions represented optimal, least stressed, environments in Eastern Europe throughout the Late Pleistocene–making these regions more akin to environments in southwestern France and Spain than to those on the nearby Eastern European Plain.

Archaeological remains in the Dnestr-Prut area come primarily from open-air stratified deposits with a number of Middle Paleolithic layers overlain by Upper Paleolithic deposits (Soffer 1989). The Crimean record is also stratified but made somewhat differently with predominantly rock-shelters.

Stratification of occupation is also in evidence at the Kostenki-Borshchevo sites along the Don. Here, however, only Upper Paleolithic materials are represented. Finally, the central part of the Plain lacks both stratified sites dating to either the Middle or to the Upper Paleolithic, and sites where these inventories are found superimposed. Middle Paleolithic sites here are extremely sparse, while the Upper Paleolithic ones all date younger than 26,000 BP (Soffer 1985). The significance of the Middle Paleolithic record from this area is that it is precisely those regions with the greatest vertical and biotic diversification that contain evidence for occupation throughout the last glaciation (Soffer 1989). The regional clustering of Middle Paleolithic sites indicates that ecotonal areas with proximal resource diversification saw more continuous occupation, while other more homogeneous loessic parts of the Plain witnessed a pattern of sporadic and discontinuous colonization and abandonment. Regions occupied during the Early Upper Paleolithic are also localized in two disparate parts–the middle Dnestr and the middle Don. The appearance of Upper Paleolithic sites in the Kostenki-Borshchevo area marks a recolonization of the open plain, which, in contrast to previous aborted attempts, expands through time.

The occupation record of eastern Eurasia mirrors this pattern also. Boriskovskii (1984, 1989) documents the presence of multilayered Middle Paleolithic sites in the Caucasus which were followed by Upper Paleolithic occupations. Davis (1987, 1990) notes the same regionalization of pre-Upper Paleolithic sites in Central Asia. Derevyanko et al. (1990), Drozdov et al. (1990), and Medvedev, Saval'ev & Svinin (1990) show that multilayered Middle Paleolithic Siberian sites are found in the southernmost diversified landscapes, while Klein (1989) and Frenzel, Pécsi, and Velichko (1992) demonstrate that the more homogeneous northern environments saw continuous human occupation only from the Upper Paleolithic on.

In summary then, the data from Eurasia suggest that these two continents were not occupied by Middle Paleolithic groups from the Atlantic to the Pacific. Human groups there were localized in discrete regional patches and were continuously present only in regions with vertical differentiation and proximal resource diversification, while Upper Paleolithic groups successfully colonized the more open biomes (Soffer 1992; Velichko 1988). In Europe this meant the presence of Middle Paleolithic groups in southwestern France and Spain, Italy, the Balkans and contiguous regions, around the Carpathians, in Crimea, and in the Caucasus. In Siberia, this meant their presence around the Altai, Saian, Yablonovy, and other southern ranges.

Faunal data from the Middle and Early Upper Paleolithic sites on the Russian Plain indicate more regionally circumscribed opportunistic subsistence strategies during the Middle Paleolithic and a change to foraging adaptations which involved mapping onto resources over wider areas and entailed seasonal group mobility in the Upper Paleolithic (Soffer 1989). Similar observations, evident only from a regional rather than a site-specific perspective, have been made for Western Europe by other researchers as well (Binford 1985, 1989; Gamble 1986; Farizy and Jaubert 1991). While some have interpreted Middle Paleolithic sites with primarily monofaunal inventories, such as Hortus, Starosel'e, Solutre, or Sukhaia Mechetka (Volgogradskaia), as evidence for the presence of specialized hunting in the Middle Paleolithic (e.g., Chase 1989), the pertinent regional records reveal instead the presence of what Farizy and Jaubert (1991) have termed "locational fidelity" and Binford (1987) has called "niche geography." Specifically, when seen in regional contexts, such sites do not show evidence for specialization in procurement but rather reflect repeated visits to locations with a specific abundant resource – a pattern of foraging behavior amply documented for a number of nonhuman primates (Garber 1987).

Evidence for raw-material procurement and use, and for the utilization of space at the sites, similarly indicates changes between the Middle and the Upper Paleolithic (Geneste 1988; Kuhn 1990; Roebroeks, Kolen & Rensink 1988; Soffer 1989, 1991). Eurasian Middle Paleolithic inventories show a redundant use of local lithic raw materials (coming from < 30 km away) regardless of their quality, suggesting a very localized and highly mobile settlement system, while Early Upper Paleolithic sites contain varying but redundant amounts of superior exotics. More significantly, Middle Paleolithic sites contain practically no art or items of personal adornment as well as a paucity of material remains we would now associate with belief systems or ideology (Conkey 1983; White 1982). Upper Paleolithic inventories, on the other hand, are replete with nonutilitarian material remains (White 1989). This patterning has generated arguments concerning limited cognitive or symboling capacities in premodern humans – an issue I address below.

Relatively small concentrations of cultural remains, sometimes with shallow unstructured hearths and often at least partially overlapping one another, are characteristic of the known Russian Middle Paleolithic sites and probably resulted from repeated palimpsest occupations. Elaborate features reported at some sites, mammoth bone dwellings at Molodova I and IV, for example, are, as I have argued elsewhere, in all likelihood fanciful present-day constructs (Soffer 1989). The organization of space and the structure of features is also far clearer at other European Upper Paleolithic sites (Binford 1985, 1989; Farizy 1988; Gamble 1986; Kuhn 1990; Simek 1987; Stiner 1990). This clear patterning, like its muted Middle Paleolithic predecessor, is redundant from site to site and more recognizable through ethnographically known examples.

In summary, disparate archaeological data also indicate significant changes between the Middle and Upper Paleolithic. Evidence from the Russian Plain, where the early Upper Paleolithic sites divide into those with strong Middle Paleolithic influence and those without, suggests that these changes involved: (1) a qualitative change in both the perception and utilization of nature, (2) a change in food management strategies and in the accompanying settlement systems, and (3) that these changes clearly did not happen overnight nor at a particular chronological or morphological Rubicon (Soffer 1989).

THE SIGNIFICANCE OF THE RECORD

In Europe, both East and West, there is a distinct absence of a neat synchronic fit between culture and morphology. In the Crimea, for example, EAMHs are found with late Middle Paleolithic inventories at Starosel'e (Alexeyev 1976; Formozov 1958). A Neanderthal with an early Upper Paleolithic industry has been found at Saint-Césaire in France (Harrold 1989). Similar juxtapositions are documented in the Near East and North Africa as well (Clark 1989; Bar-Yosef 1989). This, together with controversial but persistent evidence for some Middle Paleolithic burials (Chase and Dibble 1987; Smirnov 1989), and some, although admittedly few, pieces of nonutilitarian objects (Marshack 1988, 1989) suggest that our arguments about differential capacities of the two types of hominids are wanting. They are wanting because human behavior, as well as the behavior of higher primates, is a product not only of phylogeny but also of ontogeny and history. By focusing on questions of capacity and totally ignoring the actualization of capacity into performance (sensu Vygotsky 1986, 1987), we have ignored the social contexts crucial to understanding habitual behavior, and thereby effectively sabotaged our understanding of what happened at the transition. Primatological literature is replete with examples of paracultural behavior among primates. Strum and Mitchell (1987:97), for example, report that examples of cooperation, division of labor, strategy, and cognition all have precursors among nonhuman primates – thus, these tendencies appear primate-specific rather than human-specific. McGrew (1986), Boesch and Boesch (1984), and Kortland and Holzhaus (1987) have all documented not only tool use among chimpanzees, but one which varies between populations, and between communities and even sexes within a single population. This variability, as McGrew (1986) notes, reflects both differences in habitats and differences in social customs, leading him to conclude that the apparent limits of chimpanzee achievement may be cultural rather than organic (use of nutting stones by some groups and not others, for example). I submit that if chimps and gorillas can be taught rudimentary symboling (Linden 1981; Patterson and Linden 1981), if macaques can invent food washing and winnowing (Kawai 1963), if chimps can make fire and use inventories not too dissimilar from those of the Tasmanians (McGrew 1987), then questions of Neanderthal vs EAMH capacities are practically irrelevant while issues of performances are in dire need of research. Since primatological data are redundant in showing that the capacity for symboling, tool use, and cooperation and sharing are all well entrenched in nonhuman primates, and, as Holloway (1991) and others have argued, there are no differences in the external brains of Neanderthals and EAMHs, we should now refocus our research attention to the social contexts under which the primate capacity for such behaviors was habitually exercised and became a part of the behavioral repertoire of hominids, thus turning them into what we recognize as fully modern humans (Alexander 1989).

Explanations in the paleoanthropological literature for the differences between the Neanderthals and EAMHs pointed to changes such as from tooth use to tool use (Smith 1983), or from flakes to blades, or the advent of insert technology and long-distance kills (Frayer 1981, 1984). These explanations are clearly particularistic and insufficient to account for either all the noted changes or for the mosaic nature of these changes. Deus ex machina explanations – the arrival from elsewhere of modern humans, whether in person or through gene flow, or some combination of the two – are equally unsatisfactory, for they neither account for the nonsynchrony between morphological forms and cultural behavior nor, more importantly, specify what it is that the EAMHs did that was more adaptive than what the Neanderthals did.

Archaeological explanations which propose changes in sociocultural behavior also leave much to be desired, for, by and large, they are either nonspecific (e.g., postulating generalized cultural changes) or too particularistic (e.g., focusing on data from single sites), and, more significantly, both avoid dealing with the biological data and get mired in phenomenological arguments.

Given the total of paleoanthropological evidence, I suggest that the differences between Middle and Upper Paleolithic in northern Eurasia lay not in blade vs. teeth or in any morphological hardware, but in a dramatic change in economic and social relationships. Although each data set I presented is problematic, their patterning suggests that the time has come for us to abandon our latent but unproven assumptions that biparental provisioning of the young, division of labor, and food sharing between their progenitors goes back to the australopithecines (Isaac 1983; Lovejoy 1981). Following a number of recent models for potential social organization of premoderns (e.g., Foley and Lee 1989; Wrangham 1987), I suggest the possibility that such reorganization in both subsistence and social behavior took place much later in time – specifically arguing here that the sex-based separation and division of labor as we know it from the ethnographic record, occurred around the Middle to the Upper Paleolithic transition. Cucchiari (1981), working from social theory outward, has offered a similar argument about the Upper Paleolithic gender revolution. In contrast to him, given the primate pattern of dimorphic feeding ranges and situational food-sharing between males and females, I do not see this change as a revolution, but rather as an evolution of behavior already present but not habitually used or depended on in the repertoire of higher primates.

I argue that the small site size we see in the Middle Paleolithic, coupled with the use of local raw materials, opportunistic exploitation of biotic resources, a general absence of clear-cut site types, and a regionalization of the sites, suggests the existence of small, stable, coresidential units which moved often but within very restricted geographic space. This, together with evidence for muscle hypertrophy in both sexes, an inherited pattern of dimorphic feeding ranges present in a number of primates and postulated for the sexes during the Early and Middle Pleistocene (Frayer and Wolpoff 1985), and very equivocal evidence for food-sharing (Binford 1985), suggests to me the existence of diurnal small-sized foraging and coresidential units of mothers with their immature young plus, perhaps, a small number of related adult females, as well as small-sized all-male units. The dimorphism in feeding behavior observed among primates, and possibly present in earlier Pleistocene hominids, implies that Neanderthal females and their young may have had more restricted day ranges than males. This, in turn, suggests that males may have fed on a greater number of larger herbivores and consumed more large migratory taxa than females, who would have ingested more locally abundant small-sized species of both plants and animals. This scenario clearly carries testable propositions and predicts a greater difference in strontium isotope bone chemistry and trace element composition of bones between the Neanderthal males and females than between those of EAMHs (S. Ambrose 1991, pers. comm.).

The Neanderthal social construct outlined above is just a first approximation, and I do not preclude some situational larger sized aggregations. I base such a scenario on the social organization of hominoids inhabiting regions outside of the tropical forests (Susman 1987). I offer these scenarios not as empirically proven entities but as "something good to think" – to underscore that, given the data, we should begin thinking about Neanderthal cultural behavior in a nonnuclear family mode. Furthermore, the 74% adult to 26% juvenile ratio for the Middle Pleistocene hominids (table 1) is more like the one for EAMHs than the Neanderthals, and suggests possibilities of yet other forms of social organization further back in time such as, perhaps, combined group foraging (for possible Lower and Middle Pleistocene hominid social constructs see, e.g., Foley and Lee 1989).

THE TRANSITION SCENARIO

I now turn to considerations of some of the reasons for this social reorganization and argue that the structure of the resource base in northern environments, especially open loessic ones, presented hominids with a set of specific problems solved one way by the Neanderthals and another by EAMHs. The latitudinal increase in the patchiness and unpredictability of the food resources, especially significant in stadial times, confronted Neanderthals there with the need for much greater territories than in lower latitudes – a latitudinal imperative valid for all carnivores (Foley and Lee 1989; Gamble 1986). This need to exploit larger territories would have created special problems for the females with young. The Neanderthal solution was to occupy permanently those regions which had proximal resource diversification where female day ranges could be minimized (fig. 1). Binford (1989) has argued that it is precisely such a range shift which is reflected in the "regionalization" of the archaeological record noted by Middle Paleolithic times.

Anatomically modern humans solved these environmental problems differently (fig. 1). Confronted with an increase in temporal fluctuations in the availability of food resources, in the package size of prey resources, and an increase in the nonsynchrony of acquisition among individuals, they relied on food-sharing – a solution selected for in northern latitudes and one whose extent and inclusiveness grades latitudinally in both the ethological record of carnivores and the ethnographic record of hunter-gatherers (Foley and Lee 1989; Whallon 1989). I argue that this solution: (1) came in the form of division of labor, sharing, and biparental provisioning of the young; (2) created horizontal economic and social bonds which we recognize as uniquely human – namely, the interdependent family; and (3) linked these social units with others like them in time and space through equally uniquely human constructs of kinship and descent. As Cucchiari (1981) has noted, kinship is a sociocultural system of different distribution of rights, duties, rules, and statuses founded on the ideology of shared substances. It is a system inextricably bound up with an embedded system of gender categories reproduced by the kinship system. I submit that it is precisely this cultural definition of the

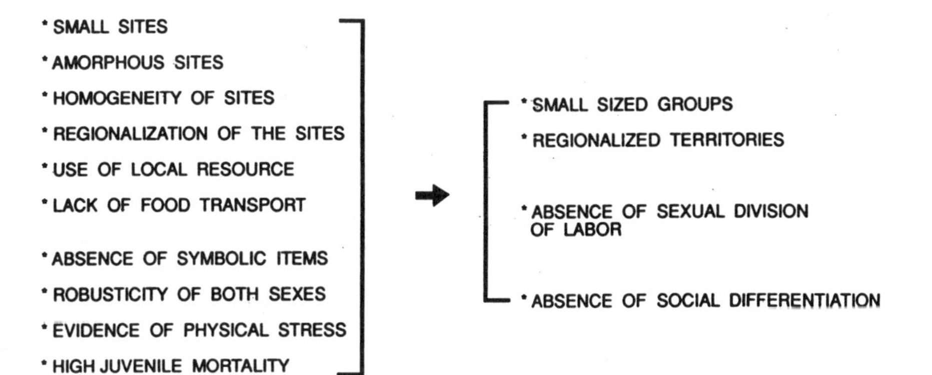

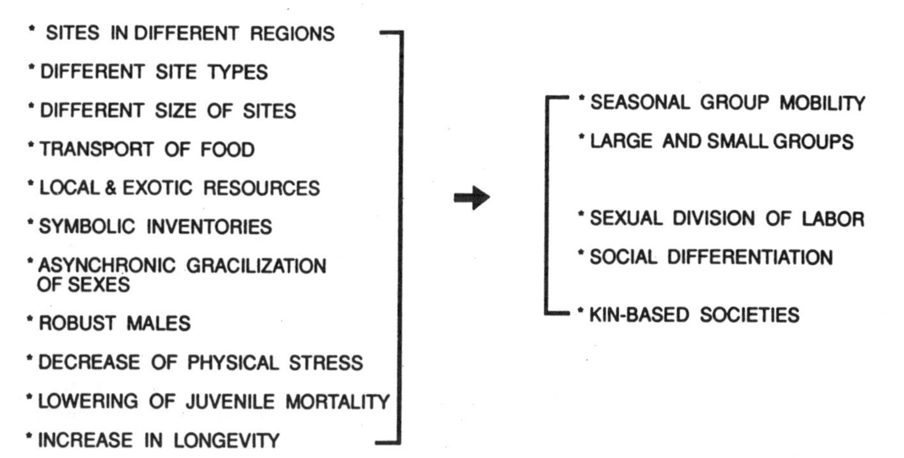

Figure 1. Paleoanthropological and archaeological data and their implications for settlement patterns and social organization.

self and its positioning in the social matrix of like and unlike others through time and across space that is reflected in the so-called "Creative Explosion" – the proliferation of items of nonutilitarian material culture characteristic of the Eurasian Upper Paleolithic.

The transition scenario offered in this chapter is one for hominids in northern latitudes, but one not arguing for a direct cause and effect relationship between the environment and behavior. Rather, it reiterates Gamble's (in press) observations that moderns are modern because of the society they construct and live in – that social context is all. Here I offer but one route to division of labor, food-sharing, and kinship, leaving alternative possible trajectories for others to explore.

Furthermore, since at the transition we are also dealing with two hominid forms, each of whom is not exclusively associated with specific cultural behavior, it remains unclear if these solutions were exclusive to a particular hominid type. Wolpoff (1979, 1980), for example, has noted an increase in deciduous tooth size among late Neanderthals and interpreted this to mean earlier weaning and a possible reduction in birth spacing. Skinner (1989) reported a continuation of this pattern of early weaning into the Upper Paleolithic. Earlier weaning in children may suggest either the advent of male provisioning or perhaps female-specific technological advances in the preparation of soft weaning foods. If the former was the case, as it may have been, since female gracilization noted for the EAMHs would not have happened overnight but would have taken time to develop, then what we may be witnessing is a strategy of reduced birth spacing begun by Neanderthals and continued by EAMHs (P. Shipman 1989, pers. comm.). In the case of the Neanderthals, such a strategy may not have been subsidized by provisioning help for the mothers, resulting in greater stress on juveniles and a continuation of high juvenile mortality. In the case of EAMHs, backed by biparental provisioning of the young, it was, and led to a dramatic reduction in stress on the subadults.

Finally, recent research by Kuhn (1990) and Stiner (1990) on Italian Late Pleistocene sites suggests that a transition from premodern to modern behavior may have occurred there around 55,000 BP. Kuhn's (1990) study of lithic inventories, from a number of Middle Paleolithic layers, notes a change at that time in the economies of stone tool manufacture and use which mirror the timing of the change from scavenging to ambush-procurement of ungulates, documented for the same layers by Stiner (1990). In Italy then, these changes occur squarely within the context of the Late Middle Paleolithic, and the lack of hominid fossil remains prevents us from assigning them to specific ancestor types. On the other hand, data from the Kostenki region in the East European Plain shows the probably coterminous presence of two distinct economic systems of resource procurement during the Early Upper Paleolithic – one more Middle Paleolithic-like and the other more Upper Paleolithic in nature (the Spitsyn and Streletskaia sites discussed above). My observations once again strongly indicate that we are not dealing with innate differences in the capacity for particular behavior between the Neanderthals and the EAMHs, but rather just with the habitual practice of that behavior – a practice which, I suggest, may have begun in the Late Middle Paleolithic but became entrenched in the Upper Paleolithic and generated the kind of an archaeological record we recognize as structurally similar to ethnographically known cases.

ACKNOWLEDGMENTS

The writing of this chapter has benefitted from friendly critiques offered by a number of colleagues, including Stanley Ambrose, Rachel Caspari, Geoff Clark, Carol Delaney, Clive Gamble, Alma Gottlieb, Linda Klepinger, Marcel Otte, Karen Rosenberg, Pat Shipman, John Speth, and Bob Whallon, not all of whom necessarily agree with the final product. My special gratitude goes to Milford Wolpoff and Dave Frayer for their generosity with unpublished data on Eurasian hominid remains and their collegial tolerance of archaeologists meddling in matters paleoanthropological. My research in Central and Eastern Europe, on which parts of this chapters are based, has over the years been supported by IREX, the National Academy of Sciences, the Fulbright-Hayes exchange programs, the National Geographic Society, as well as by the Hewlett Fund and the Research Board of the University of Illinois – their help is, once again, gratefully acknowledged.

REFERENCES

Alexander, R. D. 1989. Evolution of the human psyche, 455-513. In *The Human Revolution: Behavioural and Biological Perspectives on the Origins of Modern Humans*, ed. P. Mellars and C. Stringer. Edinburgh: Edinburgh University Press.

Alexeyev, V. P. 1976. Position of the Staroselye find in the hominid system. *Journal of Human Evolution* 5:413-21.

Bar-Yosef, O. 1989. Geochronology of the Levantine Middle Palaeolithic, 589-610. In *The Human Revolution: Behavioural and Biological Perspectives on the Origins of Modern Humans*, ed. P. Mellars and C. Stringer. Edinburgh: Edinburgh University Press.

Ben-Itzhak, S., P. Smith, and R. A. Bloom. 1988. Radiographic study of the humerus in Neanderthals and *Homo sapiens sapiens*. *American Journal of Physical Anthropology* 77: 231-42.

Binford, L. R. 1985. Human ancestors: Changing views of their behavior. *Journal of Anthropological Archaeology* 4:292-327.

Binford, L. R. 1987. Searching for camps and missing the evidence: Another look at the Lower Paleolithic, 17-32. In *The Pleistocene Old World*, ed. O. Soffer. New York: Plenum Press.

Binford, L. R. 1989. Isolating the transition to cultural adaptations: An organizational approach, 18-41. In *The Emergence of Modern Humans: Biocultural Adaptations in the Later Pleistocene*, ed. E. Trinkaus. Cambridge: Cambridge University Press.

Boesch, C., and H. Boesch. 1984. Possible causes of sex differences in the use of natural hammers by wild chimpanzees. *Journal of Human Evolution* 13:415-40.

Boriskovskii, P. I., ed. 1984. *Paleolit SSSR*. Moscow: Nauka.

Boriskovskii, P. I., ed. 1989. *Paleolit Kavkaza i Severnoj Azii*. Leningrad: Nauka.

Brennan, M. U. 1991. Health and disease in the Middle and Upper Paleolithic of Southwestern France: A bioarchaeological study. Ph.D. diss., New York University, New York, NY.

Chase, P. G. 1989. How different was Middle Palaeolithic subsistence? A zooarchaeological perspective on the Middle to Upper Palaeolithic transition, 321-37. In *The Human Revolution: Behavioural and Biological Perspectives on the Origins of Modern Humans*, ed. P. Mellars and C. Stringer. Edinburgh: Edinburgh University Press.

Chase, P. G., and H. L. Dibble. 1987. Middle Palaeolithic symbolism: A review of current evidence and interpretations. *Journal of Anthropological Archaeology* 6:263-96.

Clark, J. D. 1989. The origins and spread of modern humans: A broad perspective on the African evidence, 565-88. In *The Human Revolution: Behavioural and Biological Perspectives on the Origins of Modern Humans*, ed. P. Mellars and C. Stringer. Edinburgh: Edinburgh University Press.

Conkey, M. 1983. On the origins of Paleolithic art: A review and some critical thoughts, 201-27. In *The Mousterian Legacy*, ed. E. Trinkaus. Oxford: British Archaeological Reports International Series 164.

Cucchiari, S. 1981. The gender revolution and the transition from bisexual horde to patrilocal band: The origins of gender hierarchy, 30-79. In *Sexual Meaning: The Cultural Construction of Gender and Sexuality*, ed. S. B. Ortner and H. Whitehead. Cambridge: Cambridge University Press.

Davis, R. S. 1987. Regional perspectives on the Soviet Central Asian Paleolithic, 121-34. In *The Pleistocene Old World*, ed. O. Soffer. New York: Plenum Press.

Davis, R. S. 1990. Pleistocene climates and migration into Asia: Evidence from the loess. Paper presented at the International Symposium on the Chronostratigraphy of the Paleolithic in North, Central, East Asia and America. Novosibirsk, USSR. July.

Derevyanko, A. P., Yu. V. Grichan, M. I. Dergachev, A. N. Zenin, S. A. Laukhin, G. M. Levkovskaia, A. M. Mololetko, S. V. Markin, V. I. Molodin, N. D. Ovodov, V. T. Petrin, and M. V. Shun'kov. 1990. *Arkheologiia i paleoekologiia paleolita Gornogo Altaia*. Novosibirsk: IIPP SOAN SSSR.

Drozdov, N. I., V. P. Chekha, S. A. Laukhin, V. G. Kol'tsova, E. V. Akimova, A. V. Ermolaev, V. P. Leont'ev, S. A. Vasil'ev, A. F. Yakskikh, G. A. Demidenko, E. V. Artem'ev, A. A. Vikulov, A. A. Bokarev, I. V. Foronova, and S. D. Sodoras. 1990. *Khronostratigrafiia paleoliticheskikh pamiatnikov Srednej Sibiri*. Novosibirsk: IIPP SOAN SSSR.

Farizy, C. 1988. Spatial organization and Middle Palaeolithic open air sites. Paper presented at the International Colloquium Interpretazione funzionale dei dati i paleontologia. Rome, June.

Farizy, C., and J. Jaubert. 1991. Middle Palaeolithic specialised sites: The case of Western France-French Pyrenees. Paper presented at the 56th Annual Meeting of the Society for American Archaeology. New Orleans, April 24-28, 1991.

Foley, R. A., and P. C. Lee. 1989. Finite social space, evolutionary pathways, and reconstructing hominid behavior. *Science* 243:901-6.

Formozov, A. A. 1958. Perschernaia stoianka Starosel'e i ee mesto v Paleolite. Matieriali i Issledovaniia po Arckheologii No. 71. Moscow: Izdatel'stvo Akademii Nauk SSSR.

Frayer, D. W. 1981. Body size, weapon use, and the natural selection in the European Upper Paleolithic and Mesolithic. *American Anthropologist* 83:57-73.

Frayer, D. W. 1984. Biological and cultural change in European Late Pleistocene and Early Holocene, 211-50. In *The Origins of Modern Humans*, ed. F. H. Smith and F. Spencer. New York: Alan R. Liss.

Frayer, D. W. 1986. Cranial variation at Mladec and the relationship between Mousterian and Upper Paleolithic hominids. *Anthropos* (Brno) 23:243-56.

Frayer, D. W., and M. H. Wolpoff. 1985. Sexual dimorphism. *Annual Review of Anthropology* 14:429-73.

Frenzel, B., M. Pécsi, and A. A. Velichko, eds. 1992. *Atlas of Paleoclimates and Paleoenvironments of the Northern Hemisphere. Late Pleistocene-Holocene*. Budapest: Geographical Research Institute, Hungarian Academy of Sciences, and Stuttgart: Gustav Fischer Verlag.

Gamble, C. 1986. *Palaeolithic Europe*. Cambridge: Cambridge University Press.

Gamble, C. In press. *Timewalkers*.
Garber, P. A. 1987. Foraging strategies among living primates. *Annual Review of Anthropology* 16:339-64.
Geneste, J.-M. 1988. Systèmes d'approvisionnement en matières premières au Paléolithique moyen et au Paléolithique supérieur en Aquitaine, 61-70.In *L'Homme de Néandertal.* Vol. 8, *La Mutation*, ed. M. Otte. Liège: Etudes et Recherches Archéologiques de l'Université de Liège 35.
Gerasimov, I. P., and A. A. Velichko, eds. 1982. *Paleogeografiya Evropy za Posledniye Sto Tysyach Let*. Moscow: Nauka.
Harrold, F. B. 1989. Mousterian, Châtelperronian and Early Aurignacian in Western Europe: Continuity or discontinuity? 677-713. In *The Human Revolution: Behavioural and Biological Perspectives on the Origins of Modern Humans*, ed. P. Mellars and C. Stringer. Edinburgh: Edinburgh University Press.
Holloway, R. L. 1991. The Neandertal brain: What is primitive. *American Journal of Physical Anthropology*. Supplement 12:94 (Abstract).
Isaac, G. L. 1983. Aspects of human evolution, 509-43. In *Evolution from Molecules to Man*, ed. D. S. Bendall. Cambridge: Cambridge University Press.
Jacobs, K. 1985a. Climate and the hominid postcranial skeleton in Würm and Early Holocene Europe. *Current Anthropology* 26:512-14.
Jacobs, K. 1985b. Evolution in the postcranial skeleton of Late Glacial and Postglacial European hominids. *Zeitschrift fur Morphologie und Anthropologie* 75:307-26.
Kawai, M. 1963. On the newly acquired behavior of the natural troop of Japanese monkeys on Koshima Island. *Primates* 4:113-15.
Klein, R. G. 1989. *The Human Career*. Chicago: University of Chicago Press.
Kortland, A., and E. Holzhaus. 1987. New data on the use of stone tools by chimpanzees in Guinea and Liberia. *Primates* 28:473-96.
Kuhn, S. L. 1990. Diversity within uniformity: Tool manufacture and use in the Pontian Mousterian in Latium (Italy). Ph.D. diss. University of New Mexico, Albuquerque, NM.
Linden, E. 1981. *Apes, Men, and Language*. Middlesex: Penguin Books Ltd.
Lovejoy, C. O. 1981. The origin of man. *Science* 211:341-50.
Marshack, A. 1988. The Neanderthals and the human capacity for symbolic thought: Cognitive and problem-solving aspects of Mousterian symbol, 57-91. In *L'Homme de Néandertal*, Vol. 5, *La Pens'ee*, ed. M. Otte. Liège: Etudes et Recherches Archéologiques de l'Université de Liège 32.
Marshack, A. 1989. Evolution of the human capacity: The symbolic evidence. *Yearbook of Physical Anthropology* 32:1-34.
McGrew, W. C. 1986. Chimpanzee material culture: What are its limits, and why? In *The Pleistocene Perspective*, ed. M. Day and R. Foley. Vol. I. The World Archaeological Congress: Allen & Unwin.
McGrew, W. C. 1987. Tools to get food: The subsistence of Tasmanian aborigines and Tanzanian chimpanzees compared. *Journal of Anthropological Research* 43:247-58.
Medvedev, G. I., N. A. Saval'ev, and V. V. Svinin. 1990. *Stratigrafiia, paleogeografiia i arkheologiia iuga Srednej Sibiri*. Irkutsk: Irkutsk Gosudarstvennij Universitet.
Mellars, P., ed. 1990. *The Emergence of Modern Humans: An Archaeological Perspective*. Edinburgh: Edinburgh University Press.
Mellars, P., and C. Stringer, eds. 1989. *The Human Revolution: Behavioural and Biological Perspectives on the Origins of Modern Humans*. Edinburgh: Edinburgh University Press.
Molnar, S., and I. M. Molnar. 1985. The incidence of enamel hypoplasia among the Krapina Neandertals. *American Anthropologist* 87:536-49.
Musil, R. 1985. Paleobiography of terrestrial communities in Europe during the Last Glacial. *Acta Musei Nationalis Pragae* XLI B, no. 1-2.
Nemeskeri, J. 1989. An attempt to reconstitute demographically the Upper Palaeolithic populations of Europe and the Mediterranean region, 335-63. In *People and Culture in*

Change, ed. I. Hershkovitz. Oxford: British Archaeological Reports International Series 508(i).

Ogilvie, M. D., B. K. Curran, and E. Trinkaus. 1989. Incidence and patterning of dental enamel hypoplasia among the Neandertals. *American Journal of Physical Anthropology* 79:25-41.

Patterson, F., and E. Linden. 1981. *The Education of Koko*. New York: Holt, Rinehart & Winston.

Roebroeks, W., J. Kolen, and E. Rensink. 1988. Planning depth, anticipation, and the organization of Middle Palaeolithic technology: The 'Archaic Natives' meet Eve's descendants. *Helinium* 28:17-34.

Schoeninger, M. J. 1982. Diet and the evolution of modern human form in the Middle East. *American Journal of Physical Anthropology* 58:37-52.

Simek, J. F. 1987. Spatial order and behavioural change in the French Palaeolithic. *Antiquity* 61:25-40.

Skinner, M. 1981. Dental attrition in immature hominids of the Late Pleistocene: Implications for adult longevity. *American Journal of Physical Anthropology* 54:278-79.

Skinner, M. 1989. Dental attrition and enamel hypoplasia among Late Pleistocene immature hominids from Western Europe. *American Journal of Physical Anthropology* 78:303-4. (Abstract).

Smirnov, Yu. A. 1989. Intentional human burial: Middle Paleolithic (Last Glaciation) beginnings. *Journal of World Prehistory* 3:199-233.

Smith, F. H. 1983. Behavioral interpretation of changes in craniofacial morphology across the Archaic/Modern *Homo sapiens* transition, 141-63. In *The Mousterian Legacy*, ed. E. Trinkaus. Oxford: British Archaeological Reports International Series 164.

Smith, F. H., J. F. Simek, and M. S. Harrill. 1989. Geographic variation in supraorbital torus reduction during the later Pleistocene (c. 80 000-15 000 BP), 172-93. In *The Human Revolution: Behavioural and Biological Perspectives on the Origins of Modern Humans*, ed. P. Mellars and C. Stringer. Edinburgh: Edinburgh University Press.

Smith, F. H., and F. Spencer, eds. 1984. *The Origins of Modern Humans: A World Survey of the Fossil Evidence*. New York: Alan R. Liss.

Soffer, O. 1985. *The Upper Paleolithic of the Central Russian Plain*. New York: Academic Press.

Soffer, O. 1989. The Middle to Upper Palaeolithic transition on the Russian Plain, 714-42. In *The Human Revolution: Behavioural and Biological Perspectives on the Origins of Modern Humans*, ed. P. Mellars and C. Stringer. Edinburgh: Edinburgh University Press.

Soffer, O. 1991. Lithics and lifeways: The diversity in raw material procurement and settlement systems on the Upper Paleolithic East European Plain, 221-34. In *Raw Material Economy Among Prehistoric Hunter-Gatherers*, ed. A. Montet-White and S. Holen. University of Kansas Publications in Anthropology 19.

Soffer, O. 1992. Social transformations at the Middle to Upper Paleolithic transition: The implications of the European record, 247-59. In *Continuity or Replacement: Controversies in* Homo sapiens *Evolution*, ed. G. Bräuer and F. H. Smith. Rotterdam: A. A. Balkema.

Stiner, M. 1990. The ecology of choice: Procurement and transport of animal resources by Upper Pleistocene hominids in west-central Italy. Ph.D. diss. University of New Mexico, Albuquerque, NM.

Stoneking, M., and R. L. Cann. 1989. African origin of human mitochondrial DNA, 17-30. In *The Human Revolution: Behavioural and Biological Perspectives on the Origins of Modern Humans*, ed. P. Mellars and C. Stringer. Edinburgh: Edinburgh University Press.

Stringer, C. 1989. The origin of early modern humans: A comparison of the European and non-European evidence, 232-44. In *The Human Revolution: Behavioural and Biological*

Perspectives on the Origins of Modern Humans, ed. P. Mellars and C. Stringer. Edinburgh: Edinburgh University Press.

Strum, S. C., and W. Mitchell. 1987. Baboon models and muddles, 87-104. In *The Evolution of Human Behavior: Primate Models*, ed. W. G. Kinzey. Albany, NY: SUNY Press.

Susman, R. L. 1987. Pygmy chimpanzees and common chimpanzees: Models for the behavioral ecology of the earliest hominids, 72-86. In *The Evolution of Human Behavior: Primate Models*, ed. W. G. Kinzey. Albany, NY: SUNY Press.

Trinkaus, E., ed. 1983a. *The Mousterian Legacy*. Oxford: British Archaeological Reports International Series 164.

Trinkaus, E. 1983b. Neandertal postcrania and the adaptive shift to modern humans, 165-200. In *The Mousterian Legacy*, ed. E. Trinkaus. Oxford: British Archaeological Reports International Series 164.

Trinkaus, E. 1986. The Neandertals and modern human origins. *Annual Review of Anthropology* 15:193-218.

Trinkaus, E. 1987. Bodies, brawn, brains and noses: Human ancestors and human predation, 147-76. In *The Evolution of Human Hunting*, ed. M. H. Nitecki and D. V. Nitecki. New York: Plenum Press.

Trinkaus, E. 1988. Hominid postcrania from Krapina. Paper presented at the 12th International Congress of Anthropological and Ethnological Sciences, Zagreb, July.

Trinkaus, E. 1989. The Upper Pleistocene transition, 42-66. In *The Emergence of Modern Humans: Biocultural Adaptations in the Later Pleistocene*, ed. E. Trinkaus. Cambridge: Cambridge University Press.

Trinkaus, E. 1991. Would the real Neandertal please stand up–The search for Neandertal autapomorphies. *American Journal of Physical Anthropology*. Supplement 12:174-75.

Trinkaus, E., and F. H. Smith. 1985. The fate of the Neandertals, 325-33. In *Ancestors: The Hard Evidence*, ed. E. Delson. New York: Alan R. Liss.

Velichko, A. A. 1988. Geoecology of the Mousterian in East Europe and the adjacent areas, 179-206. In *L'Homme de Néandertal*. Vol. 2, *L'Environnement*, ed. M. Otte. Liège: Etudes et Recherches Archéologique de l'Université de Liège 29.

Vygotsky, L. S. 1986. *Thought and Language*. Translation revised and edited by A. Kozulin. Cambridge: MIT Press.

Vygotsky, L. S. 1987. Thinking and speech. In *The Collected Works of L. S. Vygotsky*. Vol. I, *Problems of General Psychology*. Translated by N. Minnick. New York: Plenum Press.

Whallon, R. 1989. Elements of cultural change in the Later Palaeolithic, 433-54. In *The Human Revolution: Behavioural and Biological Perspectives on the Origins of Modern Humans*, ed. P. Mellars and C. Stringer. Edinburgh: Edinburgh University Press.

White, R. 1982. Rethinking the Middle/Upper Paleolithic transition. *Current Anthropology* 23: 169-92.

White, R. 1989. Production complexity and standardization in early Aurignacian bead and pendant manufacture: Evolutionary implications, 366-90. In *The Human Revolution: Behavioural and Biological Perspectives on the Origins of Modern Humans*, ed. P. Mellars and C. Stringer. Edinburgh: Edinburgh University Press.

White, T. 1988. The minimum number of Neanderthals in the Krapina assemblage. Paper presented at the 12th International Congress of Anthropological and Ethnological Sciences, Zagreb, July.

Wolpoff, M. H. 1979. The Krapina dental remains. *American Journal of Physical Anthropology* 50:67-114.

Wolpoff, M. H. 1980. *Paleoanthropology*. New York: Alfred A. Knopf.

Wrangham, R. W. 1987. The significance of African apes for reconstructing human social evolution, 51-71. In *The Evolution of Human Behavior: Primate Models*, ed. W. G. Kinzey. Albany, NY: SUNY Press.

Zubrow, E. 1989. The demographic modelling of Neanderthal extinction, 212-31. In *The Human Revolution: Behavioural and Biological Perspectives on the Origins of Modern Humans*, ed. P. Mellars and C. Stringer. Edinburgh: Edinburgh University Press.

Chapter 6

New Advances in the Field of Ice Age Art

PAUL G. BAHN

Although generally regarded as a fairly static field which only comes to life when a new theory or explanation comes along, the study of Ice Age art is, in fact, constantly changing and expanding, in terms of its data base, its geographical and chronological spread, and the kinds of information that can be extracted from it. I hope to provide a brief outline here of some of the most recent advances in this field, which is currently experiencing perhaps its most exciting and revolutionary phase since the existence and authenticity of Ice Age art were recognized at the turn of the century.

The phenomenon of Upper Paleolithic art is of crucial importance in the question of the origin of Modern Humans but, strangely, it can be used to support both those who believe in a sweeping aside of archaic humans by modern people and those who instead see a considerable degree of continuity between the two groups. On the one hand, it is undeniable that the Upper Paleolithic brings the first clearly figurative depictions, which are far beyond anything produced in earlier periods. On the other hand, the very sophistication and mastery displayed in some of the earliest dated figurines, primarily the ivory carvings of Southwest Germany (Vogelherd, Geissenklosterle, Hohlenstein-Stadel), suggest very strongly that there must have been a long period of art production leading up to such masterpieces.

There have been recent attempts to scrutinize and minimize the evidence for "symbolic behavior" before the Upper Paleolithic (e.g., Chase and Dibble 1987), focusing on a few well-known claims for markings and perforations in the Middle Paleolithic, and often attributing them to natural causes or to contamination from Upper Paleolithic deposits. In fact, however, when

PAUL G. BAHN • 428 Anlaby Road, Hull, England, UK.
Origins of Anatomically Modern Humans, edited by Matthew H. Nitecki and Doris V. Nitecki. Plenum Press, New York, 1994.

one makes an effort to seek it out, there is a very wide range of evidence for nonutilitarian activity in the Middle and even the Lower Paleolithic (Bednarik 1992a; Marshack 1988, 1991), involving markings on stone and bone, use of ocher, collection of fossils, perforated objects both natural and artificial, circular objects, and even rock art. The basic problem is that very little attention has been paid to most of this evidence, and there is no consensus on how to interpret it, let alone on how to define "art" or "symbolic behavior" (Bednarik 1992a). However, as Marshack has remarked (1991:56), when one looks at the evidence, it is not its rarity but its variability that is striking, together with its chronological and geographic spread.

In addition, it should be remembered that of the numerous examples of Upper Paleolithic parietal and portable art, only a very tiny fraction come from the earliest part of the period. The vast majority are dated to the Solutrean and Magdalenian. The Aurignacian produced, as far as we can tell, a very limited range of art apart from the ivory figurines mentioned above. There are some very crude animal figures from southwest France, the varied motifs on stone blocks from seven sites in the Vezere Valley which have traditionally and unjustifiably been labelled vulvas (Bahn 1986; Delluc and Delluc 1978), beads and trinkets, and some traces of parietal art. Some parietal engravings in caves may date to this period, and examination of fallen wall fragments from even the early Upper Paleolithic layers at the abri Pataud (Dordogne) suggests that the rock-shelter was painted and engraved throughout its occupation (Delluc and Delluc 1991:206, 211). Since such shelters have been subjected to 30,000 years of weathering, it is hardly surprising that nothing remains in situ of their Aurignacian decoration. We do not know if shelter walls were also decorated in earlier periods, for the same reason of lack of preservation coupled with the general problem that archaeologists tend to look for what they expect to find, and tend to find what they look for: and cave art is not supposed to exist before the Upper Paleolithic. It is a fair bet that many more candidates for "symbolic behavior" are lurking, unnoticed or unreported, in excavated material from many pre-Upper Paleolithic sites.

In other words, Upper Paleolithic art certainly constitutes a dramatic qualitative and quantitative advance on earlier evidence for symbolic activity, but it is a change that is in some ways less radical than the proclaimers of a "cultural revolution" would have us believe, and one which must have had lengthy antecedents. For the present the only known candidates for these antecedents are the above examples from the Lower and Middle Paleolithic which, in some respects, foreshadowed Aurignacian developments.

Figure 1. The Galgenberg figurine. Drawing by R. G. Bednarik. Reprinted by permission from Bednarik 1989.

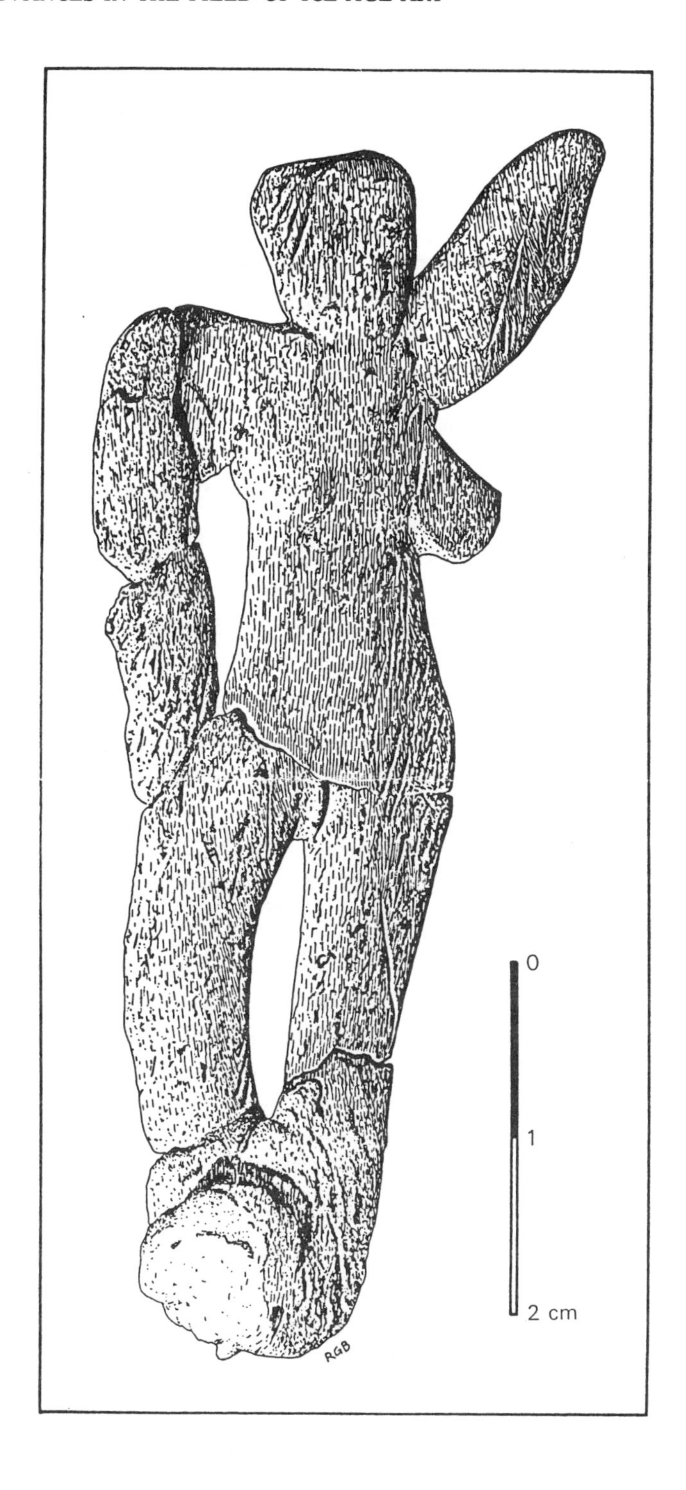
0
1
2 cm
RGB

NEW FINDS

In 1988 a new find of truly major importance was found which underlines the argument that Aurignacian art cannot have sprung ready-made from nowhere. The "Dancing Venus" of Galgenberg, a 7.2 cm-long green serpentine female figurine (fig. 1) was excavated by Christine Neugebauer-Maresch near Krems in Austria (Neugebauer-Maresch 1987; Bednarik 1989). Charcoal samples from the same layer have produced radiocarbon dates of around 30,000 ybp. The significance of the figure is that, unlike the well-known "Venus figurines" of the Ice Age, such as the Willendorf statuette from Austria found almost exactly 80 years earlier, the Galgenberg specimen is flattish rather than carved in the round: its shape was probably determined by the stone used, since serpentine often occurs in thin slabs.

The right arm and the legs are supported at both ends, while the left arm appears to be folded back at the elbow. The body is supported primarily on the left leg, while the right leg rests on a slightly higher support. The right hand is placed on the hip. This pose depicts the left breast almost in profile, while the right is in very low relief, possibly because of the stone's flatness. The vulva is depicted, a feature rare in Ice Age depictions; and, in further contrast to many of the better-known Venus figurines, there is no hint of obesity or of an emphasis on breasts and buttocks. The figure is more or less anatomically accurate, except for the thickened limbs, which are probably as thin as the artist dared carve them without making the sculpture unduly fragile.

The two most important points about the Galgenberg figurine are that, first, it is at least 5000 years older than the Gravettian period to which most Venus figurines are usually, often unjustifiably, assigned. Most of the western European specimens have no archaeological context whatsoever, and are merely assumed to be Gravettian, whereas dated specimens in eastern Europe span a period from about 23,000 to 12,000 BP. Second, the Galgenberg carving demonstrates considerable technological skill – the stone is rather delicate and brittle, and the head, left arm and breast could all easily have broken off. In addition, the two openings (under one arm and between the legs) required a delicate boring operation. The later Venuses are normally static, solid symmetrical figures with no perforations and no protruding limbs. This sophistication in technology and composition can only be the end product of a long tradition in carving (perhaps of perishable materials). Like the German ivory figurines mentioned earlier, this simply cannot represent the "first art," which must have appeared long before.

New discoveries of cave art continue to be made, including early engravings fallen from cave walls in southwest Germany (Hahn 1991). Indeed, since the publication of my survey only a few years ago (Bahn and Vertut

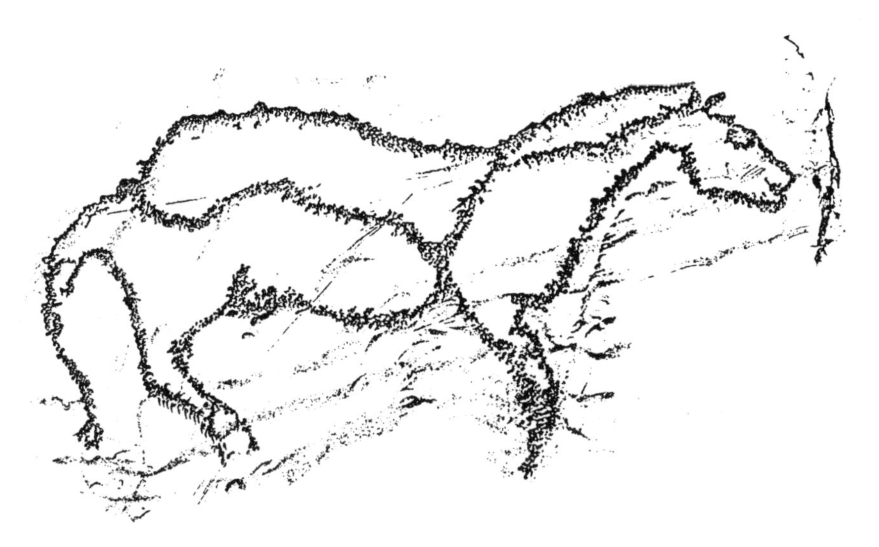

Figure 2. Horse at Siega Verde, from de Balbin Behrmann et al. 1990. Reprinted by permission of the publisher.

1988), at least half a dozen more caves in France alone have been found to contain Paleolithic parietal art, including the spectacular Cosquer Cave whose entrance is now beneath the Mediterranean (Clottes, Beltran et al. 1992). In almost all of these sites (though not Cosquer) Paleolithic occupation was already well known, but marks on the wall were suddenly noticed for the first time or, more commonly, were revealed by the removal of sediments piled up by early excavators. The most significant of these finds occurred at Le Placard cave in western France, already renowned for the richness of its portable art. The newly revealed engravings on its wall include a number of "chimney" signs, identical to painted versions in the caves of Pech-Merle and Cougnac in Quercy, 165 km away. These caves had usually been assigned on stylistic grounds to the Solutrean or early Magdalenian, and the Placard find has confirmed this, since new excavations have found fallen slabs with fragments of parietal engravings, stratified between two layers of Upper Solutrean material (Clottes, Duport & Feruglio 1990).

The relatively recent discoveries of open-air engravings in France, Spain, and Portugal, which almost certainly date to the Upper Paleolithic (Bahn and Vertut 1988:chap. 5), have also been augmented by two further finds in Spain. The first, at Siega Verde, near Ciudad Rodrigo, is only 5 km from the open-air horse engravings at Mazouco, Portugal. The new site features a number of hammered-out and engraved figures including horses (fig. 2), aurochs, and deer (de Balbin Behrmann et al. 1991). And in a totally different area, previously devoid of Paleolithic art, an open-air horse figure

Figure 3. Horse at Piedras Blancas. Modified from Martinez Garcia 1986/87.

(fig. 3) has been discovered at Piedras Blancas, Almeria (Martinez Garcia 1986/87). Situated on an inclined block at an altitude of 1400 m near the town of Escullar, the horse is made with multiple deeply incised lines. Stylistically, it has been ascribed to the final Gravettian or the Solutrean, through comparisons with the engraved plaquettes from Parpallo in eastern Spain.

PLEISTOCENE ART OUTSIDE EUROPE

There have been a few interesting developments on other continents in the last few years. The mammoth engraving on a whelk shell from Holly Oak, Delaware, always of doubtful authenticity (Bahn and Vertut 1988:26-27), has at last been radiocarbondated to 1530 ± 110 ybp (Griffin et al. 1988). Despite the often inaccurate results of radiocarbon dates from shell, it is pretty obvious that the shell in question was taken from an archaeological site of around the ninth century AD, and the engraving made on it in the 1880s.

Better authenticated art from the Piaui region of Brazil has been dated to the Late Pleistocene. The enormous decorated rock-shelter of Pedra Furada contains a number of sandstone fragments, which had fallen from the wall and become stratified in layers dating back to 50,000 BP. One such spall, bearing a clear human figure in red pigment, comes from layer XII which, in relation to dated layers above and below, has been assigned to c. 10,000 or 12,000 ybp (Guidon and Delibrias 1986; Guidon 1991). Another spall, with two parallel red lines which are very probably the legs of a human or animal, is from a layer dated to ca. 17,000 ybp. Still another spall, from a layer of c. 32,000 ybp, has a red patch, but analysis has been unable to prove whether this was made by humans.

Some of the art at the nearby rock-shelter of Toca do Baixao do Perna I is also reliably dated. A panel of small red figures was exposed by excavation of the layers covering it, and, though faded, the images have survived burial amazingly well. One fragment of charcoal still adhering to the panel was radiocarbon dated at 9650 ± 100 ybp, while charcoal from the layer touching the bottom of the panel has been dated to 10,530 ± 110 ybp (Guidon 1991:48). Hence, unless artists painted at nose level while lying on the floor, the panel must be somewhat older than the latter date. The figures correspond perfectly in size and style to those of similar age at Pedra Furada. Together with art at many other sites in the region they have been attributed to the "Serra da Gapibara" style, which is thought to be at least 12,000 years old, and probably at least 17,000.

Application of the still controversial cation-ratio dating technique to desert varnish on top of petroglyphs (bighorn sheep and abstract) in California's Coso Range has provided dates of 14,200 ± 4200, 16,600 ± 4300 and even 18,200 ± 2400 ybp (Whitley and Dorn 1988).

The Iberomaurusian (Upper Paleolithic) site of Afalou Bou Rhummel in eastern Algeria recently yielded two small terra-cotta heads of horned animals (Hachi 1985): one was found 18 cm above a layer dated to 11,450 ± 230 ybp, while the other, 210 cm beneath the first, was 30 cm below a layer dated to 12,400 ± 230 ybp, and just above a hearth containing another terra-cotta fragment. More such finds have been made at the site very recently (M. Hachid, pers. comm.).

China has produced its first definite piece of Paleolithic portable art in the form of a fragment of antler, decorated with abstract engraved motifs, from the cave of Longgu and dating to c. 13,000 years ago (Bednarik and Yuzhu 1991).

The Pleistocene art of Australia, already identified in a handful of sites (see Bahn and Vertut 1988:28-30), has now been dated in several more. At Dampier in Western Australia, Lorblanchet (1988:286) has found some very old engravings, closely associated with sea shells dated to 18,510 ± 260 ybp. At Sandy Creek, Queensland, engravings have been dated to at least 32,000 years ago by the sediments which used to cover them (Morwood 1992), and red paint has been directly dated to 24,600 ybp (Bednarik 1992b). In September 1987, at least 15 hand stencils, a roughly drawn circle, and extensive areas of wall smeared with deep-red pigment were found in a large cave known as Wargata Mina (Judds Cavern) in southwest Tasmania (Brown 1991). These were mostly in a high alcove about 35 m from the entrance, at the limit of light penetration. The red pigment has been found to contain human blood, and accelerator radiocarbon dating of it has produced dates of 9240 ± 820 ybp and 10,730 ± 830 ybp (Loy et al. 1990).

Australia has also produced the world's oldest dates for rock art, through Accelerator Mass Spectrometry analysis of organic material in varnish covering petroglyphs in South Australia. An oval figure at Wharton Hill has given a result of over 42,700 ybp, while a curved line at Panaramitec North has a minimum age of 43,140 ± 3000 ybp (Bednarik 1992b).

NEW TECHNIQUES OF ANALYSIS

These Australian researches form part of the most important advance in techniques of studying Ice Age art in recent years. Detailed pigment analysis in the French and Spanish caves has likewise been made possible by new technical methods, and for the first time it is proving possible to obtain direct dates from parietal figures, thanks to methods which require minute amounts of pigment, causing no visible damage to the art.

Previous analyses of pigment in European cave art (see Bahn and Vertut 1988:chap. 5) had consistently revealed the use of iron oxide (red) and manganese dioxide (black). But the improved analytical methods of recent years have produced far more detailed results. It has been discovered that black figures in quite a few caves in France and Spain were done with a pigment containing charcoal, which can, therefore, be dated. For example, charcoal from polychrome bison figures on the famous ceiling of Altamira Cave has produced radiocarbon dates of 14,330 ± 190, 13,940 ± 170 and 13,570 ± 190 ybp, while similar polychrome bison figures from the nearby cave of El Castillo have produced somewhat younger dates of 13,060 ± 200 and 12,910 ± 180 ybp (Valladas et al. 1992).

In the cave of Cougnac in France's Quercy region, a 100 mg sample from a black dot on the wall, between a *Megaloceros* (giant deer) done in charcoal and a "speared human figure," has produced a radiocarbon date of 14,300 ± 180 ybp (Lorblanchet 1990; Lorblanchet et al. 1990). Samples of pigment from red figures on Cougnac's walls were compared with some red ocher from the cave floor, and with ocher sources immediately outside the cave and also 15 km away. The floor deposit was probably used in production of the red figures, and the ocher was most likely obtained from local clays. Two points emerge: first, that the large red animal figures in the center of the cave's main panel were drawn with the same pigment, and were, therefore, probably a composition done in a single production event. Second, whereas stylistic studies had assigned Cougnac to the Early Magdalenian period, the dating (together with one of 15,000 ybp from a reindeer bone in the cave) points to the Middle Magdalenian. Further dating, however, will be needed for confirmation, since the engraved signs from Le Placard, mentioned earlier, have painted equivalents at Cougnac, which would place at least some of that cave's decoration in the Late Solutrean.

Interesting results have also emerged from pigment analyses in the French Pyrenees (Clottes 1993; Clottes, Menu & Walter 1990a, 1990b), especially from the caves of Niaux and Réseau Clastres. Analyses of samples by scanning electron microscopy, X-ray diffraction, and proton-induced X-ray emission have revealed four specific "recipes" of pigments mixed with mineral extenders and binders: (l) talc; (2) barite with potassium feldspar; (3) potassium feldspar; and (4) potassium feldspar with biotite. Extenders, besides making paint go further, have other advantages: adding biotite makes red paint spread easily when wet, produces a darker color than pure red ocher, improves adhesion to the wall, and prevents the paint from cracking as it dries.

In Niaux's famous "Salon Noir," most of the animal figures were first sketched in charcoal, and then covered over with manganese paint, using recipe 4. It, therefore, seems that the Salon Noir was indeed a "sanctuary," a special place where the figures were carefully planned before being executed. The figures in all other parts of this huge cave were done more spontaneously, without preliminary sketches.

These analyses are helping to establish how the caves were decorated. The Salon Noir contains all four recipes, though most of the art was done with number 4, as was a sign at the far end of the cave. The nearby Réseau Clastres, on the other hand, had only recipe 4, without preliminary sketches; this fact, together with its total lack of evidence of occupation and the clustering of its few figures near the original entrance, suggests that it was visited very briefly.

Another important application is in the detection of fakes: only one painting in Niaux, a red-painted fissure interpreted as a vulva, proved to have no extender. As the figure was not mentioned by the first researchers who studied the cave, and as there are initials nearby, it is clearly modern.

It proved possible to date recipe 4 immediately, because the occupation site of La Vache, directly across the valley from Niaux, has the same extender used with red and black paint on bones from layers dated to 12,850-11,650 ybp, the Upper Magdalenian (Buisson et al. 1989). Niaux, on the basis of style, had traditionally been assigned to the Middle Magdalenian, but this later dating has now been confirmed by a direct radiocarbon date from charcoal in one of the Salon Noir's bison figures: 12,890 ± 160 ybp (Valladas et al. 1992).

Niaux's recipe 2, on the other hand, does seem to belong to the Middle Magdalenian, as it has been found on a bead from the cave of Enlene in levels of 13,940-12,900 ybp, and also in the Middle Magdalenian of the cave of Le Mas d'Azil. However, a second bison figure in Niaux's Salon Noir has given a date of 13,580 ± 150 ybp; although this confirms that some of Niaux's decoration dates to the Middle Magdalenian, as had always been thought, the recipe used was that of the Upper Magdalenian. In other words, the recipes had a considerable duration, and are not reliable chronological markers (Clottes, Valladas et al. 1992).

One difference to emerge from the pigment analyses of Quercy and the Pyrenees is that the Quercy team believes the composition of its pigments to be entirely natural, because the mineral components discovered exist in the same proportions in local sediments. The Pyrenean team, however, insists that its compositions and recipes contain associations of components which it is quite impossible to find in nature. These claims are not necessarily contradictory – "artists" in Quercy may have used natural pigments, whereas those in the Pyrenees may have concocted new ones.

Analyses are also now being carried out to identify the binding agents used. Experiments had suggested that fatty and organic substances were unsuitable for this purpose, and failed to adhere well to humid walls. The only substance that seemed to be good at fixing and preserving pigments on the rock face was cave water, rich in calcium carbonate (Bahn and Vertut 1988: 100). However, analyses of pigments from the Magdalenian caves in the Pyrenees, using gas chromatography and mass spectrometry, have detected organic binders: at the cave of Fontanet, an oil of animal origin seems to have been used (Pepe et al. 1991), while in the two linked caves of Enlene and Les Trois Freres an oil of vegetable origin has been detected (Menu and Walter 1991:1089).

Preliminary and short-lived doubts about the authenticity of the Cosquer Cave have led to its becoming the best-dated decorated cave so far, with charcoal in its figures yielding results of around 27,000 ybp (hand stencil) and 19,000 ybp (three animals), indicating at least two periods of utilization of this cave (Clottes, Courtin et al. 1992).

In short, pigment analysis is providing a revolutionary tool, applicable, at least regionally, for dating paintings, and is revealing that behind apparent stylistic homogeneity there is technological heterogeneity. Styles of figures lasted longer than was thought, and are of very limited use in dating. Evidence is rapidly emerging for the careful preparation of pigments, combining minerals with organic materials, but with the precise "recipes" differing through space and time.

A new era in rock art studies is, therefore, opening up, which is based on more objective dating methods, and in which we shall be able to ascertain far more accurately how the decoration of panels and caves was done over time (Lorblanchet and Bahn, in press). These studies will provide the clearest insights we have ever had into the activities and thought processes of the cave artists. For thc momcnt, however, the problem of the "origins of art" and the significance of its apparent association with the arrival of the first modern humans remain as enigmatic and ambiguous as ever.

ACKNOWLEDGMENTS

I am grateful to the organizer of the symposium, Dr. Matthew Nitecki, for inviting me to this important event, and to the British Broadcasting Corporation for the opportunity of visiting the Piaui region in 1990.

REFERENCES

Bahn, P. G. 1986. No sex, please, we're Aurignacians. *Rock Art Research* 3:99-120.

Bahn, P. G., and J. Vertut. 1988. *Images of the Ice Age.* New York: Facts on File.

Balbin Behrmann, R. de, J. Alcolea Gonzalez, M. Santonja, and R. Perez Martin. 1991. Siega Verde (Salamanca). Yacimiento artistico paleolitico al aire libre, 33-48. In *Del Paleolitico a la Historia.* Salamanca: Museo de Salamanca.

Bednarik, R. G. 1989. The Galgenberg figurine from Krems, Austria. *Rock Art Research* 6(2):118-25.

Bednarik, R. G. 1992a. Palaeoart and archaeological myths. *The Cambridge Archaeological Journal* 2:27-57.

Bednarik, R. G. 1992b. Oldest dated rock art: A revision. *The Artefact* 15:39.

Bednarik, R. G., and Y. Yuzhu. 1991. Palaeolithic art from China. *Rock Art Research* 8:119-23.

Brown, S. 1991. Art and Tasmanian prehistory: Evidence for changing cultural traditions in a changing environment, 96-108. In *Rock Art and Prehistory*, ed. P. Bahn and A. Rosenfeld. Oxford: Oxbow Monograph 10.

Buisson, D., M. Menu, P. Walter, and G. Pincon. 1989. Les objets colores du Paleolithique superieur: Cas de la grotte de La Vache (Ariege). *Bulletin de la Société Préhistorique Française* 86:183-91.

Chase, P. G., and H. L. Dibble. 1987. Middle Paleolithic symbolism: A review of current evidence and interpretations. *Journal of Anthropological Archaeology* 6:263-96.

Clottes, J. 1993. Paint analyses from several Magdalenian caves in the Ariège region of France. *Journal of Archaeological Science* 20:223-35.

Clottes, J., A. Beltran, J. Courtin, and H. Cosquer. 1992. La Grotte Cosquer (Cap Margiou, Marseille). *Bulletin de la Société Préhistorique Française* 89:98-128.

Clottes, J., J. Courtin, H. Valladas, H. Cachier, N. Mercier, and M. Arnold. 1992. La grotte Cosquer datée. *Bulletin de la Société Préhistorique Française* 89:230-34.

Clottes, J., L. Duport, and V. Feruglio. 1990. Les signes du Placard. *Bulletin de la Société Préhistorique Ariège-Pyrénées* 45:15-49.

Clottes, J., M. Menu, and P. Walter. 1990a. New light on the Niaux paintings. *Rock Art Research* 7:21-26.

Clottes, J., M. Menu, and P. Walter. 1990b. La preparation des peintures magdaleniennes des cavernes ariegeoises. *Bulletin de la Société Préhistorique Française* 87:170-92.

Clottes, J., H. Valladas, H. Cachier, and M. Arnold. 1992. Des dates pour Niaux et Gargas. *Bulletin de la Société Préhistorique Française* 89:270-74.

Delluc, B., and G. Delluc. 1978. Les manifestations graphiques aurignaciens sur support rocheux des environs des Eyzies. *Gallia Préhistoire* 21:213-438.

Delluc, B., and G. Delluc. 1991. *L'Art Pariétal Archaïque en Aquitaine*. 28e Supplément à *Gallia Préhistoire*. Paris: Centre National de la Recherche Scientifique.

Griffin, J. B., D. J. Meltzer, B. D. Smith, and W. C. Sturtevant. 1988. A mammoth fraud in science. *American Antiquity* 53:578-82.

Guidon, N. 1991. *Peintures Prehistoriques du Bresil: L'art rupestre du Piauf*. Paris: Editions Recherche sur les Civilisations.

Guidon, N., and G. Delibrias. 1986. Carbon-14 dates point to man in the Americas 32,000 years ago. *Nature* 321:769-71.

Hachi, S. 1985. Figurines en terre cuite du gisement iberomaurusien d'Afalou Bou Rhummel. *Travaux du Laboratoire de Prehistoire et d'Ethnographie des Pays de la Mediterranee Occidentale*, etude 9, 6 pp.

Hahn, J. 1991. Höhlenkunst aus dem Hohlen Pels bei Schelklingen, Alb-Donau-Kreis. *Archäologische Ausgrabungen in Baden-Württemberg* 10:19-21.

Lorblanchet, M. 1988. De l'art parietal des chasseurs de rennes a l'art rupestre des chasseurs de kangourous. *L'Anthropologie* 92:271-316.

Lorblanchet, M. 1990. Etudes des pigments des grottes ornees. *Bulletin de la Societe des Etudes du Lot* 111:93-143.

Lorblanchet, M., and P. Bahn, eds. In press. *The Post-Stylistic Bra?* Oxford: Oxbow Books.

Lorblanchet, M., M. Labeau, J. L. Vernet, P. Fitte, H. Valladas, H. Cachier, and M. Arnold. 1990. Palaeolithic pigments in the Quercy, France. *Rock Art Research* 7:4-20.

Loy, T. H., R. Jones, D. E. Nelson, B. Neehan, J. Vogel, J. Southon, and R. Cosgrove. 1990. Accelerator radiocarbon dating of human blood proteins in pigments from Late Pleistocene art sites in Australia. *Antiquity* 64:110-16.

Marshack, A. 1988. The Neanderthals and the human capacity for symbolic thought: Cognitive and problem-solving aspects of Mousterian symbol, 57-91. In *L'Homme de Néandertal*. Vol. 5, *La Pens'ee*, ed. M. Otte. Liège: Etudes et Recherches Archéologiques de l'Université de Liège 32.

Marshack, A. 1991. A reply to Davidson on Mania and Mania. *Rock Art Research* 8:47-58.

Martinez Garcia, J. 1986/87. Un grabado paleolitico al aire libre en Piedras Blancas (Escullar, Almeria). *Ars Praehistorica* VJVI:49-58.

Menu, M., and P. Walter. 1991. Les premiers artistes peintres. *La Recherche* 22:1086-89.

Morwood, M. 1992. *Rock Art and Ethnography*. Melbourne: Australian Rock Art Research Association.

Neugebauer-Maresch, C. 1987. Vorbericht über die Rettungsgrabungen an der Aurignacien-Station Stratzing/Krems-Rehberg in den Jahren 1985-1988. Zum Neufund einer weiblichen Statuette. *Fundberichte aus Osterreich* 26:73-84.

Pepe, C., J. Clottes, N. Menu, and P. Walter. 1991. Le liant des peintures paleolithiques ariegeoises. *Comptes rendus de l'Academie des Sciences de Paris* 312 serie II:929-34.

Valladas, H., H. Cachier, P. Maurice, F. Bernaldo de Quiros, J. Clottes, V. Cabrera Valdés, P. Uzquiano, and M. Arnold. 1992. Direct radiocarbon dates for prehistoric paintings at the Altamira, El Castillo and Niaux caves. *Nature* 357:68-70.

Whitley, D. S., and R. I. Dorn. 1988. Cation-ratio dating of petroglyphs using PIXE. *Nuclear Instruments and Methods in Physics Research* B35:410-14.

Part III

African Center of Origin

The Out-of-Africa (or Eve) hypothesis and the Multiregional hypothesis are the two major hypotheses on the origins of modern people. Neither is robust enough to be unequivocally accepted, and either can be chosen as a working hypothesis. One of them may be eventually accepted, or our future interests and questions may make either hypothesis insufficient.

How certain are the claims based on the parsimonious analyses of phylogenetic trees is reflected in the most recent controversy over the origin of the modern people, in which the emotional issue of the Black Madonna is challenged by doubts that our pre-great-grandmother met our pre-great-grandfather in the same bed. The mtDNA, a recent Rosetta Stone for the understanding of the origin of modern people, is only a tool of population genetics. Molecular information must parallel other information from geochronology, morphology, archaeology, and behavior to support or to negate either theory–but molecular information is neither superior nor more "scientific." It only represents a different methodology, which, nevertheless, is still taxonomic!

In any case both hypotheses remain controversial in respect to the origins, emigrations, and replacement. There is, however, a gap before the appearance and spread of modern humans. Perhaps this gap is real, but not significant. After all "stone age" people are still living, long after the industrial revolution spread over most of the habitable world. And Darwin, on his voyage of the Beagle, witnessed the repatriation of three Fuegian "savage heathens" that four years earlier were taken "hostage" by Captain Robert FitzRoy. We should not expect events in the prehistory to occur simultaneously.

Rebecca Cann, Olga Rickards, and Koji Lum explain the methodology and logic of the contributions of genetics to the search for the origins of modern humans. After reviewing the history of the genetic studies of the human origins, particularly the maternally inherited genes, they strengthen the "Eve" hypothesis by application of new mtDNA techniques. The Eve hypothesis stems from research that Cann and her colleagues performed in a study which compared differences in genetic material, DNA, from contemporary people around the world. They plead for more and better collection of experimental data.

Christopher Stringer points out that the origin of modern humans was probably a two-stage morphological process. Many features shared by living humans had evolved 100,000 years ago, probably in Africa and in the Levant. However, some other features of living humans were not yet present in the earliest anatomically moderns. These latter aspects must have continued to evolve in concert with the second stage in modern human origins, the development of the regional features of the skeleton that differentiate present human populations. The large number of fossils and relatively good chronological control of the Upper Paleolithic from Europe, according to Stringer, are essential for the study of regional evolution of modern humans. Although they retain possible plesiomorphies, they also show autapomorphic features. As more fossils of early modern humans will certainly be found in non-European areas, it should become possible to reconstruct the subtle pattern of evolution of the morphological and metrical features that differentiate the extant human geographical variants. These future finds will provide independent tests for genetic conclusions about modern human origins.

Chapter 7

Mitochondrial DNA and Human Evolution: Our One Lucky Mother

REBECCA L. CANN, OLGA RICKARDS, and J. KOJI LUM

Anthropologists who study human evolution today, especially the later stages, have many advantages over scientists working on human origins 50 years ago. Biotechnology and the discovery of new fossils *in situ* help us deduce that some lineages of humans survived and some perished without leaving modern descendants. This information is crucial for understanding the major transitions in human evolution, whether the problem at hand is the origin of bipedalism, or in our case in this volume, the origin of anatomically modern people. This chapter will discuss the contribution that genetics and specifically, studies of maternally inherited genes, have made to furthering our understanding of where and when modern humans arose. It will summarize the insights of many scientists. In the end, we will make a plea for better experimental models which allow us to get at the actual numbers, e.g., how many generations does a morphological transition take, how many ancestors (original founders) were involved, how much genetic isolation was needed, how many survivors were there, and what were the rates of subdivision, rates of extinction, and rates of at which populations expanded? This focus derives from our work on Polynesian migration to Hawaii, lineage extinction and stability, and demographic change due to the introduction of infectious diseases by colonizers to aboriginal peoples. In the Hawaiian example, some of the numbers are known, some are ball-park estimates, and some are just wild

REBECCA L. CANN • Department of Genetics and Molecular Biology, University of Hawaii at Manoa, Honolulu, HI 96822. OLGA RICKARDS • Dipartimento di Biologia, Universita di Roma, Rome, Italy. J. KOJI LUM • Department of Genetics and Molecular Biology, University of Hawaii at Manoa, Honolulu, HI 96822.
Origins of Anatomically Modern Humans, edited by Matthew H. Nitecki and Doris V. Nitecki. Plenum Press, New York, 1994.

speculation. Still, we think they will drive discussion in a more fruitful manner than the current impasse which now exists.

A New Era of Human Evolutionary Studies

Genetics today is part of a revolution in biological thought that is driven by advances in biotechnology. Of the 100,000 or so genes estimated to exist in the human body, a few are known in exquisite detail. Most are known only in the broadest outlines, if at all. It is possible to pick up a standard textbook about the new genetics and read "current ignorance is vaster than current knowledge" (Singer and Berg 1991). Experimental scientists do not find this statement either depressing or demoralizing, but use the knowledge built on over time to extract new principles and ask new questions.

The study of human mitochondrial genes, in an evolutionary sense, began with Wesley Brown when he was a Ph.D. student at California Institute of Technology about 20 years ago. For the first ten years of human mitochondrial population genetics, it was only practical to compare populations at the DNA level indirectly using restriction enzymes. The earliest studies of human genetic diversity concentrated on describing patterns of mutations revealed using purified mitochondrial DNA (mtDNA) and restriction endonucleases.

At first, many molecular biologists and biochemists were shocked that humans varied in their mitochondrial genes at the DNA level, because they were essential for life and, therefore, the bias was that they should be extraordinarily conservative in their mode and rate of change. Brown (1980) demonstrated that there was indeed variation, it could be easily detected, and it contained implications for thinking about how human populations evolved.

This work was later extended by students and postdoctoral fellows working with Allan Wilson at the University of California at Berkeley. They refined the precision of restriction site studies to catalogue particular mutations in maternal lineages after one complete human mtDNA sequence was published. In a survey of about 140 donors from mostly urban populations in California, the Berkeley report described a pattern of human evolution suggesting two things. First, that our origins were African, and second, that all human mtDNA coalesced to a single mother who probably lived about 200,000 years ago (Cann, Stoneking & Wilson 1987).

FURTHER ADVANCES WITH HUMAN mtDNA

Now, Russel Higuchi, Mark Stoneking, Linda Vigilant, Anna Di Rienzo, and Tom Kocher have made three significant advances in human mtDNA studies while working with Wilson at Berkeley. They perfected the application of a cell-free cloning technology, the polymerase chain reaction, or PCR, to make use of human hairs instead of blood or other tissues as sources of mtDNA (Higuchi 1989). Using direct DNA sequencing, they estimate the time of divergence from common ancestors within human populations from the chimpanzee perspective (Kocher and Wilson 1991). Finally, they demonstrate one way to think about which specific African populations may have been involved in the spread of anatomically modern people (Vigilant et al. 1989; Vigilant et al. 1991; Di Rienzo and Wilson 1991).

Work on mtDNA in modern humans suggests that there was widescale replacement of many archaic people by African lineages with limited gene flow. Our characterization of the transition from anatomically archaic to modern people as one of replacement by African populations is supported by nuclear gene data as well (Bowcock et al. 1987). We will explain the assumptions of the models, discuss the limitations of current knowledge, and lay out the plans for future research directed at resolving this debate. We are confident that new technology for extracting DNA from bone will eventually resolve many mysteries, including exactly what became of Neandertal populations in western Europe.

Such information is crucial to getting at the question we posed in 1987, namely, understanding the mechanism by which anatomically modern people arose. It could have entailed a smooth, gradual, world-wide transition from widespread but semi-isolated archaic populations. Alternatively, it might have been a relatively sudden change involving the replacement of archaic peoples by modern groups immigrating from elsewhere with little gene flow between the various groups. Regional continuity or replacement, with various gradations in between, describe this problem (Mellars and Stringer 1989). Various interpretations of the transition are possible if researchers concentrate on only fossil evidence, while the mtDNA studies more strongly support replacement. The statistical certainty of the earliest branches is not as strong as we originally thought (Templeton 1992), but the best approximation of the process still appears to be an African-based spread (Hedges et al. 1992). It is, therefore, useful to return to the reasons why the mitochondrial approach at solving the problem of modern human origins yields this answer.

MtDNA TRANSMISSION

MtDNA is such a decisive biological tool for phylogenetic systematics in part due to its mode of transmission (maternal). Some difficulties that people have had in the past with understanding the mitochondrial evidence for human origins stem from confusion about the transmission patterns for mtDNA lineages in populations and their evolutionary implications. Usually, we describe and quantify mitochondrial diversity in population, based on sampling individual lineages, and then attempt to reconstruct the history of the population based on the genetic relationships of people carrying those maternal lineages.

Evolutionary biologists often show complicated pedigrees to depict the connections between genes and ancestors. Fortunately, mtDNA is simple to follow, because there is no genetic recombination between parental types. A man inherits the mtDNA of his mother, but fails to transmit it to his children because his sperm's mtDNA is destroyed or diluted upon fertilization of an egg. His children will have their mother's mtDNA, and his daughters will transmit it to their children. If we go back five generations in time, this man will have 32 potential genetic ancestors, but a single mitochondrial ancestor. We can speak of his having a single mitochondrial source for his family, but this in no way implies that there was only a single female in the population five generations ago. Also, the one mitochondrial mother of this family was real, not an artifact of a computer program that requires hypothetical ancestors.

In the same way, the human population 10,000 generations ago contained many genetic ancestors. The coalescent of mtDNA to a single maternal ancestor 200,000 years ago easily accommodates an effective population size of 4000 females (Hartl 1988). A nuclear gene picture, if we could account for recombination, would give 4 times this number, or 16,000 breeding adults. Mathematic theory predicts that in time, most mitochondrial lineages will go extinct because many families will fail to have female children. The rate by which this happens depends on the number of lineages in the population, the rate of mutation, and the number of generations elapsed. John Avise and his students at the University of Georgia have described this phenomenon as lineage sorting.

The evolutionary process of genetic descent with modification applied to mtDNA does not require that a single human female gave rise to all modern people. There has been confusion about this notion, brought on by the general misunderstanding surrounding genetic bottlenecks and genetic revolutions in producing adaptation, especially in the process of speciation. No revolution or bottleneck need have taken place, to accommodate the numbers described either by restriction site analysis or direct sequencing.

Mitochondrial mutations tend to accumulate in the third positions of codons, and so are silent in a functional sense when it comes to considering what consequence mutations have for the protein encoded. We tend to think that the bulk of mitochondrial variation at the DNA level is uncoupled from adaptation, in a strict sense, and is largely independent of population size under neutral population genetic theory (Nei 1987).

Evolutionary biologists fall between two general extremes today when it comes to considering population size, genetics, and adaptation. Sometimes, they prefer to think in terms of small populations, and fixation of mutations around adaptive peaks, as Sewall Wright stressed in his shifting balance models. Others, like R. A. Fisher, tend to emphasize the potential for adaptation that exists in large populations, due to the potential for observational learning and the large number of genetic variants as a consequence of recombination and mutation. This debate is inappropriate for mtDNA, because it is uncertain how many particular mitochondrial mutations support the notion of adaptation.

Mitochondrial variants can act as markers for population movement, and some mutations do appear to be restricted in a geographic sense. However, the most common markers are unlikely to convey increased reproductive fitness because most are at third positions of codons or in non-coding regions of the genome. Some scientists believe that enhanced language abilities may have been due to particular mitochondrial mutations, but the linkage would probably be indirect and this idea is purely speculative at present.

Mitochondrial variation would always be less than nuclear variation in a population if the two genomes had the same rate of mutation because of its transmission as a haploid genome. Owing to little if any repair of DNA damage, variation accumulates faster in mitochondrial genes, but individual alleles are sorted through the population quicker than nuclear gene alleles. This can give the appearance of a population bottleneck, when, in fact, there might not have been one. It is possible that low levels of mtDNA variation in some groups could be due to purifying selection or adaptation driving to fixation lineages with higher fitness, but it is most probable that low variation simply reflects a recent descent from a common mitochondrial mother (Hartl and Clark 1989).

MtDNA VARIABILITY

PCR has changed the rate at which new information about human genetic variation accumulates. It is now possible to gather in two afternoons the same amount of information that it took a whole year to accumulate by restriction endonuclease digestion of mtDNA isolated from human donors. For comparisons of variation within our species, two approaches are now common. The first is to focus on a well-characterized mutation that can be shown to have a nonrandom distribution by continents. In simple terms, the mutation should be confined to single ethnic, political, language, or religious group. In 2000 years of cultural isolation, the few mutations that naturally arise over time might spread by marriage between cousins, so that they come to characterize an entire group. When the molecular anthropologist encounters the mutation, it provides a key to a past often lost in mythology and conflicting oral tradition. A second approach requires brute-force sequencing, where the investigator focuses on the control region's 1000 base-pairs of mtDNA that change the fastest. In primates, this noncoding part of the mitochondrial genome has an even higher rate of change than the rest of the mtDNA circle combined, up to perhaps 100 times as much as some portions of the ribosomal regions, and this property allows us to use it as a fast clock. The control region is most useful for looking at recent expansion and migration of populations.

An illustration of the first approach is shown by the mutation due to a small deletion of 9 base-pairs in a tiny, non-coding region separating the Cytochrome Oxidase subunit 2 gene and a tRNA Lysine gene. The deletion occurs sporadically in humans, and is commonly restricted to a few Asian lineages and their descendants. (Linda Vigilant also finds it in some African pygmies.) The deletion hits its highest frequency in Oceania, and has reached fixation in particular parts of Polynesia (Hertzberg et al. 1989; Lum et al. submitted), presumably due to genetic founder events in colonization of remote islands of the Pacific Ocean. For the same reason, it can also be detected today at high frequency in some Native American tribal groups. By assaying for the presence or absence of the deletion with the polymerase chain reaction, one can trace the migration of specific lineages over short periods of time, on the order of 2000 to 30,000 years.

This approach gains resolving power when combined with sequence information from the hypervariable control region. Mutations commonly found among Native Hawaiian donors for this part of the mitochondrial genome are clustered into two major portions of the noncoding region (fig. 1), as had been described previously for African donors (Vigilant et al. 1989).

New insights into the identification of this clustering of mutations have led to the way well-defined regions can be quickly assayed in populations, without laborious and costly direct sequencing. Synthetic DNA probes can be designed to pick up specific substitutions using DNA hybridization and dot-blots instead. The oligonucleotide probe approach was recently described for human mtDNA control regions by Stoneking et al. (1991).

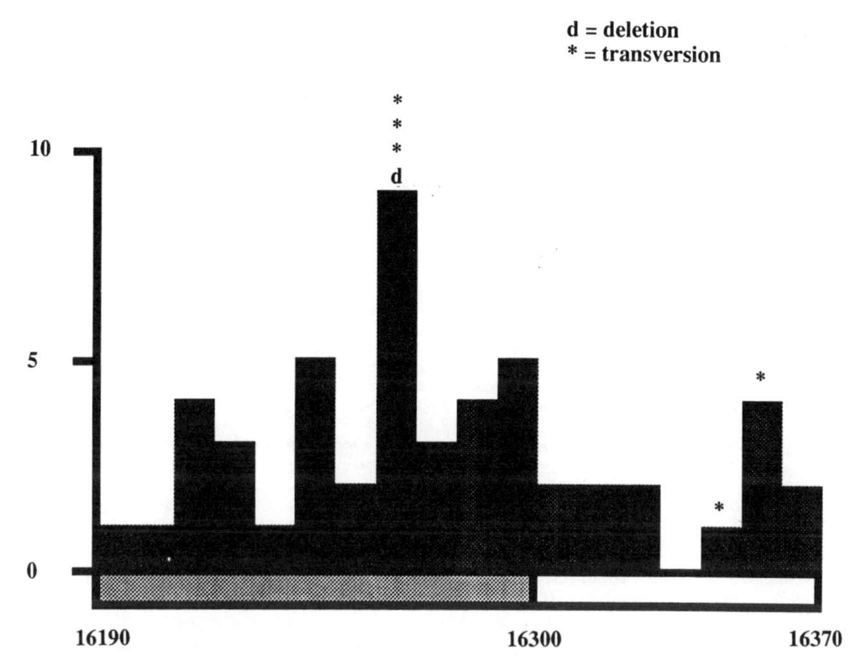

Figure 1. Hypervariable regions of human mitochondrial DNA in the major noncoding portion of the molecule, the control region. Mutations are clustered in 10 bp intervals, and the number of mutations within the subregion are stacked vertically. The line at the base refers to a piece of mtDNA spanning nucleotides 16190-16300, apparently unique to higher primates. Most of the substitutions are transition mutations, and are clustered within this region. Rarer transversions are also indicated.

APPLICATIONS OF NEW mtDNA TECHNIQUES

Human geneticists today sometimes study populations which have become admixed in the last 500 years as a consequence of expanding European and Asian trade. Hawaii is a good example of this mixing, and can serve as a model for anthropologists to study biological as well as cultural contact between societies with different morphologies, beliefs, practices, and germs. For the resident population of Native Hawaiians, the results have been devastating. Epidemic diseases brought by outsiders resulted in horrendous mortality during the 1820s and 1830s, and hastened the social collapse of entire valleys (Stannard 1989).

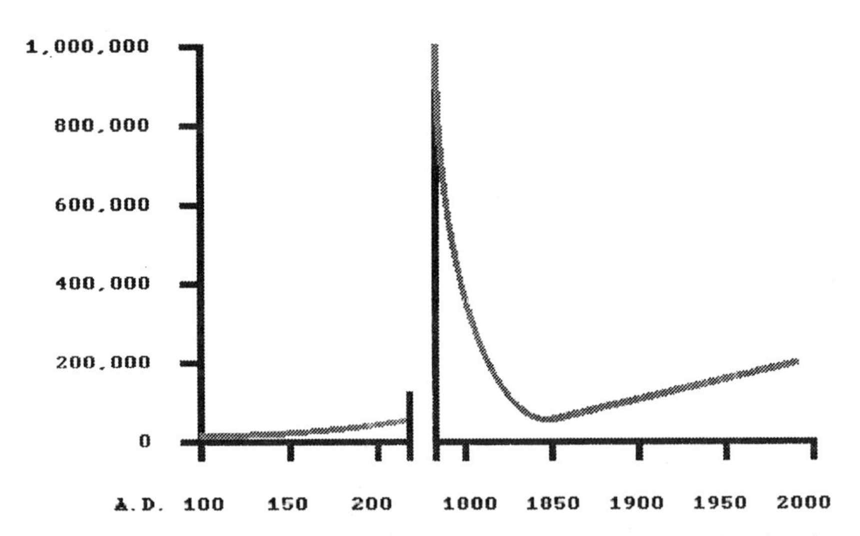

Figure 2. Population decline is attributed to social collapse brought on by epidemics of measles, small pox, influenza, tuberculosis and venereal diseases. Native Hawaiians experienced a catastrophic loss upon contact with Europeans in the late 1700s. Many aboriginal human populations underwent similar declines on other continents, and this process is still continuing in South America. The Hawaiian case is instructive for illustrating that just 5000 years of biological isolation from other populations can bring about incredible biological changes in infectious disease resistance.

Diachronic comparisons between ancient and modern DNA donors help us identify and estimate the number of mtDNA lineages segregating in the Native Hawaiian population for the time range examined. They can also help us understand which lineages carry loci that may be the basis of genetic-mediated disease resistance. Some aboriginal people object to genetic studies using human bone, on the grounds that it violates their sense of religion and ethics about treatment of the bones of their ancestors. Not all Native Hawaiians agree on this point. Genetic research utilizing skeletal material from Hawaiians has stopped in our laboratory, although it continues in other laboratories around the world.

We are interested in these comparisons for Polynesians because they allow us to measure directly the effects of infectious disease on population size and genetic diversity. Hawaii was initially settled by only a few founders, perhaps as few as 100 individuals, although oral traditions tell of two-way voyages between Tahiti and Hawaii, that probably resulted in gene flow, and which augmented the original genetic diversity of the Hawaiians. The Polynesian population in Hawaii grew in isolation for about 1000 years, until it was "found" by Captain James Cook in 1778 during his third voyage in the Pacific.

Original census estimates at the time of European contact were made by Cook and his men based on sightings of villages, counting of canoes, and observations of settled valleys during short trips on shore to the dry, leeward shores of islands. These, and later estimates based on correspondence of European and American missionaries form the basis of a growing debate about systematic underestimates of the aboriginal population. If the conservative estimate of 200,000 to 400,000 is too low, and as Stannard (1989) suggests, might more accurately be really upwards of 1 to 2 million, the epidemics in Hawaii were truly genocide. In less than 2 generations, the population might have crashed 2 orders of magnitude, from 2 million to 20,000 survivors (fig. 2). Today, census estimates place the number of "pure" Hawaiians at less than 10,000, and most of these people are elderly. There is a large population, 100,000-200,000, who are admixed Hawaiians. We find now 4 major maternal lineages in Hawaii, and most "pure" Hawaiians can be identified by 3 single nucleotide substitutions and a small deletion. The control region nucleotide substitutions and region 5 deletion mutations will help identify mitochondrial lineages in donors who may be unaware of their true family history, due to an informal type of adoption called *hanai* that was and is practiced by Hawaiian families. Social, medical, and economic programs designed for Hawaiians would experience a flood of applicants if all eligible by birth were to apply.

APPLICATION TO FORMAL MODELS FOR REPLACEMENT

Could the extraordinary tragedy of largely European-borne infectious diseases on Native Hawaiians be an example of what happened around the world 200,000 to 100,000 years ago, when another migration seems to have begun? Infectious diseases could explain the appearance of population replacement, even if gene flow was occurring at relatively high rates. There could be admixed populations, or hybrids between geographic groups where the two came into contact. Judging from data accumulating in the Pacific, the hybrid population could even be quite large, and contain a number of resident and colonist mtDNAs. But the question we have to ask is, what about the evolutionary stability of that mixed population?

Regional continuity implies that social and biological stability must be maintained over 20,000 generations, not just 10 or 20, in societies with predominantly oral traditions. We know from conservation biology literature that taxa with effective population sizes (N_e) less than 500 face a precarious future, and most probably, extinction, due to demographic catastrophes (Soule 1987). For regional continuity to be supported by genetic data, either the rates of mutation or fixation of mitochondrial genes are wrong. All estimates for mtDNA mutation rates in humans must be 5 times too fast, or all reconstructions by demographers of human population growth until 10,000 years before present are in error.

Two tests can be applied to this problem. First, we can use biogeographical, archaeological, and linguistic studies to fix some internally consistent data points in human population expansion, and use this time estimate to evaluate rate linearity estimates. This type of argument was first introduced by Mark Stoneking for highland New Guinea versus coastal and Australian mainland populations (Stoneking, Bhatia & Wilson 1986).

One important date in such internally-driven rate calculations is the separation of Australians from the Asian mainland and island South East Asian populations. Using the date of 50,000 plus or minus 10,000 years ago for a general estimate of Sahul colonization (Roberts, Jones & Smith 1990), we can estimate the rate at which genetic diversity accumulates in the control regions of Australian Aborigines. This has been measured by pairwise sequence comparisons as averaging 2.9%, while the same measurement in Native Hawaiians with the "Lapita" mtDNA haplotype is 1.6% (fig. 3). Lapita refers to the pottery and the cultural complex associated with Polynesia's original colonizers (Bellwood 1987; Green 1989). This would place the last common maternal ancestor of Native Hawaiians with 3 specific control region mutations that are single base substitutions, and the region 5 deletion mutation as having lived about 27,000 years ago. Our estimate agrees well with hypotheses which place the proto-Polynesian peoples in a region of Asia, perhaps around the Philippine Islands. It allows their genotypes to spread to Polynesia, to Micronesia, and to remain also on the Asian mainland. The Lapita cluster so far contains Hawaiians, Samoans, a Marshall Islander, a Native American from Vancouver Island, as well as Malaysian, Japanese, and Chinese donors. The molecular data is, therefore, in reasonable agreement with dental models as well as archaeological estimates for the spread of Asian mainlanders into outlying areas of the Pacific.

The second method of calibration relies on DNA sequence comparisons using an outside reference point, a closely related species or group of species. Tom Kocher's DNA sequences for chimpanzees and humans reinforce the view that the divisions in the human mitochondrial tree are relatively recent, compared to those in chimpanzees as a species. His tree also shows that the most divergent human lineages, those with the most change relative to each other, are found among African donors, not Asians, as predicted by the regional continuity model. Using chimpanzees as an outgroup,

Haplotype I

	N	Region V deletion	16217	16247	16261
Hawaiian	1	1	0	1	1
Hawaiian	7	1	1	1	1
Ad. Haw.	4	1	1	1	1
Samoan	7	1	1	1	1
Tongan	2	1	1	1	1
Micronesian	3	1	1	1	1
Ad. Haw.	1	1	1	0	1
Indonesian	7	1	1	0	1
Malaysian	1	1	1	0	1
Japanese	1	1	1	0	1
Chinese	1	1	1	0	1
Samoan	1	1	1	0	0
Indonesian	2	1	1	0	0
Amerindian*	2	1	1	0	0
Indonesian	1	1	0	0	0
	42	100%	95%	57%	86%

* Ward et al. 1991.

Figure 3. The most common maternal genotype among Hawaiians today can be characterized by 4 mitochondrial mutations. We know that this genotype is shared by other Polynesians from Samoa and Tonga, as well as non-Polynesians. We infer that this genotype was present and probably a common one in the mainland Asian population which later spread to Polynesia, Micronesia, and North America.

the entire control region is aligned and compared for 4 chimps and 14 ethnically diverse humans. Humans are about 1/40 to 1/50 as diverse as chimpanzees. If human and chimp mtDNA sequences diverged from a common ancestor 7 million years ago, the basal node for humans as a species is 125,000 to 175,000 years ago. This estimate fits fairly well with the appearance of the first anatomically modern people in Africa about 120,000 to 100,000 years ago. It also fits the pattern of dispersion of human populations seen in the analysis of Vigilant et al. (1991). Because of difficulties in analysis of a data set now so large, the problem of solving the deepest branches of the tree will likely remain for some time (Hedges et al. 1992). Researchers need new methods which may not be based in parsimony analysis, which start from a different set of assumptions and can handle all the computations without a bias to order of entry. There is an additional problem associated with such reconstructions, because the deepest branches of the tree constrain maximum likelihood estimates drawn from pairwise sequence estimates (Felsenstein 1992).

HOW DOES THE NUCLEAR GENOME FIT THE mtDNA PICTURE?

In the last year, a number of new nuclear genetic studies have reported data that reinforce our view that all modern humans share a relatively recent African past. These include reports of hemoglobin diversity (Long et al. 1990), Y chromosome comparisons (Dorozynski 1991, reporting about the 1991 analysis of G. Lucotte and S. Hazout using the p49 probe), and apolipoprotein B surveys (Rapacz et al. 1990). Due to the possibility of undetected or uncontrolled recombination events, a time scale is harder to establish for sequence change in the Y genes using a molecular clock model (Zuckerkandl and Pauling 1962). However, they all find that genetic diversity is highest in modern African populations, compared to donors from other regions of the world. This is consistent with predictions from mtDNA.

Now, Michael Hammer will soon report on his newest study of change in humans and chimps for a 2.6 kilobase single-copy DNA sequence on the Y chromosome. Human genetic diversity again is the most extreme among African donors, and humans as a group are 1/40-1/50 as diverse as the chimpanzees. The time needed to accumulate mutations on the Y is different for this DNA in comparison to mtDNA, but the relative distance predicted from mitochondrial comparisons fits, as does the diversity pattern among human geographic groups (M. Hammer, pers. comm.).

SUMMARY

We remain puzzled by exactly what caused the break in the transition from our archaic ancestors to anatomically modern people. Our working hypothesis is now that it was due to infectious disease, which will require gathering more nuclear genetic data to evaluate. Specifically we will examine genes for cell-surface protein components and circulating enzymes. If Anna Di Rienzo and colleagues are correct in having identified the migration corridor out of Africa as the Nile valley (Di Rienzo et al. submitted), North African populations would have a quite different genetic profile 150,000 years ago, in comparison to Chinese and European populations of the same age. With all our uncertainty about technology to extract DNA from fossils this old, we are still confident that human geneticists and human paleontologists will have many areas of fertile discussion for the next 50 years.

ACKNOWLEDGMENTS

We thank past and present members of Allan Wilson's laboratory for advice and discussion. We also thank H. Carson, B. Feldman, B. Finney, L. Freed, B. Griffon, P. Kirch, B. Kirk, J. Linnekin, C. Reeb, S. Serjeantson, M. Spriggs, and D. Yen for their stimulating input.

REFERENCES

Bellwood, P. 1987. *The Polynesians*. Rev. ed. London: Thames and Hudson.

Bowcock, A. M., C. Bucci, J. M. Herbert, J. R. Kidd, K. K. Kidd, J. S. Friedlaender, and L. L. Cavalli-Sforza. 1987. Study of 47 DNA markers in five populations from four continents. *Gene Geography* 1:47-64.

Brown, W. 1980. Polymorphism in mitochondrial DNA of humans as revealed by restriction endonuclease analysis. *Proceedings of the National Academy of Sciences,* USA 77: 3605-9.

Cann, R. L., M. Stoneking, and A. C. Wilson. 1987. Mitochondrial DNA and human evolution. *Nature* 325:31-36.

Di Rienzo, A., and A. C. Wilson. 1991. Branching pattern in the evolutionary tree for human mitochondrial DNA. *Proceedings of the National Academy of Sciences,* USA 88:1597-1601.

Dorozynski, A. 1991. Was Adam very, very short? *Science* 251:379.

Felsenstein, J. 1992. Estimating effective population size from samples of sequences: Inefficiency of pairwise and segregating sites as compared to phylogenetic estimates. *Journal of Genetic Research* 50:139-47.

Green, R. C. 1989. Lapita people: An introductory context for skeletal materials associated with pottery of this cultural complex. *Records of the Australian Museum* 41:207-13.

Hartl, D. 1988. *A Primer of Population Genetics*. 2d ed. Sunderland, MA: Sinauer Associates.

Hartl, D., and A. G. Clark. 1989. *Principles of Population Genetics*. 2d ed. Sunderland, MA: Sinauer Associates.

Hedges, S. B., S. Kumar, K. Tamura, and M. Stoneking. 1992. Human origins and analysis of mitochondrial DNA sequences. *Science* 255:737-39.

Hertzberg, M., K. N. P. Mickleson, S. Serjeantson, J. F. Prior, and R. J. Trent. 1989. An Asian-specific 9-bp deletion of mitochondrial DNA is frequently found in Polynesians. *American Journal of Human Genetics* 44:510-40.

Higuchi, R. 1989. Simple and rapid preparation of samples for PCR, 31-38. In *PCR Technology*, ed. H. Erlich. New York: Stockton Press.

Kocher, T. D., and A. C. Wilson. 1991. Sequence evolution of mitochondrial DNA in humans and chimpanzees: Control region and a protein-coding region. In *Evolution of Life*, a symposium volume from Kyoto, Japan, held March, 1990.

Long, J. C., A. Chakravarti, C. D. Boehm, S. Antonarakis, and H. Kazazian. 1990. Phylogeny of human B-globin haplotypes and its implications for recent human evolution. *American Journal of Physical Anthropology*. 81:113-30.

Mellars, P., and C. Stringer, eds. 1989. *The Human Revolution: Behaviuoral and Biological Perspectives on the Origins of Modern Humans*. Edinburgh: Edinburgh University Press.

Nei, M. 1987. *Molecular Evolutionary Genetics*. New York: Columbia University Press.

Rapacz, J., L. Chen, E. Butler-Brunner, M.-J. Wu, J. O. Hasler-Rapacz, R. Butler, and V. N. Schumaker. 1991. Identification of the ancestral haplotype for apolipoprotein B suggests an African origin of *Homo sapiens sapiens* and traces their subsequent migration to Europe and the Pacific. *Proceedings of the National Academy of Sciences, USA* 88:1403-6.

Roberts, R. G., R. Jones, and M. A. Smith. 1990. Thermoluminescence dating of a 50,000-year-old human occupation site in northern Australia. *Nature* 345:153-56.

Singer, M., and P. Berg. 1991. *Genes and Genomes*. Mill Valley, CA: University Science Books.

Soule, M., ed. 1987. *Viable Populations for Conservation*. New York: Cambridge University Press.

Stannard, D. E. 1989. *Before the Horror*. Honolulu, HI: Hawaii Press.

Stoneking, M., K. K. Bhatia, and A. C. Wilson. 1986. Rate of sequence divergence estimated from restriction maps of mitochondrial DNAs from Papua New Guinea. *Cold Spring Harbor Symposium on Quantitative Biology* 51:433-39.

Stoneking, M., D. Hedgecock, R. G. Higuchi, L. Vigilant, and H. Erlich. 1991. Population variation of human mitochondrial DNA control region sequences detected by enzymatic amplification and sequence-specific oligonucleotide probes. *American Journal of Human Genetics* 48:370-382.

Templeton, A. R. 1992. Human origins and analysis of mitochondrial DNA sequences. *Science* 255:737.

Vigilant, L., R. Pennington, H. Harpending, T. D. Kocher, and A. C. Wilson. 1989. Mitochondrial DNA sequences in single hairs from a southern African population. *Proceedings of the National Academy of Sciences, USA* 86:9350-54.

Vigilant, L., M. Stoneking, H. Harpending, K. Hawkes, and A. C. Wilson. 1991. African populations and the evolution of human mitochondrial DNA. *Science* 253:1503-8.

Ward, R. H., B. L. Frazier, K. Dew-Jager, and S. Pääbo. 1991. Extensive mitochondrial diversity within a single Amerindian tribe. *Proceedings of the National Academy of Sciences, USA* 88:8720-24.

Zuckerkandl, E., and L. Pauling. 1962. Molecular disease, evolution, and genic heterogeneity, 189. In *Horizons in Biochemistry*, ed. M. Kasha and N. Pullman. New York: Academic Press.

Chapter **8**

Out of Africa – A Personal History

CHRISTOPHER B. STRINGER

To the general public and the media (but certainly not for all palaeoanthropologists) two evolutionary models dominate current debate about the origins of modern humans. These are the model of multiregional evolution, which traces the origins of modern human anatomy and racial diversity to variation developed over the period 1,000,000 to 100,000 years ago during the evolution of *Homo erectus* and so-called "archaic *Homo sapiens*," and the recent African origin ("Out of Africa") model, which argues for an African origin probably within the last 200,000 years, and a subsequent radiation and diversification of early modern humans around the world during only the last 100,000 years. However, when I began my research over twenty years ago on recent human evolution, the present versions of these models did not exist and to my knowledge the Out of Africa model had not been proposed in any form. It is instructive to examine the way in which the Out of Africa model has developed since 1970, both from my own perspective and more generally, because there have been many misunderstandings and misrepresentations of the idea, both in the media, which perhaps can be absolved of some blame through ignorance, and also by some expert opponents. It is seen as retrograde and antievolutionary by some, and a return to a "pre-*sapiens*" model of modern human origins. Others see it as implying the dispersal of "killer Africans," who wiped out other human populations in a Pleistocene holocaust. Some opponents have particularly targeted the concept of a "mitochondrial Eve," or have attacked the view that the appearance of modern humans represented a speciation event, believing (incorrectly) that these concepts are fundamental to the Out of Africa model, and that by undermining them, the whole model

---

CHRISTOPHER B. STRINGER • Department of Palaeontology, The Natural History Museum, London SW7 5BD England, UK.
Origins of Anatomically Modern Humans, edited by Matthew H. Nitecki and Doris V. Nitecki. Plenum Press, New York, 1994.

can be falsified. However, the model has had a complex gestation period, and relies on several independent lines of evidence for its support, as I hope to show in this paper.

1969-1974

In 1970, when I began my doctoral research, I addressed four main phylogenetic hypotheses (fig. 1) about the origin of modern humans (Stringer 1974). The first was the unilinear, or "Neanderthal phase," hypothesis of Hrdlicka, with its multiregional variants proposed by workers such as Weidenreich, Coon and Brace. Here, populations around the inhabited world, throughout the Middle and Late Pleistocene, were evolving inexorably through a "Neanderthal phase" towards modern *Homo sapiens*. Orthogenesis and contact between the regions was emphasized in Weidenreich's formulation, whereas cultural innovation was the predominant force for change in Brace's version. However, the basic proposition was consistent – groups represented by fossils such as Swanscombe and Neanderthal (Europe), Tabun and Skhul (Levant), Broken Hill (Africa), Zhoukoudian (China), and Ngandong (Java) were ancestral to present-day populations in the same regions of the world. The second model, the "spectrum hypothesis" of Weiner and Campbell, resembled the Neanderthal phase model except that there was a much greater blending of features between the different populations and regions, so that separate evolutionary lines did not really exist. This implied open genetic network meant that all descendant populations of *Homo erectus* could potentially have contributed to recent human evolution, although not all necessarily actually did so, and the Neanderthal and Ngandong peoples may have been partly ancestral to modern groups. The third model, associated with workers such as Sergi, Breitinger, and Clark Howell, and often called the pre-Neanderthal or early Neanderthal model, proposed the existence of an early or generalized Neanderthal or Neanderthal-like population, in Europe and/or the Levant. This population later diverged into a European lineage leading to the late Neanderthals and a lineage (probably centered in the Levant) which gave rise to modern humans. In the middle of the last glaciation, the latter group migrated from its place of origin to replace the more specialized European Neanderthals. The fourth model was the "pre-*sapiens* hypothesis" of Boule and Vallois (with variations from workers such as Keith and Louis Leakey). Here it was envisaged that there was an ancient split between the true *sapiens* lineage leading to modern people, and the lineage leading to other groups such as *Homo erectus* and the Neanderthals. The lineage of modern humans must have coexisted with that of the Neanderthals, even within one continent, so that in Europe the pre-*sapiens*

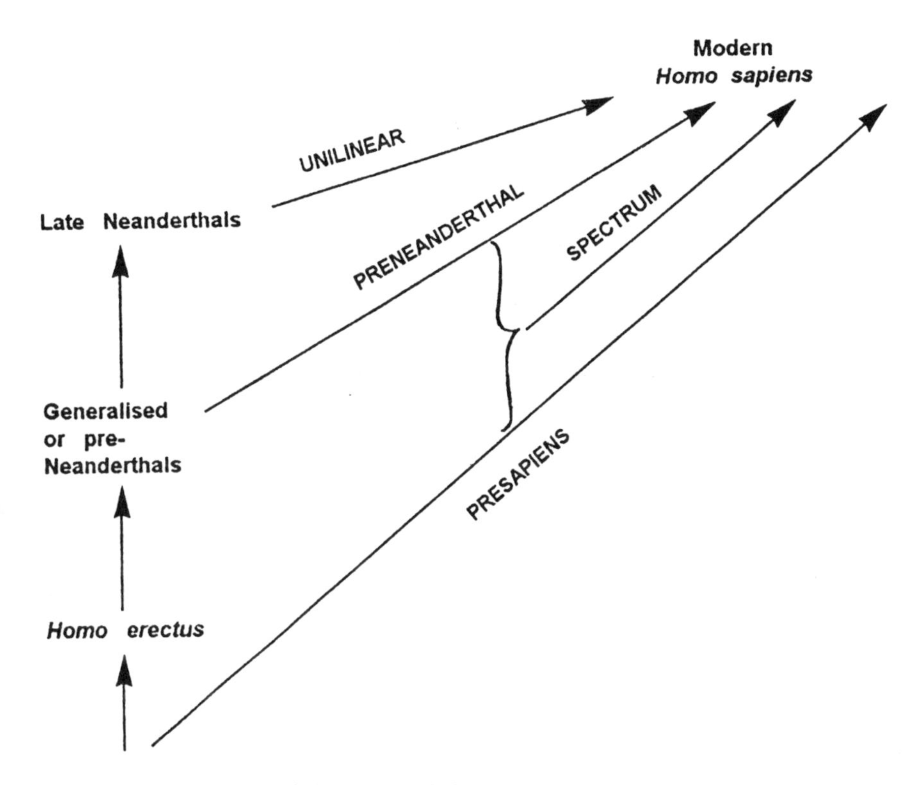

Figure 1. Pre-1970 evolutionary models for the evolution of modern *Homo sapiens.*

line was represented by supposedly advanced fossils such as Swanscombe and Fontéchevade, which were more or less contemporaneous with more primitive groups such as Steinheim and the early Neanderthals. One model which I did not address in 1970 was the idea of a single center of origin for modern humans, discussed in a general way by workers such as Howells (1967), but articulated only in rather vague terms because of the lack of clear support at that time from interpretations of the fossil record and chronological evidence.

These last two factors were as important in limiting research in 1970 as were the straitjackets of the prevailing "expert" views. Europe had by far the best fossil evidence, but some of it had not received comprehensive study or restudy for many years, and work was hampered by uncertainty about the pattern of Middle Pleistocene human evolution in the region. Important fossils such as Petralona were still virtually unstudied, and the evolutionary positions of others such as Swanscombe, Steinheim, and Fontéchevade were unclear. Although it was apparent that there had been a major period of morphological and behavioral change in the middle of the last glaciation, the exact chronological relationship between the last Neanderthals and the first Cro-Magnons was very unclear, while the manufacturers of industries such as the Late Mousterian and the Châtelperronian were still unknown. Interpretations of human evolution in the Middle East were still dominated by the Mount Carmel remains from Tabun and Skhul, since the Amud skeleton was still in the process of publication, the Shanidar and Qafzeh discoveries were not published in any detail, nor, in fact, had some of the Qafzeh discoveries yet been made. Here again, much guesswork had to be used to assess the chronological relationships of the different fossil samples, and a commonly expressed opinion was that the Skhul and Qafzeh samples either represented Neanderthals, or populations morphologically intermediate between Neanderthals and modern humans, either through their intermediate evolutionary position or through hybridization (Brace 1964; Brose and Wolpoff 1971).

Unfortunately, interpreting the Far Eastern record was also hampered by both the lack of fossil material and its uncertain chronology, with doubts surrounding the status of the Chinese Maba discovery (Howells 1967) and the enigmatic Ngandong sample from Java (Weidenreich 1951). At this stage, too, some of the most important Australian discoveries from the Willandra Lakes and Kow Swamp remained to be made or published in detail or dated. The situation in Africa was hardly any better, and views of primitive "relict" populations lingering at the northern and southern ends of the continent predominated. Thus the Broken Hill skull ("Rhodesian Man") was generally believed to be only about 50,000 years old (Clark 1970), and claims for the presence of earlier, more advanced forms, such as at Omo-Kibish, were only just being made (Leakey, Butzer & Day 1969; Day 1972). Important discoveries at sites such as the Klasies River Mouth Caves and Jebel Irhoud had already been made, but remained to be published in detail or made more widely known. For each continent then, the fossil record required clarification, expansion or both. A final straitjacket was the conviction of many workers that technological change was the only possible mechanism for the evolution of modern humans, and thus the origin of modern people must have been intimately linked with the Middle to Upper Palaeolithic transition. This particular straitjacket still seems to be with us.

As I chose my research topic I was influenced by several factors, one of which was my long-standing fascination with the Neanderthals. Second, I spent several months working at the Natural History Museum in London during 1969 and 1970 where I came under the influence of Don Brothwell, who had a healthily skeptical approach to received wisdom about the human fossil record. Third, I was inspired by the liberal sentiments of the late 1960s and by Brace's polemical article in *Current Anthropology* (Brace 1964) to feel that the Neanderthals had indeed been unfairly treated by much previous research. Lastly, through the library of the Natural History Museum, I also learned about the comprehensive cranial measurement system devised by Howells, and the multivariate methods of analysis he was using (subsequently published as Howells 1973). Such phenetic methods of study were in the ascendancy at that time, before cladistics had made any impact on palaeoanthropology. I felt, along with many others, that multivariate analyses could give us the objectivity we were seeking in interpreting fossil hominids. I had only to feed the data into a computer, and the answers would rapidly emerge! Thus, "in 1970 I embarked on a study with the aims of measuring the available fossil cranial material from the Middle and Upper Pleistocene and attempting to quantify the extent of variation in cranial measurements to be found in the fossils themselves compared to that found in recent populations" (Stringer 1974).

The research was undertaken at the University of Bristol under the supervision of Jonathan Musgrave, who had already used multivariate methods in a study of Neanderthal hand bones (Musgrave 1973). The problems in assembling reasonable samples from the fossil material, even for the Upper Palaeolithic and Neanderthal groups, were formidable. I spent nearly four months in 1971 traveling around Europe measuring fossil material. In most places I was received with great courtesy, but, nevertheless, I found some important material completely unavailable for study, including fossils such as Steinheim, Qafzeh 6, and even La Ferrassie 1, a specimen found over 60 years earlier! But, by good fortune, I was able to circumvent one particular problem with the help of Yves Coppens and make a study of the Jebel Irhoud 1 skull, and this chance event was later to prove quite important to the development of my ideas about recent human evolution.

For my thesis I analyzed the cranial data in various ways. There were bivariate plots of cranial angles and indices, there were canonical variates analyses, and there was the computation of Mahalanobis D^2 matrices, and secondary analyses of these matrices. Because of the inconsistent nature of preservation in the fossils, it was necessary to use samples of specimens which possessed parts in common (e.g., frontal bones, parietals, faces, cranial vaults), which could then be combined into larger units until most of the skull was covered, but with decreasing sample sizes. Table 1 shows extracts from two D^2 matrices for different combinations of 25 cranial measurements. The close relationship between Amud and the European Neanderthals is clear from their small or even nonsignificant D^2 values. The relatively close relationship between Skhul 5 and the Upper Palaeolithic group is also indicated. Finally, the very distant relationship between the Upper Palaeolithic and the Neanderthal samples is clear. What are much less clear are the relationships between Amud, Jebel Irhoud, and Skhul 5, and their D^2 values are clearly dependent on the particular suite of measurements used.

I was able to reach reasonably firm conclusions about the status of cer-

Table 1. Mahalanobis D^2 distances (data from Stringer 1974). These two matrices are for different combinations of 25 cranial measurements. N.S. = nonsignificant D^2 distance; NEA = European Neanderthal; IRH = Jebel Irhoud; SK5 = Skhul 5; UP = Upper Palaeolithic.

D^2 DISTANCES - 25 VARIATIONS								
	NEA	AMUD	IRH	SK5	NEA	AMUD	IRH	SK5
AMUD	22	-			N.S.	-		
IRH	67	66	-		58	36	-	
SK5	147	114	42	-	82	53	46	-
UP	134	110	100	43	127	98	121	42

tain fossils I had measured. Swanscombe and Fontéchevade seemed archaic or Neanderthal-like in their metrical features, and I could find no support for the classical pre-*sapiens* hypothesis of Boule and Vallois. Equally, I felt I could reject simple unilinear evolution from the European Neanderthals to the European Upper Palaeolithic "population" on the grounds that, not only were there consistently large D^2 distances between these samples, but other recent human samples were often actually closer to the Neanderthals than were the Upper Palaeolithics. However, it was much less easy for me to decide on the relative merits of the spectrum and preneanderthal theories, and my 1974 conclusions, therefore, incorporated elements of both in my favored model of recent human evolution. I attempted to summarize in a complex diagrammatic network (fig. 2) the main relationships between the fossil crania I had studied. There appeared to be a "spectrum" of approximately contemporaneous populations around the Mediterranean ranging from a seemingly near-modern form represented by Skhul 5, through Jebel Irhoud and Amud to the European Neanderthals. Some form of "early Neanderthal" (not necessarily European) might well have been ancestral to both Neanderthal and modern humans, but I considered that Jebel Irhoud 1 was probably geologically too young to occupy such a position, despite its "generalised" nature. On the limited evidence available, Omo-Kibish 1 represented a plausible early modern specimen, but was an isolated find from which far-reaching conclusions could not be drawn.

Looking back at my thesis research now, it is obvious that it suffered from a certain naivety about the power of multivariate methods to resolve evolutionary problems, but also from factors which were partly or largely beyond my control at that time. Thus my "Upper Palaeolithic" sample included a number of specimens which are now known to be of terminal Pleistocene or even Holocene age, and my "Mesolithic" sample can be seen to be a confusing amalgam of genuine European Mesolithics and largely more ancient north Africans. Further beyond my control was the paucity of Middle

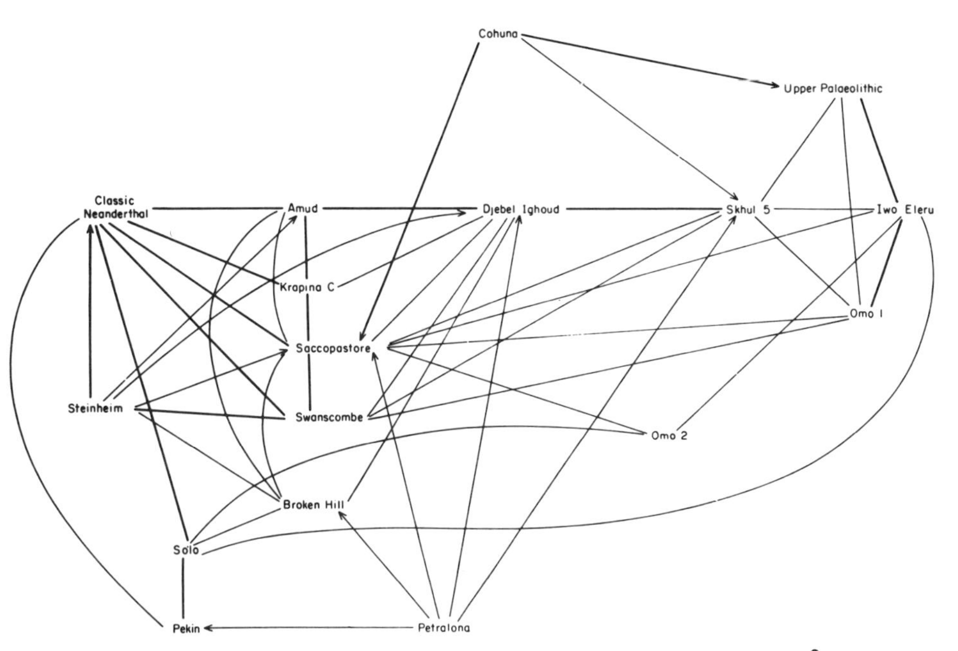

Figure 2. Diagrammatic network summarizing relationships of fossil crania from D^2 analyses (Stringer 1974). Mutually closely related fossils are joined by lines, with an arrow indicating where the nearest neighbors are actually relatively distant. Reproduced by permission of Academic Press.

Palaeolithic-associated material sampled from the Middle East and Africa, and the complete absence of such material from the Far East (although the Javanese Ngandong crania and a cast of the Cohuna skull, now regarded as related to the Kow Swamp sample, were included).

One positive aspect of my emphasis on metrical characters was the computation of the "coefficients of separate determination," which allowed an examination of the variables which were contributing most to the D^2 distances generated. In table 2, I have given some examples of these to show how patterns of similarities and differences between the fossils could be built up. When I later turned to a more cladistic approach, these tables were to be a valuable source of information about, for example, features in common between the Neanderthal crania, and between anatomically modern crania.

1975-1978

Following the completion and partial publication of my doctoral research, I began to move away from a primary emphasis on the multivariate

Table 2. Coefficients of separate determination for the most significant contributing variables to the D^2 analyses of 25 cranial measurements (data from Stringer 1974). A recent sample has been added (86 Tasmanian crania; data from Howells 1973) for comparison, and variables which reoccur with the same size factor are underlined. Patterns of similarities and differences are thus readily apparent (e.g., large NOL is particular to Neanderthals, − XCB is shared in the moderns, + PRR is probably a plesiomorphy retained in Skhul 5).

COMMON DISCRIMINATING VARIABLES (+ = LARGER, − = SMALLER)						
	AGAINST EUROPEAN NEANDERTHALS					
For:						
EUR. NEAS			---------------			
IRHOUD	−NOL	+EKB	−NAS	+PAC	−NLH	−OBB
SKHUL 5	−NOL	+EKB	−NAS	−SOS	−XCB	−NAR
UP. PAL.	−NOL	−PRR	−NLB	−SOS	−XCB	−NPH
TASM.	−NOL	−GOL	−ZMB	−SOS	−XCB	−NPH
	AGAINST EUROPEAN UPPER PALAEOLITHIC					
For:						
EUR. NEAS	+PRR	+SOS	+NOL	+XCB	+NLB	+NPH
IRHOUD	+PRR	+SOS	+FMB	−PAS	+ZYB	+NLB
SKHUL 5	+PRR	−NAR	+FMB	−NOL	+EKB	+NLH
UP. PAL.			---------------			
TASM.	−ZOR	−NPH	−MDB	−NOL	−VRR	+NLB

Legend: NOL = nasio-occipital length; EKB = biorbital breadth; NAS = nasal subtense; PAC = parietal chord; NLH = nasal height; OBB = orbital breadth; SOS = supraorbital projection; XCB = maximum cranial breadth; NAR = nasion radius; PRR = prosthion radius; NLB = nasal breadth; GOL = glabello-occipital length; ZMB = zygomaxillary breadth; NPH = nasion-prosthion height; FMB = frontomalare breadth; PAS = parietal subtense; ZYB = zygomatic breadth; ZOR = zygoorbitale radius; MDB = mastoid breadth; VRR = vertex radius.

distances generated between the specimens and groups, and towards the factors that lay behind such distances. This change of emphasis was paralleled in the work of Howells (1974, 1975) and was encouraged by contact with the biometrician Michael Hills when I returned to work at the Natural History Museum in 1973. He advised me to spend more time looking at the raw cranial data and at the relationships between different elements of the data, and advised against complicated multivariate procedures where interpretation of the results was actually much more difficult than working with simple metrical comparisons through bivariate plots, ratios, etc. He also emphasized the need to separate factors of size from those of shape, and recommended the use of the relatively simple Penrose Size and Shape statistic (Penrose 1954). This advice was put to use in a paper prepared in 1976 and published two years later (Stringer 1978). Here I employed the same basic data set as in my thesis, but condensed by a more rigorous selection of specimens for the Upper Palaeolithic sample, and expanded by the addition of data from fossils such as Arago 21 and Qafzeh 6, which I had been able to study in 1975. Penrose analyses were combined with bivariate plots to produce some confirmation of my earlier results, but also clarification of the relative roles of size and shape in producing the phenetic distance measures obtained between crania. In addition, through contact with workers such as Peter Andrews, Eric Delson, Jeff Schwartz, and Ian Tattersall, I was becoming aware of the potential of cladistics to contribute to my understanding of the data I was trying to analyze. The crania began to fall into more clearly defined groups on the basis of shared derived characters: Neanderthals (early and late in Europe, and Asian), moderns (recent, Upper Palaeolithic, Skhul and Qafzeh, and Omo 1), and a more primitive third group consisting of specimens like Petralona, Broken Hill and Arago. If both Neanderthals and moderns were to be regarded as *Homo sapiens*, this third group could be regarded as "archaic *Homo sapiens*."

However, the position of specimens such as Jebel Irhoud 1 was still unclear, as can be seen from my 1978 diagram summarizing the results of the size and shape analyses (fig. 3). It appeared that the Middle Pleistocene ancestral condition of a long broad and low cranial vault of small size, coupled with a broad flat face of large size, had given rise to three main evolutionary trends. One, leading to modern humans, had modified the cranial shape to be higher, narrower and shorter, but had mainly retained the primitive facial shape, only reducing its size. In contrast, the Neanderthal lineage had maintained a primitive cranial shape, but made it larger, and had mainly modified facial shape, producing a narrower upper face with pronounced midfacial projection. The third trend was apparently mainly a retention of the primitive cranial proportions into the Late Pleistocene, as found in the enigmatic Jebel Irhoud 1 specimen. The Skhul/Qafzeh sample and the Asian Neanderthals seemed to be trending in opposite directions towards modern humans and European Neanderthals, respectively.

1979-1984

In a paper which marked a transition from a predominantly phenetic to a predominantly cladistic approach, I published a collaborative and quite detailed study of the Petralona cranium in 1979 (Stringer, Howell & Melentis 1979), but two subsequent chance events were to have significant effects on my views on recent human evolution. The first was the invitation to collaborate with Erik Trinkaus on a study of the Shanidar crania in 1979, and the second was a similar invitation to collaborate with Michael Day on a new

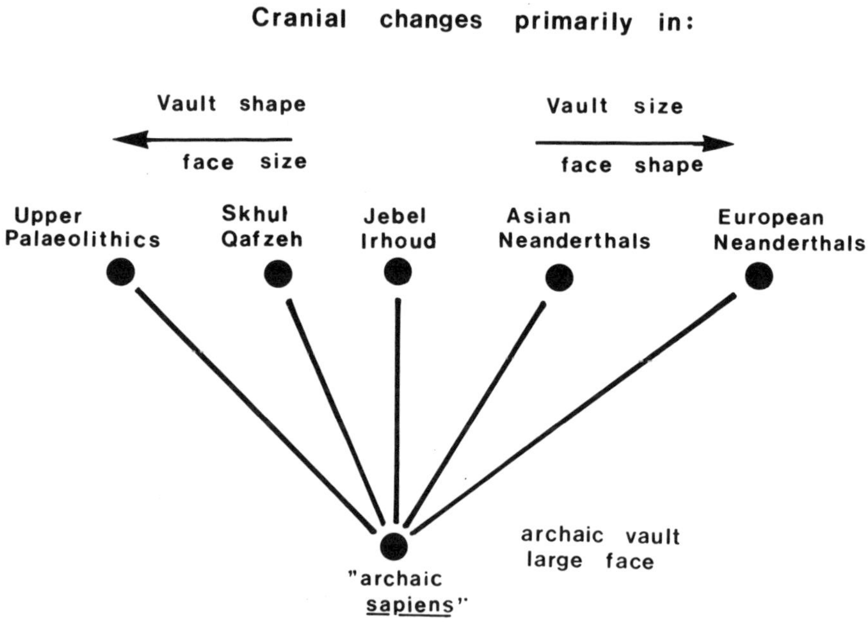

Figure 3. Summary diagram of Penrose Size and Shape Analyses (modified from Stringer 1978).

reconstruction and study of the Omo-Kibish remains in 1981. In these studies I moved over to a predominantly cladistic approach, and attempted to use my metrical data base cladistically. I was also able to utilize cladistic approaches to Neanderthal fossils pioneered by Santa Luca (1978) and Hublin (1978).

The main aim of the Shanidar study was to compare the Shanidar crania with my more extensive data base of the European Neanderthals to address the problem of whether the Shanidar specimens were merely geographical variants of the Neanderthal type, or whether they were in fact "progressive" Neanderthals, perhaps evolving towards an early modern form. Here I was able to make full use of my data set on the Qafzeh hominids for the first time, and was able to refer to the parallel work of my coauthor Trinkaus on the postcranial anatomy of the same groups (Trinkaus 1981). We concluded that "it seems most likely that the Shanidar sample represents a geographical or clinal variant within what we are calling here the subspecies *Homo sapiens neanderthalensis* or *Homo sapiens grade* 3a" (Stringer and Trinkaus 1981). By separating out primitive and derived characters or metrical tendencies, it was possible to see that whereas the European and Shanidar Neanderthal similarities were based on both symplesiomorphies (shared primitive features) and the more significant synapomorphies (shared derived

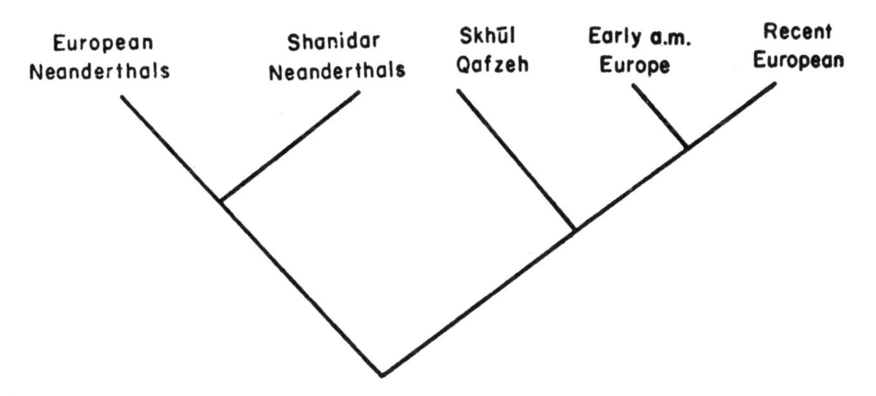

Figure 4. Cladogram of western Eurasian hominid relationships (Stringer 1982). Reproduced by permission of Academic Press.

features), those between the Neanderthal groups and the Skhul-Qafzeh sample were almost entirely symplesiomorphies. When synapomorphies were examined, the latter group was closely allied to anatomically modern humans rather than to the Neanderthals (fig. 4).

The Omo-Kibish study enabled me to address the problem of identifying an incomplete fossil as anatomically modern or not. Omo-Kibish 1 had been characterized as a predominantly archaic or Neanderthal-like form by some workers, despite Day's clear presentation of modern features in his 1969 and 1972 preliminary reports (Leakey, Butzer & Day 1969; Day 1972). Thus, using our reconstruction, we proceeded to list in a "prétirage" manuscript some metrical and morphological features derived from large studies of modern human variation which might be able to discriminate an anatomically modern fossil from others, such as Neanderthals (Day and Stringer 1982). On applying these, Omo-Kibish 2 could not be classed as "modern," whereas Omo-Kibish 1 certainly could, particularly as the postcranial material also fell within the range of modern humans. These criteria for the recognition of fossil *Homo sapiens* were necessarily strict in order to provide good discrimination, and it was never intended that they be used to exclude recent crania from our species, as has been done by Wolpoff (1986) and Brown (1990). Even in these last cases, the criteria used were not those I published in 1984, but our preliminary version from the preconference manuscript of 1982 (now revised as Day and Stringer 1991). Nevertheless, I recognize the inadequacy of some of these criteria when applied to certain large recent cranial samples, and hope to publish further revisions after more research. When it came to discussing the dating and evolutionary significance of the Omo remains, Day and I looked at the dating evidence critically, but felt that an early Late Pleistocene age at minimum was quite feasible for Omo 1 and 2. Thus, bearing in mind the generally accepted date for the Skhul and Qafzeh remains of 40,000 to 50,000 years, it seemed quite possible that Omo 1 was the oldest anatomically modern specimen yet found.

When coupled with my previous conclusions concerning the absence of evidence for regional continuity in Europe and the Middle East, I was now predisposed to favor an origin for modern humans outside of those regions. I was becoming more aware of claims for the existence of very early modern human remains at the southern African sites of Klasies and Border Cave, and realized that late archaic African specimens such as Irhoud and Omo 2 showed a rather generalised cranial anatomy which could not exclude them from an ancestral position for modern humans. There were also signs of possible modern synapomorphies such as in the supraorbital morphology of Omo 2, and the frontal morphology of the Irhoud 2 cranium, which I had not yet been able to study directly. From the Far East there was evidence of possible late archaic forms at sites such as Maba (Wu and Wu 1985) and Ngandong (Santa Luca 1980), in which I could see no obvious links to modern humans, but there was no evidence of a comparably early presence of anatomically modern features. Instead, it seemed to me that the known early modern specimens from sites such as Zhoukoudian Upper Cave (China) and Mungo and Keilor (Australia) were similar enough to each other and to western samples to indicate a probable recent common origin. Thus, taking the negative evidence from most regions of an absence of transitional forms and an absence of evidence for very early modern humans (including the assumption that the Skhul-Qafzeh samples were rather late in age), I presented an explicit African origin viewpoint for the first time in public at Nice in 1982 (Day and Stringer 1982) and in print at the end of the same year (Stringer 1982).

The years 1982-1984 saw me preparing papers which dealt with *Homo erectus*, Middle Pleistocene hominids, Neanderthals and modern human origins (Stringer 1982, 1984, 1985; Stringer, Hublin & Vandermeersch 1984). All of these incorporated cladistic procedures, and I began to see how the diagrammatic network of my thesis conclusions could now be redrafted on the basis of symplesiomorphies and synapomorphies (fig. 5). There were apparently five main groups within that network: a *Homo erectus* group, an archaic or primitive *Homo sapiens* group, a Neanderthal group, an African archaic group which apparently retained primitive features into the later Pleistocene, and an early modern group. There were also two rather enigmatic specimens: Steinheim appeared to have affinities to both the Neanderthal and the archaic *sapiens* groups, and the Australian Cohuna skull was apparently a very peculiar anatomically modern specimen, characterized by a massive faee

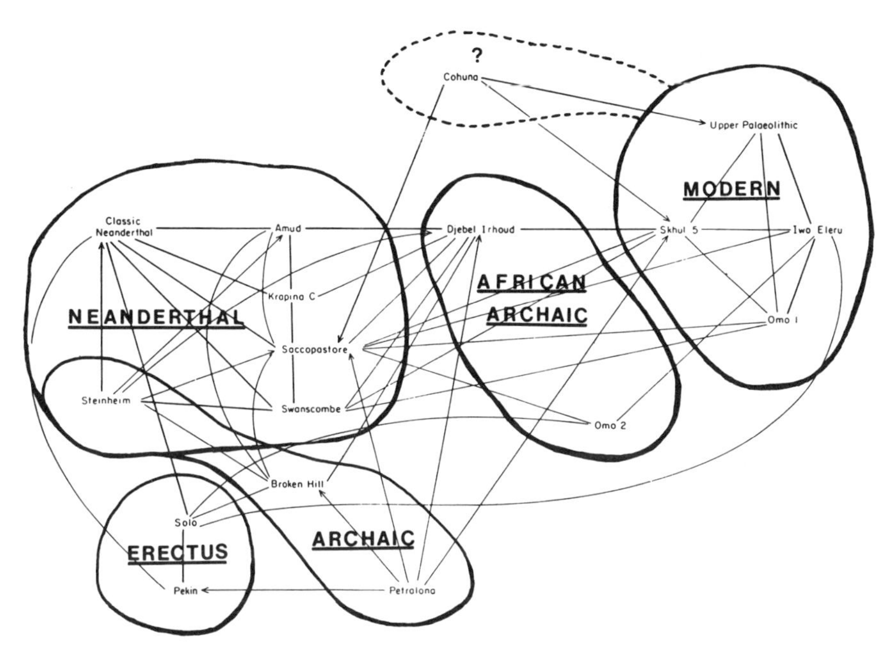

Figure 5. A revised version of figure 2, based on a cladistic rather than phenetic approach.

and a very flat frontal bone (now thought to have been artificially deformed, Brown 1987).

At this point it would be worth emphasizing that my progress towards an Out of Africa viewpoint by 1981 was not unique. Even during the 1960s, workers such as Brothwell (1963) had unfashionably considered Africa as a possible homeland for modern humans, and during the 1970s, archaeological and chronological evidence accrued that the Middle Stone Age of Africa was not the rather retarded poor relation of the European Upper Palaeolithic that it had come to be considered. Instead, new evidence suggested that it was much more likely to be time equivalent to the Middle Palaeolithic of Eurasia, and contained variants (such as the Howieson's Poort) which, in fact, looked rather "advanced" for the Middle Palaeolithic. Thus, Desmond Clark answered his 1975 question "Africa in prehistory: peripheral or paramount?" with an answer favoring the paramount (Clark 1975). At about the same time, Reiner Protsch advanced an early version of an Out of Africa model, but this was flawed by an unreliable chronology and a very idiosyncratic taxonomic approach, including the proposition that the archaic "*Homo sapiens rhodesiensis*" had actually evolved from the early anatomically modern "*Homo sapiens capensis*" (named from the Boskop skull) about 50,000 years ago (Protsch 1975). This ignored research by Klein (1973) which suggested that the Broken Hill skull was not a late African relict form, but was instead a probable Middle Pleistocene "archaic *Homo sapiens*."

A rather better argued Out of Africa viewpoint came from Peter Beaumont and colleagues, based on the study of the Border Cave hominids by de Villiers and on his and Vogel's work at Border Cave and other South African sites (Beaumont, de Villiers & Vogel 1978). An archaic *sapiens* type comparable to the South African Elandsfontein cranium or Swanscombe and Steinheim had given rise to two Middle Pleistocene lines of human evolution. One, north of the Sahara, lead to the Neanderthals, the other to the south, gave rise to modern humans such as those from Omo-Kibish, Klasies and Border Cave by about 100,000 years ago. However, Beaumont, de Villiers, and Vogel (1978) also tried to connect the spread of early modern humans with the spread of novel technologies, for example, in north Africa and the Levant, and this aspect of the model is now untenable. Two other researchers had also identified Africa as an early source for modern human anatomy by the early 1980s. One of these, Philip Rightmire (1984), remained very cautious about extrapolating from sub-Saharan African evidence to events elsewhere, whereas Günter Bräuer (1984) had formulated the "Afro-European *sapiens* hypothesis," proposing a direct evolutionary connection between early modern humans in Africa and the subsequent appearance of modern humans in the Middle East and Europe. However, this particular Out of Africa interpretation applied only to the western half of the Pleistocene world and included hybridization with "neanderthaloids" in North Africa and with Neanderthals in the Levant, followed by the spread of these new populations to Europe, where they took over from the remaining Neanderthals. But Bräuer was cautious about the course of human evolution in Asia and felt that the evidence was too inadequate to indicate a monocentric or polycentric origin for modern humans generally.

As my views on a recent African origin for modern humans crystallized, I came increasingly to the realization that the standard classification of later Pleistocene hominids into chronologically and geographically based subspecies of *Homo sapiens* was inappropriate. Instead, it appeared that there were more likely to have been three recognizably distinct morphologies which could correspond to phylogenetic species – *Homo heidelbergensis* in the Middle Pleistocene of Europe and Africa, *Homo neanderthalensis* in the Late Pleistocene of western Eurasia, and *Homo sapiens*, originating in Africa and then dispersing to the other regions. The taxonomy of Far Eastern hominids was problematic, but there seemed no evidence for the presence of *Homo neanderthalensis* or *Homo sapiens* in the early Late Pleistocene of China or Indonesia. If the Ngandong hominids were, in fact, Late Pleistocene, as suggested by some interpretations of the Ngandong stratigraphy fauna and dating, then these might represent the continuation of *Homo erectus* very late in this region. If *Homo erectus* was actually a distinct species from *Homo neanderthalensis* and *Homo sapiens*, then a Middle Pleistocene cladogenesis would be implied. This would have occurred at the appearance of *Homo heidelbergensis*, if the Zhoukoudian hominids were really penecontemporaneous with those known from European sites such as Petralona and Arago. However, the uncertain status of Chinese fossils such as those from Dali (Wu and Wu 1985) and Jinniu Shan (Lü 1990) means that it is still unclear whether *Homo heidelbergensis* was, in fact, present in China during the later Middle Pleistocene (Stringer, in press).

1985-1988

The final phases of development of the Out of Africa model came with the attention of geneticists and geochronologists. Up to 1984 it had seemed that genetic data could not clarify the story of modern human origins, despite early attempts by workers such as Bodmer and Cavalli-Sforza (1976) and Nei and Roychoudhury (1974). This was because the data base was still relatively small and generally indirect, with insufficient resolving power or potential for reliable calibration. However, when I presented a paper at a meeting in Oxford in 1984 with an Out of Africa perspective (eventually published as Stringer 1989a), I was approached after the meeting by James Wainscoat, an Oxford geneticist, who told me that his recent collaborative research on beta-globins had produced a pattern of variation in modern humans which was consistent with a recent African origin. This work was published in *Nature* in 1986, and at about that time I began to hear of the new conclusions of Berkeley researchers working with mitochondrial DNA (mtDNA). I had been aware of the preliminary conclusions of Rebecca Cann, which were rather misrepresented in a book by Gribbin and Cherfas (1982), but the conclusions from further sampling and analyses were quite remarkable. Reconstruction of the ancestral mtDNA pattern suggested that it was African in origin, and, moreover, the hypothetical last maternal ancestor for the entire sample of over 150 individuals from around the world may have lived only about 200,000 years ago. Even more remarkable was the claim that, because outside of Africa there were apparently no more ancient mtDNA lineages, this indicated that the more ancient lineages inferred for Neanderthals and for Asian *Homo erectus* had not come through to modern populations. Hence, there had been a recent African origin for modern human mtDNA, followed by an inferred complete replacement of other more ancient mtDNA lineages. I was able to introduce these conclusions into discussions at the School of American Research seminar at Santa Fe in 1986, and I made some preliminary comments on how these results might be integrated with the fossil record in the published version of my paper (Stringer 1989b). Other developments followed rapidly, including the publication of very extensive analyses of the products of nuclear DNA or of nuclear DNA coding itself (Cavalli-Sforza et al. 1988; Long et al. 1990; Nei and Ota 1991; Bowcock et al. 1991), which now pointed clearly to a primary African/non-African split and a hierarchical pattern of population relationships more consistent with dispersal than long term multiregional evolution. Peter Andrews and I reviewed the fossil and molecular evidence for recent human evolution in 1988, and laid out several tests for the multiregional and Out of Africa models (Stringer and Andrews 1988). We concluded that the majority of the evidence favored an Out of Africa interpretation. This paper has been criticized for choosing two extreme models, and then being biased in favor of one of them. In retrospect there may be some basis for both of these criticisms, but I think that the debate which has arisen because of the way we focused our presentation has been healthy and productive.

1988-1992

The emerging chronologies for the archaic-modern "transitions" or "replacements" have recently undergone considerable revision. Thermoluminescence (TL) and Electron Spin Resonance (ESR) datings have confirmed the persistence of Neanderthals to less than 40,000 years ago in western Europe, when early modern humans were apparently also present there. These dating techniques also suggest that morphologically modern hominids at Klasies and Border Cave in South Africa could be 65-100,000 years old. They have produced even more remarkable results in the Levant, where the primitive moderns from Skhul and Qafzeh are now dated at about 80-120,000 years ago, whereas some Neanderthal fossils in the region appear to be only 60,000 years old or less (Grün and Stringer 1991).

A comparison of the European and Levantine dates clearly shows the inappropriateness of statements about "holocausts" and "killer Africans." Whatever processes led to the replacement of archaic morphologies by early modern ones, such processes operated over many millennia, and probably at different rates and times in different areas. However, the new age estimates for early modern humans both confirm the independence of the Neanderthal and modern evolutionary linages, and question the assumption that sub-Saharan Africa must be the homeland of modern *Homo sapiens*. North Africa and the Levant should also be considered as possible areas of origin. The old idea that northern Africa was an isolated retreat for relict archaic populations now seems dubious. Mousterian levels at the Irhoud site have been dated by ESR to over 80,000, and perhaps as much as 150,000 years old.

Table 3. Matrix of Penrose shape distances for 39 cranial variables. ASNEA = Asian Neanderthal; AF = African late archaic; SQ = Skhul-Qafzeh; UP = Upper Paleolithic.

SHAPE DISTANCES - 39 VARIABLES				
	NEA	ASNEA	AF	SQ
ASNEA	0.33	-		
AF	0.64	0.63	-	
SQ	1.50	1.00	0.77	-
UP	1.97	1.55	1.30	0.90

Fauna from the Singa hominid site in Sudan have also been dated to a comparable antiquity, with the association of the dated mammal teeth and the hominid discussed in Grün and Stringer (1991) now supported by still unpublished mineralogical analyses.

Recently, I have returned to the phenetic methods I previously used to see if I can make a clearer interpretation of the multivariate distance results in the light of the improved samples and absolute chronology now available. Using enlarged samples for the Asian Neanderthals (Shanidar 1 and 5, as well as Amud), Levantine early moderns (Qafzeh 3, 6 and 9 and Skhul 9, as well as Skhul 5) and for late archaic Africans (Florisbad, Laetoli 18, Omo 2, Eliye Springs and Irhoud 2, as well as Irhoud 1), and improved computing power, I have carried out size and shape analyses for 39 cranial variables. The matrix of shape distance is shown in table 3. As can be seen, the results are quite comparable with those of table 1, in that the closest relationships are again between the European and Asian Neanderthals, and there is a large distance between these and the Upper Palaeolithic sample (now restricted to specimens believed to date prior to 20,000 years ago). The African archaic and Skhul-Qafzeh early modern samples, like Irhoud and Skhul 5 earlier on, remain less clear in their relationships. The African sample is somewhat closer to the Neanderthals than to the Skhul-Qafzeh moderns, and the latter are in turn somewhat closer to the African archaic than to the Upper Palaeolithic sample. It looks rather clearer when plotted in a phenogram which approximates the multidimensional relationship in only two dimensions (fig. 6). The Skhul-Qafzeh sample has to be somewhat compressed to occupy its appropriate relative position, but is about midway way between either the Asian Neanderthals or Africans, and the Upper Palaeolithics. The archaic Africans are phenetically more closely allied to the nonmodern than the modern groups, to which they are indirectly linked through being the nearest neighbors of Skhul-Qafzeh.

However, when the implications of the revised chronologies are taken into account, it is possible to look again at figure 3 and redraft it to take ac-

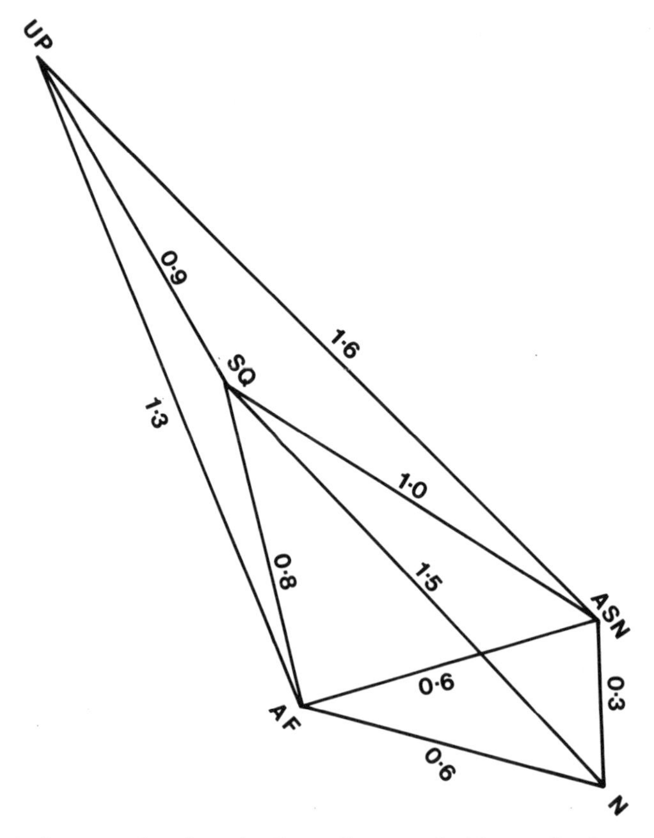

Figure 6. A phenogram based on the Shape distances of table 3. All values are indicated except for UP-N(eanderthal) which would be 2.0. The distances cannot all be represented exactly, especially in the case of SQ (Skhul-Qafzeh).

count of the new data. Placing the archaic African sample at about 150,000 years, it is possible to make more sense of the evolutionary position of specimens such as Jebel Irhoud, which were so equivocal previously (fig. 7). They are fundamentally plesiomorphous, and hence show similarities in cranial vault form to the Neanderthal groups, and in facial form (and to a lesser extent cranial form) to the Skhul-Qafzeh group. The degree of retained cranial shape plesiomorphies in the Skhul-Qafzeh group is still sufficient to maintain an intermediate position vis-à-vis the Asian Neanderthals and African archaics. But when the time dimension is added to the phenetic distance measures, the route to the Upper Palaeolithics via the African and Skhul-Qafzeh groups will always require a more moderate rate of change (and less alterations in the direction of shape change) than the route via the Neander-

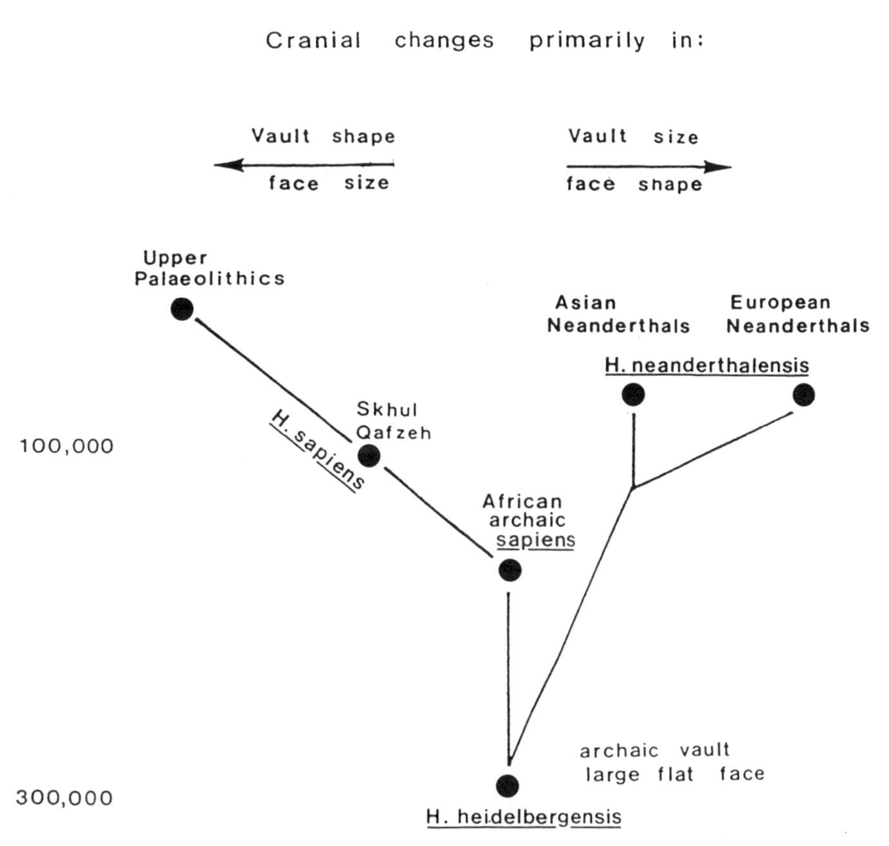

Figure 7. A revised version of figure 3, taking account of taxonomic revisions, figure 6, and actual or possible chronological revisions. The Middle Pleistocene African and European "archaic *sapiens*" group of specimens such as Arago, Petralona, Broken Hill and Elandsfontein, is now classified as *H. heidelbergensis*, and the Eurasian Neanderthal clade is recognized as a separate species *H. neanderthalensis*. The clade of anatomically modern humans (*H. sapiens*) can be extended to include the more primitive Skhul-Qafzeh group and, less certainly, the late archaic Africans such as Irhoud, Ngaloba and Florisbad. They are chronologically positioned using the linear uptake ESR results from Irhoud and Singa, but most of the sample remains to be dated satisfactorily.

thal groups. Of course, this phenetic analysis still does not, and cannot, prove which evolutionary route was taken, only indicate which is the most plausible one at present. It should also be borne in mind that the data presented here are cranial, but can be supported in some instances from the postcranial evidence as well (Trinkaus 1989; Kennedy 1992). The African archaic sample used here is heterogeneous geographically and also rather heterogeneous morphologically, and it remains to be seen how heterogeneous it is chronologically. Further samples are required, not only from Africa but also from western Asia, and indeed Asia generally. For example, the little that is known about Chinese probable late Middle Pleistocene hominids indicates that, like Irhoud 1, they are rather plesiomorphous facially and cranially, and data from these specimens is now being added to the comparisons. Data from the Dali cranium indicate that while it is similar to European and African "archaic *sapiens*" (= *Homo heidelbergensis*) fossils such as Broken Hill, Petralona, and Bodo, it is also rather like late archaic Africans in cranial shape (Stringer, in press).

SOME CONCLUDING THOUGHTS ON OUT OF AFRICA

I have tried to show how the development of an Out of Africa perspective was a gradual one in my case, where several lines of evidence converged to reinforce my original view that Eurasian Neanderthal populations were unlikely ancestors for modern humans. Instead the evidence points to late Middle Pleistocene Africa (or perhaps the Levant) as the most likely source area. The probability that a speciation event occurred seems to me to be high, but this is not a necessity for the Out of Africa model, nor does such a view preclude gene flow between *Homo sapiens* and other species. One could have had intraspecific population replacements with negligible gene flow and equally, interspecific replacements with more gene flow. However, the reality of the situation is that, over a period of time from 100,000 to 30,000 years ago, archaic morphologies outside of Africa were replaced by early modern ones. Those who demand to know what allowed the replacement to occur and what factors drove the dispersal event before they will accept that it occurred at all, are asking important put premature questions. For over a century, scientists have accepted that the dinosaurs rapidly became extinct, with the real causes only now becoming apparent. Equally, we accept that bipedalism evolved in the hominid lineage over 4 million years ago, without any clear idea of how and why this happened. The Out of Africa model has only been under serious general discussion for the last five years, so it is a little early for all its implications to have been considered and for its research potential to have been realized. But at least the right sorts of questions can now be asked in the various fields of palaeontology, genetics, archaeology, linguistics, palaeoclimatology, etc.

One of these questions concerns the factors behind the evolution of the modern skeletal pattern, but this question is not unique to the Out of Africa model, and may be easier to answer from an African (or Levantine) perspective than from the global one required by the multiregional model. The pivotal position of the Skhul-Qafzeh samples in current arguments means that a whole series of new questions can be asked about their evolutionary position using their relatively complete preservation. Attempts to get the australopithecines accepted as hominids undoubtedly led to an overstatement of their hominid nature and the creation of a false human-ape dichotomy which obscured (and still obscures) their real evolutionary significance. Similarly, I have probably been overemphatic about the modernity of the Skhul-Qafzeh and other comparable early modern samples in reaction to claims that they are closely related to Neanderthals. Now that morphological and chronological data seem to have established them in the eyes of most workers as a primitive form of modern human, we should be able to examine how "modern" they actually are, and in answering that question we will illuminate evolutionary patterns for modern humans generally (Stringer, in press).

Finally, the concept of complete global replacement of archaic by early modern hominids with an African origin is one which has arisen from the special case of the mitochondrial DNA data, and interpretations of these data are now undergoing considerable reevaluation (Maddison 1991; Wilson, Stoneking & Cann 1991; Templeton, in press). While I have been impressed by this evidence, I have never argued that complete genetic replacement was a requisite of the Out of Africa model, nor do I see how the fossil evidence could ever be adequate to address this question properly on a global basis. Where the fossil evidence does score highly is that it can reveal populations, both anatomically archaic and modern, which may *not* have modern descendants, and their morphologies are in fact more important to the construction of realistic phylogenies and evolutionary scenarios than the morphologies of the sole "survivors" – that is, us. The marriage of genetics and palaeontology which has recently occurred in the study of *Homo sapiens* origins may seem to some to have been an unhappy one – even a shotgun affair (Wolpoff 1989; Thorne and Wolpoff 1992) – but after a difficult honeymoon, I think this will eventually prove to be a long and very fruitful union.

ACKNOWLEDGMENTS

I would particularly like to thank Matthew Nitecki from the Field Museum for organizing and running the conference so effectively, and for his hospitality. I would also like to thank colleagues, especially Erik Trinkaus, for their helpful comments about the manuscript, and Robert Kruszynski for his help in drafting some of the figures.

REFERENCES

Beaumont, P. B., H. de Villiers, and J. C. Vogel. 1978. Modern man in sub-Saharan Africa prior to 49000 years BP: A review and evaluation with particular reference to Border Cave. *South African Journal of Science* 74:409-19.

Bodmer, W. F., and L. L. Cavalli-Sforza. 1976. *Genetics, Evolution and Man*. San Francisco: Freeman.

Bowcock, A. M., J. R. Kidd, J. L. Mountain, J. M. Herbert, L. Carotenuto, K. K. Kidd, and L. L. Cavalli-Sforza. 1991. Drift, admixture and selection in human evolution: A study with DNA polymorphisms. *Proceedings of the National Academy of Sciences, USA* 88:839-43.

Brace, C. L. 1964. The fate of the "Classic" Neanderthals: A consideration of hominid catastrophism. *Current Anthropology* 5:3-43.

Bräuer, G. 1984. The "Afro-European *sapiens*-hypothesis" and hominid evolution in East Asia during the Late Middle and Upper Pleistocene. *Courier Forschungsinstitut Senckenberg* 69:145-65.

Brose, D. S., and M. H. Wolpoff. 1971. Early Upper Paleolithic man and late Middle Paleolithic tools. *American Anthropologist* 73:1156-94.

Brothwell, D. R. 1963. Where and when did Man become wise? *Discovery* 24:10-14.

Brown, P. 1987. Pleistocene homogeneity and Holocene size reduction: The Australian human skeletal evidence. *Archaeology in Oceania* 22:47-71.

Brown, P. 1990. Osteological definitions of "anatomically modern" *Homo sapiens*: A test using modern and terminal Pleistocene *Homo sapiens*, 51-74. In *Is Our Future Limited by Our Past?* ed. L. Freedman. Nedlands: Proceedings of the Third Conference of the Australasian Society of Human Biology. Centre for Human Biology, University of Western Australia.

Cavalli-Sforza, L. L., A. Piazza, P. Menozzi, and J. Mountain. 1988. Reconstruction of human evolution: Bringing together genetic, archaeological and linguistic data. *Proceedings of the National Academy of Sciences, USA* 85:6002-6.

Clark, J. D. 1970. *The Prehistory of Africa.* New York: Praeger.

Clark, J. D. 1975. Africa in prehistory: Peripheral or paramount? *Man* 10:175-98.

Day, M. H. 1972. The Omo human skeletal remains, 31-35. In *The Origin of* Homo sapiens, ed. F. Bordes. Paris: Unesco.

Day, M. H., and C. B. Stringer. 1982. A reconsideration of the Omo Kibish remains and the *erectus-sapiens* transition, 814-46. In *L'*Homo erectus *et la Place de L'Homme de Tautavel Parmi les Hominidés Fossiles.* Vol. 2, ed. H. de Lumley. Nice: Centre National de la Recherche Scientifique.

Day, M. H., and C. B. Stringer. 1991. Les restes crâniens d'Omo-Kibish et leur classification à l'interieur de genre *Homo. L'Anthropologie* 95:573-94.

Gribbin, J., and J. Cherfas. 1982. *The Monkey Puzzle: A Family Tree*. London: Bodley Head.

Grün, R., and C. B. Stringer. 1991. Electron spin resonance dating and the evolution of modern humans. *Archaeometry* 33:153-99.

Howells, W. W. 1967. *Mankind in the Making*. London: Penguin Books.

Howells, W. W. 1973. Cranial variation in man: A study by multivariate analysis of patterns of differences among recent human populations. *Papers of the Peabody Museum of Archaeology and Ethnology* 67:1-259.

Howells, W. W. 1974. Neanderthals: Names, hypotheses and scientific method. *American Anthropologist* 76:24-38.

Howells, W. W. 1975. Neanderthal man: Facts and figures, 389-407. In *Paleoanthropology, Morphology and Paleoecology*, ed. R. H. Tuttle. The Hague: Mouton.

Hublin, J. J. 1978. Quelques caractères apomorphes du crâne néandertalien et leur interpretation phylogénique. *Compte Rendu Hebdomadaire des Seances de l'Academie des Sciences, Paris* (D) 287:923-26.
Kennedy, G. E. 1990. The evolution of *Homo sapiens* as indicated by features of the postcranium, 209-18. In *Continuity or Replacement: Controversies in* Homo sapiens *Evolution*, ed. G. Bräuer and F. H. Smith. Rotterdam: A. A. Balkema.
Klein, R. G. 1973. Geological antiquity of Rhodesian man. *Nature* 244:311-12.
Leakey, R. E. F., K. Butzer, and M. H. Day. 1969. Early *Homo sapiens* remains from the Omo River region of south-west Ethiopia. *Nature* 222:1132-38.
Long, J. C., A. Chakravarti, C. D. Boehm, S. Antonarakis, and H. H. Kazazian. 1990. Phylogeny of human ß-Globin haplotypes and its implications for recent human evolution. *American Journal of Physical Anthropology* 81:113-30.
Lü Zun'e. 1990. La decouverte de l'homme fossile de Jing-Niu-Shan. Première étude. *L'Anthropologie* 94:899-902.
Maddison, D. R. 1991. African origin of human mitochondrial DNA reexamined. *Systematic Zoology* 40:355-63.
Musgrave, J. H. 1973. The phalanges of Neanderthal and Upper Palaeolithic hands, 59-85. In *Human Evolution*, ed. M. H. Day. London: Taylor and Francis.
Nei, M., and T. Ota. 1991. Evolutionary relationships of human populations at the molecular level, 415-28. In *Evolution of Life: Fossils, Molecules and Culture*, ed. S. Osawa and T. Honjo. Tokyo and Heidelberg: Springer.
Nei, M., and A. Roychoudhury. 1974. Genic variation within and between the three major races of Man: Caucasoids, Negroids, and Mongoloids. *American Journal of Human Genetics* 26:421-43.
Penrose, L. S. 1954. Distance, size and shape. *Annals of Eugenics* 18:337-43.
Protsch, R. 1975. The absolute dating of Upper Pleistocene sub-Saharan fossil hominids and their place in human evolution. *Journal of Human Evolution* 4:297-322.
Rightmire, G. P. 1984. *Homo sapiens* in sub-Saharan Africa, 295-325. In *The Origins of Modern Humans: A World Survey of the Fossil Evidence*, ed. F. H. Smith and F. Spencer. New York: Alan R. Liss.
Santa Luca, A. P. 1978. A re-examination of presumed Neandertal-like fossils. *Journal of Human Evolution* 7:619-36.
Santa Luca, A. P. 1980. *The Ngandong Fossil Hominids*. Yale University Publications in Anthropology 78:1-175.
Stringer, C. B. 1974. Population relationships of later Pleistocene hominids: A multivariate study of available crania. *Journal of Archaeological Science* 1:317-42.
Stringer, C. B. 1978. Some problems in Middle and Upper Pleistocene hominid relationships, 395-418. In *Recent Advances in Primatology*. Vol. 3, *Evolution*, ed. D. J. Chivers and K. Joysey. London: Academic Press.
Stringer, C. B. 1982. Towards a solution to the Neanderthal problem. *Journal of Human Evolution* 11:431-38.
Stringer, C. B. 1984. The definition of *Homo erectus* and the existence of the species in Africa and Europe. *Courier Forschungsinstitut Senckenberg* 69:131-44.
Stringer, C. B. 1985. Middle Pleistocene hominid variability and the origin of late Pleistocene humans, 289-95. In *Ancestors: The Hard Evidence*, ed. E. Delson. New York: Alan R. Liss.
Stringer, C. B. 1989a. *Homo sapiens*: Single or multiple origin? 63-80. In *Human Origins*, ed. J. Durant. Oxford: Clarendon Press.
Stringer, C. B. 1989b. Documenting the origin of modern humans, 67-96. In *The Emergence of Modern Humans: Biocultural Adaptations in the Late Pleistocene*, ed. E. Trinkaus. Cambridge: Cambridge University Press.

Stringer, C. B. In press. Reconstructing recent human evolution. *Philosophical Transactions of The Royal Society*, London.

Stringer, C. B., and P. Andrews. 1988. Genetic and fossil evidence for the origin of modern humans. *Science* 239:1263-68.

Stringer, C. B., F. C. Howell, and J. K. Melentis. 1979. The significance of the fossil hominid skull from Petralona, Greece. *Journal of Archaeological Science* 6:235-53.

Stringer, C. B., J. J. Hublin, and B. Vandermeersch. 1984. The origin of anatomically modern humans in Western Europe, 51-135. In *The Origins of Modern Humans: A World Survey of the Fossil Evidence*, ed. F. H. Smith and F. Spencer. New York: Alan R. Liss.

Stringer, C. B., and E. Trinkaus. 1981. The Shanidar Neanderthal crania, 129-65. In *Aspects of Human Evolution*, ed. C. B. Stringer. London: Taylor and Francis.

Templeton, A. R. In press. The "Eve" Hypothesis: A genetic critique and reanalysis. *American Anthropologist*.

Thorne, A. G., and M. H. Wolpoff. 1992. The multiregional evolution of humans. *Scientific American* 266:76-83.

Trinkaus, E. 1981. Neandertal limb proportions and cold adaptation, 187-224. In *Aspects of Human Evolution*, ed. C. B. Stringer. London: Taylor and Francis.

Trinkaus, E., ed. 1989. *The Emergence of Modern Humans: Biocultural Adaptations in the Later Pleistocene*. Cambridge: Cambridge University Press.

Wainscoat, J. S., A. V. S. Hill, A. L. Boyce, J. Flint, M. Hernandez, S. L. Thein, J. M. Old, J. R. Lynch, A. G. Falusi, D. J. Wetherall, and J. B. Clegg. 1986. Evolutionary relationships of human populations from an analysis of nuclear DNA polymorphisms. *Nature* 319:491-93.

Weidenreich, F. 1951. Morphology of Solo Man. *Anthropological Papers of the American Museum of Natural History* 43:205-90.

Wilson, A. C., M. Stoneking, and R. L. Cann. 1991. Ancestral geographic states and the peril of parsimony. *Systematic Zoology* 40:363-65.

Wolpoff, M. H. 1986. Describing anatomically modern *Homo sapiens*: A distinction without a definable difference. *Anthropos* (Brno) 23:41-53.

Wolpoff, M. H. 1989. Multiregional evolution: The fossil alternative to Eden, 62-108. In *The Human Revolution: Behavioural and Biological Perspectives on the Origins of Modern Humans*, ed. P. Mellars and C. Stringer. Edinburgh: Edinburgh University Press.

Wu Xinzhi, and Wu Maolin. 1985. Early *Homo sapiens* in China, 91-96. In *Palaeoanthropology and Palaeolithic Archaeology in the People's Republic of China*, ed. Wu Rukang and J. W. Olsen. Orlando: Academic Press.

Part IV

Multiregional Hypothesis

The present debate in evolutionary biology is between the principle of continuity and discontinuity. In human anthropology this means whether the modern humans are monophyletic or polyphyletic. The answer depends on the focus and the position of the observer. From a distance we may see a phylogenetic tree and speciation events; from nearby we may see a net and not a tree.

Milford Wolpoff, Alan Thorne, Fred Smith, David Frayer and Geoffrey Pope concentrate on the refutation of the Eve hypothesis. Their Multiregional Evolution explains the formation of races by the early humans' occupation of the world outside of Africa, and by their adaptations to those regions. The evolution of human races they see as continuous and worldwide, and characterized by gene flow, continuity, selection, and drift at the peripheries. To Wolpoff and his coauthors, as to most contributors to our volume, the story of human evolution is in the fossil record. The retention of ancient features in the modern humans outside of the African continent supports the continuity between ancient and modern populations, and, thus, the multiregional evolution. The most recent publications on the mtDNA also fail to support the Eve theory. Therefore, they conclude that the living people do not have a unique, recent origin.

Fred Smith points out the many difficulties associated with the human origins. He shows the great complexities of the problem, and the incompleteness of the fossil record. He urges moderation in the formulation of definitive theories of human origin. He suggests that since modern humans have a greater antiquity outside Europe, a tendency arose to believe that modern humans migrated to Europe where they replaced archaic populations. Thus, Neanderthals' contributions to the emergence of modern people around the Mediterranean has been considered small or nonexistent. But Smith argues that a total replacement model for the origin of modern Europeans cannot be supported. He documents the continuity in morphology in Neanderthals and in the earliest post-Neanderthal populations in central Europe. In addition, possible genetic influences of Neanderthals to the non-Neanderthals appear in North Africa. Therefore, the modern humans in the circum-Mediterranean area originated by clines balancing gene flow, continuity and selection, rather than replacement.

Tal Simmons is concerned with the influences of evolutionary theories upon the formulation of models of the origin of modern humans. The Out-of-Africa hypothesis is based on the punctuation model, and the regional continuity hypothesis on the traditional concepts of the modern synthesis. The North African and the Levant contact zone of the anatomically modern and the premodern humans clearly delineates the constraints that these models impose on the theories of the origin of modern people. This is the reason why the Out-of-Africa model excludes transitional forms, and advocates the speciation events for the Neanderthals, and the Multiregional hypothesis depends on the continuity of the gene flow, and a nonspeciation. Simmons argues that archaic and modern *H. sapiens* coexisted in the circum-Mediterranean region for tens of thousands of years, and that the Levant provides an excellent sample from which to evaluate the theories of the modern human origins. The Out-of-Africa model is untenable and *natura non facit saltum.*

Chapter **9**

Multiregional Evolution: A World-Wide Source for Modern Human Populations

MILFORD H. WOLPOFF, ALAN G. THORNE, FRED H. SMITH, DAVID W. FRAYER, AND GEOFFREY G. POPE

One of the great advances of twentieth-century biology has been the demonstration that all living people are extremely closely related (Lewontin 1984). Genetic research has provided what for some is the surprising result that our DNA similarities are far greater than the much more disparate anatomical variations of humanity might suggest. These variations, the object of systematic studies for over 150 years, involve both the visible external features of our bodies seen across the world, and their underlying skeletal structures. The detailing of this variation across the world, and for skeletal features over time as well, created a broad spectrum of theories about the human races–their relationships to each other and their origin. These genetic advances have rendered virtually all of them obsolete.

While much of our understanding of genetics has come from the analysis of our nuclear genes, the recent explosion of information from studies of DNA in the mitochondria (mtDNA) has reinforced the notion of similarity and provided many additional details of its pattern. The proliferation of new human fossil remains over the last 20 years, combined with these mitochondrial studies, has reduced a plethora of different theories to two distinct views about the origins of those differences, and at the present time

---

MILFORD H. WOLPOFF • Department of Anthropology, University of Michigan, Ann Arbor, MI 48109. ALAN G. THORNE • Department of Prehistory, Australian National University, Canberra, ACT 2601, Australia. FRED H. SMITH • Department of Anthropology, Northern Illinois University, DeKalb, IL 60115. DAVID W. FRAYER • Department of Anthropology, University of Kansas, Lawrence, KS 66045. GEOFFREY G. POPE • Department of Anthropology, University of Illinois, Urbana, IL 61801.
Origins of Anatomically Modern Humans, edited by Matthew H. Nitecki and Doris V. Nitecki. Plenum Press, New York, 1994.

there is a major dispute among anthropologists and other scientists about the mechanics and timing of the origin of modern human populations.

THREE THEORIES AND THE MIDDLE GROUND

Two decades before the analysis of mtDNA became focused on modern human origins, extreme theories about modern human origins had developed within the paleoanthropological community. Each of these was an outgrowth of ideas proposed earlier in the century but involved reformulations that reflected the recent "New Synthesis" and increasing genetic sophistication. One theory emphasized variation due to genic exchanges (meaning both gene flow and migrations) and common cultural adaptations to the virtual exclusion of geographic variation. All humans were seen as one large population that evolved through a series of world-wide stages from the most ancient to the modern forms. Arambourg (1958) was the first in recent times to develop this approach. The other two theories were quite the opposite, in that they emphasized geographic variation to the virtual exclusion of any role for genic exchanges. One of these, promoted by Coon (1962), posited that the human races developed independently, largely in geographic isolation, from a common source in what was then considered the parental species *Homo erectus* (today, it is more reasonable to include *Homo erectus* within the species *Homo sapiens* as these paleospecies are not separated by cladogenesis, or even a distinct, definable boundary). The other theory was first described by Sarich (1971) as the "Garden of Eden" interpretation. It posits a single recent origin for modern people, with replacement as the main mechanism of their spread around the world. This was subsequently elaborated as Howells' (1976) "Noah's Ark" hypothesis. Prior to this elaboration Africa had been proposed as the place of origin by Protsch (1975).

The new source of mitochondrial information further separates these extremes, but at the same time allows focus on middle ground theories developed from the insights of Weidenreich (especially 1939, 1943, 1946). Multiregional Evolution encompasses three seemingly contradictory elements of the earlier extremes: (1) *long-term geographic variation,* (2) *significant genic exchange between geographic variant*, and (3) *common evolutionary trends throughout the human range*. Coon's parallelism, where all the major divisions of living humanity were subspecies of what he called *Homo erectus* before they somehow all independently moved over to become subspecies (or races) of *Homo sapiens*, has been widely, and in our view correctly, rejected. The "stages" approach, however satisfying to those who posit that no human races exist, has proved to be old fashioned and incapable of dealing with the obvious geographic elements of human variation, let alone their origin. These two frameworks have not survived the implications of the new genetic information. What remains of the earlier theories is the combination of Howells's Noah's Ark and Protsch's claim of an African origin for modern humans, which developed into the concept of evolution through replacement by Africans or their descendants. Ultimately joined with claims made in some interpretations of mtDNA variation this became the "Eve" theory (see Stringer this volume, and Cann, Rickards & Lum this volume) of *total replacement*, held among paleontologists most prominently, if not uniquely, by Gould (1987, 1989) and Stringer (1990, 1992, this volume).

The middle ground between the three earlier theories thereby ultimately became the only surviving alternative to the Eve theory, making it appear to be much more extreme than it actually is. It is represented by ourselves and by colleagues from a number of countries. The Multiregional Evolution explanation (Wolpoff, Wu & Thorne 1984) posits a role for both local genetic continuity and genetic exchanges between regions. Multiregional Evolution begins with the precept that some of the features distinguishing major human groups such as Asians, Australian Aborigines or Europeans, appeared when these regions were first inhabited, with modern expressions that evolved over a long period in approximately the same geographic regions where these traits are found in high frequency today. This view traces all modern populations back to what was ultimately an African source, to a time when people lived only there, through a web of ancient lineages whose genetic contributions to each other and to the present were significant. These contributions persisted throughout human evolution, although they were not constant, and varied from region to region, and from time to time. Regional variants first became distinct at the peripheries, a consequence of the colonization events that limited peripheral variability, primarily because of drift and bottlenecking. The regional differences were maintained through a series of balances between genic exchange (often, but certainly not always from the center toward the peripheries), and selection, which in some cases was more intense at the peripheries and in any event tended to be unique from one peripheral area to another (Wolpoff 1989a). Thus, many of the differences established during the process when our species became polytypic and widespread geographically were maintained throughout human evolution, even as cultural innovations and gene combinations spread widely and guided all populations of *Homo* down a common evolutionary path (Smith, Falsetti & Donnelly 1989).

This dynamic view of the evolutionary process accounts for how human populations both *developed* and *maintained* regional differences, while at the same time continuing to exchange genes so that selectively advantageous gene frequency changes (and the occasional new allele) could spread rapidly throughout the human range. Thus, we find that just as in the world today where very distinctive populations maintain their physical differences even though they are linked by genic exchange, this situation has existed within humanity ever since humans first spread out of Africa and colonized the Old World.

Several different patterns of *modern* human origins are compatible with the Multiregional Evolution model. For instance, it is possible that modern humans originated as a single population that spread and hybridized with other populations, with the most successful of the "modern" traits spreading furthest, perhaps even beyond the range over which the hypothesized first population can be identified. This would certainly fit the modeling of evolutionary change across the polytypic species in which the changes are caused by shifting balances of genic exchange and selection (or drift) magnitudes. One of us (FHS) is inclined toward this explanation. The rest of us think this is most unlikely, and believe that the best reading of the fossil record suggests a different compatible pattern–that modern humans evolved through the coalescence of a series of modern traits that appeared independently in various areas at different times. Gene flow, rather than migration, would appear to be the more important mode of genic exchange in the process of modern human origins. This idea that traits, rather than populations, "originated" was perhaps best described by Leakey and Lewin:

> . . . one can think of taking a handful of pebbles and flinging them into a pool of water. Each pebble generates outward-spreading ripples that sooner or later meet the oncoming ripples set in motion by other pebbles. The pool represents the Old World with its basic *sapiens* population. (1977:137)

The landing place of each pebble would be the point of origin for a unique modern feature, in this analogy. The ripples would denote the spread of the modern features and where the wave fronts meet each other unique interactive patterns are formed, contributing to the already geographically distinct natures of the populations. With continued pebble throwing, more and more modern features coalesce in this manner, which suggests that there is no specific morphological rubicon to be crossed to mark the appearance of modern humans. This view also implies that the other part of the constellation of features that distinguish some modern groups was developed very early in our history, after the exodus from Africa. Clearly, not all of our features emerged at precisely the same time (Pope 1992; Thorne 1981; Wolpoff 1989a, 1992; Wu 1988, 1990).

COMPLETE REPLACEMENT–THE EVE THEORY

The replacement theory proposes that modern humans are a new species that arose as recently as between 50,000 and 500,000 years ago (Stoneking and Cann 1989), a range usually reduced to a single date of 200,000 years ago. This Eve theory postulates a unique African origin for all modern humans, and ultimately the descent of all races from a single African woman. The theory relies on an interpretation of genetic data which concludes that over a period as short as 150,000 years there was a complete replacement of all pre-existing hunter-gatherers in Africa and the rest of the then inhabited world. In the analogy above, the Eve theory would account for modern human origins by dropping a very large rock into the pool.

While it is not uncommon to find a specific geographic area where one animal species replaced another in a fairly short time period, what is unusual about this claim is the worldwide panclimatic scope of the rapid replacement. Moreover, this replacement is unlike examples from the animal kingdom as it involves the worldwide extinction of human hunter-gatherers by other human hunter-gatherers. Even the best documented case of competitive exclusion in the hominids, the replacement of *Australopithecus boisei* by *Homo*, took place over a period of more than one million years.

The impetus for the Eve theory comes from the study of mtDNA. Some biochemists believe that the evolutionary record of this extranuclear organelle can be traced back to a single female, the putative Eve. Because Eve lived long after the initial habitation of the Old World, according to this theory, she carried not only the last common ancestor for all mtDNA but must also be the last common ancestor for nuclear DNA, and Stringer's (1990) replacement theory is thereby incorporated into Wilson's Eve theory (Cann, Stoneking & Wilson 1987). This logic is required by the explanation developed for the observed mtDNA variation, tracing all existing mtDNA to a single recent source. If Eve's descendants had hybridized with other peoples as their populations expanded, we would expect to find the remnants of other mtDNA lines, especially outside Africa where Eve is said to have originated (Cann 1988). This is because it is normal for men of successful invading people to take on wives from the people they invade, thereby perpetuating and even proliferating the mtDNA of the people being replaced. In modern examples of this kind we find, paradoxically, that wherever major invasions have taken place, the resulting people of mixed descent would often show a diminution of the physical features of the original inhabitants but an expansion of their mtDNA.

However, the mtDNA of Eve's contemporaries cannot be found among the people of today–only Eve's and those of her female descendants who replaced their less fortunate contemporaries throughout the Old World

(Cann 1987, 1988; Cann, Stoneking & Wilson 1987). The only credible explanation for the absence of other ancient mtDNA lines in modern humans is that none of the other people who lived in the distant past (or, specifically, none of their women) mixed with these invaders. Therefore, this mtDNA interpretation requires the assumption that there was no admixture between Eve's descendants and the people they replaced. It is reasonable, if not necessary, to interpret the assumption of no mixture to mean that Eve founded a new species, since, by definition, members of different species cannot have fertile offspring.

SIMILARITIES

Even though the concepts in Multiregional Evolution and the Eve theory offer very different interpretations of modern human origins, there are several points on which both sides agree. This is because we all acknowledge the validity of the mtDNA data (although not necessarily of the tree analyses used to interpret it) and the techniques used to develop it. First, both accept this evidence as strong confirmation that all human beings are closely related. Second, both point to Africa as the original source of humanity, regardless of the length of time over which they see modern human variation developing. Third, we agree that only one mtDNA lineage, with a single origin, is found in the mtDNA of all living humans. Fourth, when Eve was supposed to have lived, some 200,000 years ago, there were other people spread all over the Old World, from the tropics to the temperate zones.

CONTRADICTIONS

In spite of these points of agreement, Multiregional Evolution and the Eve theory are largely contradictory (Wolpoff 1992; Wolpoff and Thorne 1991; Thorne and Wolpoff 1992). The most important consequence of the mtDNA studies is the unavoidable implication that one of the two views of our origins must be incorrect. There is no compromise that incorporates elements of both because the genetic interpretation underlying the Eve theory requires that if there was a replacement, it must have been complete, without admixture. Thus any evidence of admixture precludes the possibility of the Eve theory, which is based on the claim that there are no surviving ancient non-African mtDNA lineages.

The details of the fossil record should allow us to discover which theory is incorrect. The role played by mtDNA has been useful in theory formation, but theories must be tested and only fossils can provide the basis for refutation of one or the other. This is because the genetic information, at best, provides a *theory* of how modern human origins *might have happened* if the assumptions used in interpreting the genes are correct. One theory about the past, however, cannot be used to disprove another. As Stoneking and Cann (1989) once wrote, "it should be stressed that the transformation from archaic to modern humans is defined by the fossil and archaeological evidence." We would add to this only that those who would reject the remnants of the very past they hope to understand (cf. Wilson and Cann 1992) are in the position of historians who begin their study of an historical event by discarding what was written about it when the event took place. The scientific method requires that we try to incorporate all sources of data in an explanatory theory because the power of a theory is measured by how much it can explain.

The fossils are the direct evidence for human evolution, and their role in the analysis of modern human origins theories is in their unique potential to refute. Their potential for refutation is high since the evidence for human evolution over the past million years is rich, in both human skeletal material and archaeological remains. Moreover, unlike the genetic data, fossils can be matched to the predictions of theories about the past without relying on a long list of assumptions. We take the Eve theory to be the null hypothesis and propose here to examine five major predictions that are unique to the Eve theory and which diametrically oppose predictions drawn from Multiregional Evolution. We will use a small sample taken from the extensive record of human fossil remains in this attempt to determine whether the predictions are fit by the evidence. If the Eve theory is correct:

(1) In a complete replacement of one worldwide human hunter-gatherer species by another, species change must have been highly competitive, presumably, if the present is any guide to understanding the past, ranging from passive coexistence to violent extermination of the resident archaic groups by the new species. Moreover, we would expect the complete replacement to have involved a technological advantage at least as dramatic as the European expansions of the past several hundred years, even though these were not as successful in replacing native populations. We predict that there should be *evidence of a spreading technology with this advantage in the archaeological record.*

(2) The evidence of the *earliest modern people should be found in Africa.*

(3) Everywhere *outside Africa the earliest modern people should resemble their African ancestors* and not the local people who lived there first. Here, we are testing the Stringer and Andrews (1988), and Bräuer (1989) contention that modern people developed African features first and then spread. It is not possible to test Bräuer's latest suggestion that the first

moderns were a "less differentiated anatomical form" (1992) because it is not clear what this means. Especially given the fact that several anatomical definitions of "modern *Homo sapiens*" fail to include all recent and living people (P. Brown 1990; Kidder, Jantz & Smith 1992; Wolpoff 1986), the features diagnosing their presence probably cannot be defined if none of them are unique to Africans.

(4) There should be *no anatomical evidence of mixing* between the earliest modern people from Africa and the populations they replaced.

(5) In regions outside of Africa there should be *no evidence of anatomical continuity* spanning the period before and after the replacement time.

BEHAVIORAL EVIDENCE

We are troubled by a theory that must assume there was a total replacement of the hunter-gatherer populations of 200,000 years ago by another (presumably rapidly expanding) group of hunter-gatherers, because it would have involved the demise of people who had successful long term occupations and adaptations to many different environments (Pope 1988; Jones 1989). We would expect natives to have an adaptive and demographic advantage over any newcomers who were allegedly able to overwhelm them. This expectation arises from the fact that historically there are many hunter-gatherer groups that have survived in Australia and in both North and South America after the arrival of Europeans, in spite of the fact that they came in large numbers with a series of technologies that were vastly more complex and destructive than those of the original inhabitants. This does not mean that invasions and complete replacements could not have happened, but rather suggests the magnitude of adaptive difference they would have to involve – certainly more than the European colonizations as these generally did not result in complete replacements.

The fact is that, quite to the contrary, a model of technological superiority associated with the first appearance of (what is regarded as) modern human form is directly at odds with the archaeological evidence virtually everywhere. In east Asia, where the persistence through the Pleistocene of distinct non-Acheulean lithic assemblages provides no evidence for the introduction of novel technology. The unique nature of the Asian Paleolithic record compels one to consider the implications of the fact that six decades of research have failed to show any indication of intrusive cultures or technologies. Artifact types first recorded in Asia's earliest Paleolithic assemblages continue into the latest Pleistocene (Pope et al. 1990). If invading Africans replaced local Middle or Late Pleistocene *Homo sapiens* populations, or their descendants, they must have adopted the cultures and technologies of the people they replaced, even as their own (presumably superior) cultures vanished without trace.

If such a worldwide invasion and complete replacement of all native peoples actually took place, we should expect to find at least some archaeological traces of the behaviors that made it successful. But are there any? In western Asia, where some researchers believe the earliest modern humans outside of Africa can be found at the Qafzeh site, this evidence is lacking. Indeed the superb archaeological record at Qafzeh shows that these people had the same stone industries. They made the same tool types, manufactured these tools using the same technology, utilized the same stylized burial customs, hunted the same game, and even used the same butchering procedures as their Neandertal contemporaries (Bar-Yosef 1987, this volume; Klein 1992; Lindly and Clark 1990b; Speth, pers. comm.). By any measure that can be applied to the archaeological record they participated in the same culture. Moreover, at the time when Eve's descendants are supposed to have left Africa, there is no evidence for the emergence of any new African stone industry, or other local African technology (Klein 1983; Lindly and Clark 1990a).

Thus, technological difference is not the explanation provided for replacement by the Eve theorists. For instance, in his review of the archaeology of modern human origins, Klein (1992) recognizes the disjunction of biological form and behavior (as it is reflected in the archaeological record), but unlike many of his colleagues who attempt to explain these behavioral changes in ecological, adaptive, and social contexts (Lindly and Clark 1990a; Harrod 1992; Simek 1992; Soffer 1992) he posits that the biological changes underlying modern human origins must be skeletally invisible because they were neurological. The idea of a major worldwide neurological change in the last 40,000 years has escaped the attention of those who study brain evolution (Holloway 1983; Falk 1987; Deacon 1989), and for good reason – there is absolutely no evidence for it. The idea that after the transformation in human behavior (Klein perceives to have occurred 40,000 years ago) "evolution of the human form all but ceased" (Klein 1992:12) ignores the voluminous literature focused on the worldwide skeletal changes in the last 40,000 years. The Neolithic changes alone exceed the magnitude of any changes that can be associated with Klein's hypothesized behavioral revolution, and occurred very much faster, and independently in a number of regions.

As an alternative to technological superiority, there are several "not with a bang, but a whimper" theories of native demise (cf. Stringer and Grün 1991). One of these is based on the fact that a population with a higher death rate can be replaced by a population with a lower death rate. This theory would envisage natives pushed into marginal environments where they were marginalized to extinction – a possibility that would have more credibility if it had not been tried and failed as the unsuccessful national policy of several colonial nations. Another "explanation" relies on unknown diseases presumably carried by the invaders–unknown because there is no evidence suggesting that natives anywhere had unusual diseases at the time they were supposed to have been replaced. The problem here is that *something like this actually happened historically*. The urban diseases that evolved in Europe during the Middle Ages decimated but fell far short of extinguishing natives when Europeans became colonial powers, even when they were purposefully spread. Diseases of similar potency that could have been carried to a susceptible world of low density hunter-gatherers by invading African hunter-gatherers have never been identified and the species distinction for the Africans suggested by some authors make this entire scenario even more unlikely. Moreover, the evidence from the Upper Pleistocene Levant, where populations who are supposed to represent natives and their replacers lived in the same areas for over 60,000 years, argues against these stories as there was neither marginalization nor banging and whimpering there.

Finally, the late Alan Wilson turned to an unusual behavioral advantage he posited for the invading Africans, asserting in a talk before the American Association for the Advancement of Science that the invasion was successful because Eve and her descendants carried a gene for language ability in their mtDNA (M. H. Brown 1990). This proposal is yet to be widely accepted, and little wonder. It is in direct disagreement with evidence from paleoneurology (Falk 1987; Holloway 1983; Deacon 1989), and would provide the strongest possible disproof for Wilson's own hypothesis of neutrality for mtDNA evolution.

In all, while the Eve theory clearly predicts certain behavioral consequences, the data as we understand them refute these predictions and we conclude that there is no archaeological support for the Eve theory.

ANATOMICAL EVIDENCE

The remaining predictions of the Eve theory relate to the question of anatomical evidence for abrupt change and the question of whether the earliest recognizably "modern" humans resemble earlier regional populations or Africans. With the fossil evidence known at this time, these questions can be unambiguously resolved in at least two, and possibly three, regions of the world. The most convincing data are from east and southeast Asia.

The hominid fossils from Australasia (Indonesia, New Guinea, and Australia), one of the most peripheral areas of long-term Pleistocene occupation, show a clear anatomical sequence (Wolpoff, Wu & Thorne 1984) uninterrupted by African migrants at any time. The distinguishing features of the earliest "Java Man" remains, dated to about one million years ago, show that these regional characters were developed at the time the region was first inhabited. The characteristic features of these Javan people that are unique to the region lie in the pattern of their overall cranial robusticity and the detailed morphology of their faces.

Compared to other regional samples from the Middle Pleistocene, the Javan people have a unique combination of features (Thorne and Wolpoff 1981; Habgood 1989a) that includes thick cranial bone, with strong continuous brow ridges forming an almost straight bar of bone across their eye sockets (orbits), and a second well-developed shelf of bone (nuchal torus) at the back of the skull for the neck muscles. Behind the brows the forehead is flat and retreating. These early Indonesians have large projecting faces with massive rounded cheek bones, especially the men. Their teeth are among the largest known from the Middle Pleistocene. A series of small but important features can be found in the most complete face known from the region (Sangiran 17) and in other facial fragments that are preserved. These include such things as a "rolled edge" on the lower margin of the orbits, a distinctive ridge on the zygomatic, and a nasal floor that flows out smoothly onto the face. In the case of this nasal floor anatomy, other later populations, such as some modern Africans also possess the feature. What makes it important here is the fact that it characterizes the Australasian peoples both in the past and today in combination with the other features we discuss – a point that serves to emphasize our contention that it is the frequencies of combinations of features, and not their presence or absence, that provides evidence for or against regional continuity. The combined presence of these features together distinguishes these early *Homo sapiens* people from other early *Homo sapiens* groups in other areas.

This unique combination of regional features of the Javan morphology was stable for at least 700,000 years, while, of course, other characteristics continued evolving. More recent Javan remains have expanded brains that have reached the modern range, but the crania from Sambungmachan (Jacob 1976) and the later Ngawi and Ngandong people (Weidenreich 1951), are otherwise remarkably similar to the earlier individuals. The large series from Ngandong, which recent evidence suggests may be about 100,000 years old (Bartstra, Soegondho & van der Wijk 1988), is striking proof of this.

The earliest Australians arrived by sea, probably before 60,000 years ago (Jones 1989), beginning an immigration process that has continued to the present-day "Boat People." The first inhabitants, by their behavior and anatomy, are clearly modern *Homo sapiens*. Their skeletons show the continuing presence of the Javan complex of features, along with further braincase expansions and other modernizations (Wolpoff, Wu & Thorne 1984). Several dozen well-preserved Late Pleistocene-Early Holocene fossils demonstrate three things: the continued regionality of the combination of Australasian skeletal features described above, the geographic source area for the majority of Australia's first colonists, and the fact that the combination of features which distinguishes modern Australoids from other living human populations is precisely that which earlier distinguished their regional predecessors from their contemporaries in east Asia, Africa and Europe (Thorne 1980; Wu 1987). If the earliest Australians were descendants of Africans, as the Eve theory predicts, and this if continuity was merely apparent, it would mean that all the features evolved a second time, independently, even as the expected characteristically African features disappeared without trace. Evolution of an individual feature a second time would be rare, but the double evolution of a whole set of features is improbable to the extreme.

In northern Asia there is also evidence of anatomical links between the area's modern and ancient inhabitants. What makes this evidence a convincing rebuttal of the Eve theory is that the links involve different features than those in the south, discussed above (Wolpoff, Wu & Thorne 1984; Wu 1987). Similar patterns based on a different set of features of the face and forehead show the different southern and northern Asian links are unlikely to be fortuitous. In China, there is a continuous sequence of human fossil remains, both earlier and later than the Zhoukoudian Lower Cave, which is widely known for the unfortunate loss of its fossils during the war with the Japanese. The very earliest Chinese fossil crania, perhaps older than a million years (Pope 1988, 1991), differ from their Javan counterparts in many ways that parallel the differences between north Asians and Australians today. The Chinese fossils are less robust, have smaller faces and teeth, flatter flaring cheeks in a very anterior position, and rounder foreheads separated from their arched brow ridges. Their noses are much less prominent and flattened at the top.

This combination continues at Zhoukoudian where a dramatic fossil sample was recovered (and subsequently lost) including six partial crania (Weidenreich 1943; Chiu et al 1973). These crania have larger brains and other features that show the ancient Chinese to be evolving in a modern direction. Moreover, as Wu Rukang of the Chinese Academy of Sciences pointed out in a review of the site (Wu and Lin 1983), even within the 150,000 or more years spanned by the Zhoukoudian occupation, evolutionary changes in the modern direction can be established for the crania. Similar trends toward modernity are demonstrable within the Zhoukoudian dental sample as well (Zhang 1991).

The origin of the complex of multiple characteristics that distinguishes the modern Asian face is to be found in the Chinese fossil record of the last million years. In addition to the features described above that characterize the earliest fragmentary specimens, the most distinctive midfacial features include relatively horizontal and anteriorly facing cheek bones, anteriorly facing frontal process of the maxilla with a distinct paranasal inflation, and the frequent presence of a distinct notch along the lower border of the cheek. Perhaps the most undebatable indication of morphological continuity is the high frequency in living Asians for the strongest manifestations of maxillary incisor shoveling (Cadien 1972), that also characterizes every fossil Asian hominid preserving the teeth. Moreover, the morphology of Asian shoveling is unique (Crummett 1991).

Toward the end of the Middle Pleistocene, crania from Dali (Wu 1981) and Jinniushan (Lü, in press) in north China, and the two specimens from Yunxian (Li and Etler 1992) – all dated to 200,000-300,000 years ago – illustrate both the continuing modernization of the Chinese cranium and provide the strongest evidence of a unique regional craniofacial complex that link the earliest Chinese remains to modern Chinese populations (Pope 1991, 1992; Etler 1990). In paleoanthropology, hypotheses are retested every time new specimens are found. The newest Chinese discoveries from Yunxian confirm more than a half-century of contentions that

> Middle Pleistocene hominids were highly polytypic and regionally differentiated, . . . the events leading to the emergence of modern humans were not restricted to one region of the world alone. (Li and Etler 1992:407)

Our careful examinations of the Middle Pleistocene Chinese specimens discovered before Yunxian show no anatomical evidence of African features, even though such evidence is predicted by the Eve theory for specimens from this time span, and the new material confirms this observation as well. The later remains from China, right down to the Zhoukoudian Upper Cave specimens in the north and the Liujiang skeleton in the south, reflect the consequences of a smooth transition into the living peoples of east Asia. While we believe there was genic exchange in and out of the region throughout the Pleistocene, this evidence unequivocally demonstrates the absence of a complete invasion by modern (or other) Africans.

Even in Europe, long thought to be the best source of evidence for a replacement, the fossil record shows that any influx of new people was neither complete nor without mixture, as the Eve theory demands, and did not involve any obvious behavioral advantages. The Eve theorists traditionally held that this influx was of modern Africans (Stringer and Andrews 1988). Bräuer, for instance, asserted that

> the ancestors of the inhabitants of Europe and Western Asia . . . consisted of *modern Africans* with some admixture of Neanderthals, (1989:139; italics ours)

although in more recent publications it has now been argued that these invaders were "generalized moderns" without regional affiliation. This position, a response to the repeated demonstrations that the earliest "modern humans" found out of Africa lack any unique African features, makes it difficult to test any "Out-of-Africa" model. Our experience (cf. Frayer et al., in press) is that the earliest of the post-Neandertal Europeans are neither Africans nor are characterized by a unique pattern of *either* modern *or* archaic African features. Behavioral evidence in Europe has also proven to be a poor source of support for Eve. For instance, the youngest known Neandertals, from Arcy-sur-Cure and Saint Césaire in France (Vandermeersch 1984), show that the people allegedly about to be replaced had already taken on the behavioral characteristics of the invaders.

To the contrary, convincing morphological evidence of mixture or in situ evolution has been demonstrated by Frayer (1992), Smith (1985, 1991, 1992), and Wolpoff (1989b), who have reported that many of the "unique" native Neandertal features are found in the Europeans who followed the Neandertals – the Upper Paleolithic, Mesolithic and later peoples. In fact, it is now recognized that only a few Neandertal features completely disappeared from the later European skeletal record. These characteristics lowered in frequency so that their appearance in combination became rare and Neandertals disappeared as a distinct morphology. Yet, Neandertal features persist well into the Upper Paleolithic, ranging from highly visible structures, such as the shape and size of the prominent nose of both Neandertal and later Europeans, to much more minute traits such as the bunned form of the back of the skull and the details of its surface such as the suprainiac fossa. A good example, reported by Frayer (1992) as one of many, is the shape of the mylohyoid foramen, the opening for the mandibular nerve. In many Neandertals, the lower portion of the opening is covered by a broad bony bridge, while in other Neandertals the bridge is absent so that the bottom rim of the canal opening has a "V-shape" form. In European Neandertals the bridged form occurs in 53% of known individuals, while in the earliest Upper Paleolithic people its incidence is 44%. In later Upper Paleolithic, Mesolithic and recent groups the incidence drops to levels below 6%.

What makes this small bony structure so important in confronting the predictions of the Eve theory is that while it is common in the Neandertals and the later populations of Europe, it is seen only very rarely in fossil or modern people from Asia and Australia. In Africa the few jaws that date from the suggested Eve period do not have it. This mandibular trait and a number of others like it on the skull and rest of the skeleton, must have evolved twice in Europe for the Eve theory to be correct. A better and much less tortuous explanation is that these characters indicate a measure of continuity that is feasible if we accept some genetic contribution of Neandertals to later European populations.

In summary, the evolutionary patterns of three different regions show that the earliest "modern" humans are not Africans and do not have the unique complex of features that characterize the Africans of that time or any other. In fact there is no evidence of specific admixture with Africans at any time, let alone replacement by them. Contrary to the Eve theory's expectations, there is indisputable evidence for the continuity of distinct unique combinations of skeletal features in different regions, connecting the earliest human populations with recent and living peoples.

EVIDENCE FROM AFRICA

The final prediction of the Eve theory is that Africa was the Garden of Eden for all living people. One would expect that only there can evidence of an archaic to modern transition be found, and that the earliest modern humans will be African. Beginning with Protsch (1975), some biological anthropologists reasoned that modern *Homo sapiens* originated first in Africa because they believed the earliest modern-looking humans were found there, and that modern African racial features can be seen in these fossils (Stringer and Andrews 1988:1263; Bräuer 1984, 1989:139 – although Bräuer [1992] now argues differently, that no unique African features can be found in these fossils, which, if true, would make it impossible to identify the place of origin for the earliest modern populations out of Africa). This would leave the African fossil record as the only valid source of information to test the Eve theory, or for that matter any Out-of-Africa theory. Unfortunately, the African evidence is sparse and fragmentary, and for the most part poorly provenienced and/or dated – the criticism of the first Out-of-Africa formulation as "flawed by an unreliable chronology" (Stringer, this volume) is truly ironic when the best proveniences and dates accepted by the Eve theorists are reviewed. Exacerbating these problems, the African data include materials which do not seem to fit the Eve theory. Moreover, the most experienced of the African specialists (Rightmire 1986; Habgood 1989b; Morris 1992) have questioned the evidence for continuity of unique regional features in Africa (the prediction of Multiregional Evolution is that regional distinctions will be found last at the center of the human range (Thorne 1981)).

Our research suggests that transitional fossils confirm the expectation that modernization occurred in Africa, as it did everywhere else. However, there is no pattern that indicates the early emergence of a new set of skeletal features that uniquely characterize modern humans generally, let alone Africans in particular. Transitional fossils from the sites of Djebel Irhoud (Morocco), Ngaloba (Tanzania), and Florisbad (South Africa) demonstrate that humans were evolving in the modern direction by in the early Upper Pleistocene. These specimens, regarded as "anatomically modern humans" by some workers (Delson 1988; Stringer 1988; and, for the first two, Bräuer 1992) and late archaics by others (cf. Smith 1985; Smith, Falsetti & Donnelly 1989; Wolpoff 1980), are most like modern humans in their faces – especially reduced height and midfacial gracilization (zygomaxillary notch, canine fossa). The mosaic nature of their morphology is demonstrated by the fact that the form of their frontal bones and other aspects of cranial vault morphology remain rather archaic (Simmons and Smith 1991; Smith, Falsetti & Donnelly 1989)

The evidence generally cited in support of a great antiquity for modern-looking people in Africa comes from three sites – Omo (Kibish Formation site KHS) in Ethiopia, Border Cave, and Klasies River Mouth Cave, both in South Africa. Fragmentary crania and mandibles from the first two sites exhibit some morphological details that could support the idea of early representatives for modern people (Day 1972; Rightmire 1979, 1986). We concur with the contention that Omo 1 has some modern features, although we do not accept the validity of the multivariate analysis linking the specimen with recent Norse (Day and Stringer 1982) as it is based on the all-plaster reconstruction of the missing face. Ethiopia was not a Viking colony, or *vice versa.* Moreover, this Omo calvarium retains certain archaic features (Corruccini 1992; Smith 1985). The mixture of archaic and modern features, however, is less of a problem for their interpretation than the context or the dating for these two sites. The Omo 1 (KHS) incomplete skeleton was found partly in situ in Member 1 of the Omo Kibish Formation. A uranium-thorium date on Etheria shell of 130,000 BP was obtained from just above the specimen and a standard radiocarbon of > 37,000 BP (also from shell) was derived from the overlying Member III. However, both techniques give generally unreliable results when used on shell and have yet to be confirmed by additional dating. Even if the radiocarbon estimate was verified on another material, it is unreasonable to interpret it to mean anything more than "> 30,000." The faunal remains associated with Omo 1 are, according to Howell, "unconvincing of any remote antiquity" (1978:216).

The Border Cave partial cranium and mandible are claimed to derive from a stratum dated from 70,000 to 80,000 BP by electron spin resonance on presumably associated nonhuman remains (Grün, Beaumont & Stringer 1990). Yet, the context of the human material relative to the dated nonhuman remains is uncertain, since, in fact, Border Cave 1 was recovered by workers digging for fertilizer in the cave (Morris 1992). Even the second adult mandible with a well-developed chin and mental trigone (both characteristic of modern humans), supposedly excavated in situ from a 50,000 to 65,000-year-old level, may actually be from an intrusive burial (Rightmire 1979). This is suggested by the fact that the human and animal bones from this level exhibit a strong contrast in the state of preservation (Klein 1983). The specimens from neither of these sites, therefore, can be said to be dated, in any conventional meaning of the term.

The best excavated remains are from Klasies, securely dated to between 80,000 and 100,000 years ago (Deacon and Geleijnse 1988; Thackeray 1989). We believe that the claims that they are modern are highly influenced by expectations developed from the morphology of the Omo 1 and Border Cave specimens, which are much more complete, but may be of considerably less antiquity than that of the Klasies people. Some of the Klasies fragments are small and delicate, but a comparative analysis shows that others are not at all modern-looking (Wolpoff and Caspari 1990; Smith, this volume). Two of the four lower jaws do not have chins, and the only distinct chin development is in KRM 41815, an aged individual with significant alveolar resorption above the trigone that emphasizes its apparent projection. Facial morphology, critical to the interpretation of modernity, is represented by only four fragmentary specimens. The single cheek bone from the site is robust and has an unusual development of the orbital pillar (the bony area forming the outer orbital rim) which is columnar-like, a condition common to Neandertals and other archaic humans, but quite rare in modern humans from anywhere in the world. This zygomatic is not only larger than those of living African people but larger and more robust than both the earlier transitional and the archaic humans found in Africa. There are two very fragmentary pieces of left maxilla. The larger of these is quite heavily constructed, while the smaller shows enough difference to suggest "a substantial level of sexual dimorphism" (Rightmire and Deacon 1991:145). KRM 16425 is a portion of a right frontal with attached superior nasal bones. While the specimen exhibits weakly developed supraciliary arches, and, therefore, is used as the centerpiece of the "anatomically modern" description for the site, the age at death is uncertain, which undermines the significance of this taxonomically important feature. Thus, the reduced brows in this specimen (like the projecting chin of KRM 41815) may be more a function of age than an indication of taxonomic modernity.

Finally, even if the KRM 16425 frontal and 41815 mandible each possess a modern feature, the question of how to define modernity in a sample must be addressed. Does the earliest appearance of modernity in one or two features of an otherwise archaic sample mean that the sample is modern (Caspari and Wolpoff 1990)? If so, there are earlier cases for its first appearance. Modern midfacial form first appears in the Middle Pleistocene of China (Pope 1991), and Jinniushan is the earliest specimen to combine a very large cranial vault with thin vault bone (Pope 1988). Vérteszöllös is the first specimen to show a distinct supreme nuchal line. Tabun B may have the first prominent chin. The point is that these specimens are not regarded as modern because of a few modern features, and we wonder why the same criteria should not be applied to the superciliary arches of the KRM 16425 frontal fragment. In summary, the claim that this sample is modern, or modern African, is highly dubious. For these reasons, the Protsch-Bräuer proposal, based on the claim that the earliest modern humans are to be found in Africa, is quite unjustified.

GENETIC EVIDENCE

With the disproof of the unique African ancestry theory for the living people of most areas, and the lack of evidence showing that modern people first appeared in Africa, we conclude that the five predictions of the Eve theory cannot be substantiated. Therefore, we wonder why an analysis of mtDNA could suggest a theory so contrary to the known facts. Overwhelming evidence suggests that it is the mtDNA that has been misinterpreted.

Beginning with the work of the late A. Wilson and his colleagues and students (Cann, Stoneking & Wilson 1987), one interpretation of mtDNA variation has been dominant–indeed, it provided the genesis of Eve theory as it is understood today (Cann 1987, 1988; Stoneking and Cann 1989; Wilson and Cann 1992). However, this is not the *only* interpretation of mtDNA variation, and, as of this date, a number of geneticists have questioned and in many cases rejected every assumption required by the Eve theory. The most damaging criticisms are accurately and thoroughly reviewed by Templeton (in press) in a paper whose careful reading might profitably be recommended. We review two of the faulty linchpins for the theory below.

One assumption critical to Eve is the demonstration that all human mtDNA can be traced to an *African* source. The computer analyses that led to this reading of the data are now recognized to be incorrect (Barinaga 1992; Maddison 1991; Templeton 1992), even by some of the Eve supporters (Hedges et al. 1992). Maximum parsimony relationship trees are extraordinarily numerous, and many of them coalesce to non-African ancestries. There is no practical way to calculate all of the most likely of these trees, let alone choose which one may have the best chance of being correct. Therefore, there is no internal mtDNA evidence for African, or any other, mtDNA origins (Excoffier and Langaney 1989; Maddison, Ruvolo & Swofford 1992).

A second critical assumption is that there is a valid "molecular clock," providing a date estimate for when the last common mtDNA ancestor lived. This clock was recognized as extraordinarily inaccurate, even by its supporters; Stoneking and Cann (1989), for instance, estimated an order-of-magnitude error range. Three reasons have emerged that undermine the clock's validity, at least as far as evolution within a species is concerned:

(1) Any clock assumes evolutionary neutrality for its assumption of a constant rate of change, but this assumption is incorrect for within-species mtDNA variation where considerable evidence suggests that mtDNA is not neutral (MacRae and Anderson 1988; Palca 1990; Fos et al. 1990). Selection acting on part of the molecule must affect the survivorship of the whole because the entire molecule is only a single genetic locus (Spuhler 1988), inheriting as a single unit. The role of mtDNA in aging and degenerative diseases (Wallace 1992) shows that the molecule is under considerable selection and the mode of inheritance (cloning, along female lines) destroys an assumption of neutrality for any part of it.

(2) Small sample effects make stochastic loss a major evolutionary force in subdivided populations (Avise, Neigel & Arnold 1984; Avise, Ball & Arnold 1988). The effect of random mtDNA losses alters the reconstruction of the tree of human mtDNA branching by pruning off the evidence of many past divergences. Each of these overlooked divergences is a mutation that was not counted when the number of mutations is used to determine how long ago Eve lived. The exact number of uncounted mutations depends on how active the pruning process was, and this in turn is a direct consequence of the exact history of population size fluctuations for each population, which can never be known. Thus, considering human populational history, this factor alone would invalidate the use of mtDNA variation to "clock" past events (Thorne and Wolpoff 1992).

(3) Finally, mistakes have been identified in the mathematical underpinnings of the mtDNA "clock." For instance, even if the assumptions necessary for a valid clock were met, Lovejoy and his colleagues propose that in the most recent calculation of Eve's age (Vigilant et al. 1991) the wrong transition/transversion ratio was used (Lovejoy et al., n.d.). The correct ratio places Eve's age at least 1.3 myr ago. We believe with others (Lewin 1990; Wayne et al. 1990) that accurate calibration of any mitochondrial "clock" has major problems that probably can never be resolved. As far as human evolution is concerned, there is no molecular clock.

Finally, recent developments in DNA analysis have come to provide support for the Multiregional Evolution model. Xiong et al. (1991) report on two rare nucleotide polymorphisms that, the authors argue, would have disappeared had there been significant bottlenecking during human evolution (a necessary consequence of the single-origin Eve theory). The rare mutant allele sequences differ from the common form in a number of bases and appear to have a common ancestry (in a single phase-shift deletion) between one-half million and two million years ago, the estimate depending on which stochastic assumptions are made. These allele sequences were found, respectively, in a Japanese and Venezuelan individual. Because the allele sequences are not independent, this distribution could derive from the common Mongoloid ancestry, or less probably from an (unknown) common Caucasoid ancestry. They could not, however, reflect a shared African ancestry. Therefore, this evidence both argues against the validity of a key element of the Eve theory, and supports the interpretation of regional continuity, most probably in Asia.

In summary, there have been dramatic changes in the interpretation of mtDNA variation between the presentation of the Spring Symposium and the finalization of this paper. As recently as this year, authors were able to turn to the mtDNA interpretations for the best evidence supporting their unique African origins models. This is no longer possible. Stringer concludes his paper in this volume by suggesting that:

> The marriage of genetics and palaeontology . . . may seem to some to have been an unhappy one – even a shotgun affair . . . – but after a difficult honeymoon, I think this will eventually prove to be a long and very fruitful union.

Indeed the union *has* been unhappy, with the discovery that one of the mates was never faithful from the start, but with proper (genetic) counseling the underlying compatibility of the marriage can now be seen by all.

CONCLUSIONS

We have undertaken to refute the Eve theory as the null hypothesis explaining modern human origins. There is more than sufficient evidence, behavioral, anatomical, and genetic, to show that this theory is wrong. This does not prove Multiregional Evolution carries the "correct" explanation, but means that it is here that we must focus our subsequent attempts at refutation *because Multiregional Evolution is the best explanation for the pattern of human evolution we have at present.*

The dramatic mtDNA similarities throughout the human race do not reflect a recent common ancestry for all living people. Instead they reflect both the considerable genetic linkage between the populations of today, with genic exchange across the world, and a similar pattern of genic exchange since our ancestors first populated the old world, more than a million years ago. They are the consequences of an ancient history of population connections and mate exchanges that characterized the human race from its inception.

Thus, the human fossil record and the interpretation of mtDNA variation can be synthesized to form a view of human origins that fits all of the data now known. This synthetic model combines the best sources of evidence about human evolution, making sense of both the archaeological and fossil records, as well as the information locked up in the genetic variation of living people all over the world. The richness of the breadth and scope of human diversity, contrasted with the closeness of human relationships, is a direct consequence of the evolutionary process. There is something very literal about the precept that we are most alike where it matters, under the skin.

REFERENCES

Arambourg, C. 1958. Les stades evolutifs de l'humanité. *The Leech* (Raymond A. Dart Commemorative Number) 28:106-11.

Avise, J. C., R. M. Ball, and J. Arnold. 1988. Current versus historical populations sizes in vertebrate species with high gene flow: A comparison based on mitochondrial DNA lineages and inbreeding theory for neutral mutations. *Molecular Biology and Evolution* 5:331-44.

Avise, J. C., J. E. Neigel, and J. Arnold. 1984. Demographic influences on mitochondrial DNA lineage survivorship in animal populations. *Journal of Molecular Evolution* 20:99-105.

Barinaga, M. 1992. "African Eve" backers beat a retreat. *Science* 255:686-87.

Bartstra, G.-J., S. Soegondho, and A. van der Wijk. 1988. Ngandong man: Age and artifacts. *Journal of Human Evolution* 17:325-37.

Bar-Yosef, O. 1987. Late Pleistocene adaptations in the Levant, 219-36. In *The Pleistocene Old World. Regional Perspectives*, ed. O. Soffer. New York: Plenum Press.

Bräuer, G. 1984. A craniological approach to the origin of anatomically modern *Homo sapiens* in Africa and implications for the appearance of modern Europeans, 327-410. In *The Origins of Modern Humans: A World Survey of the Fossil Evidence*, ed. F. H. Smith and F. Spencer. New York: Alan R. Liss.

Bräuer, G. 1989. The evolution of modern humans: A comparison of the African and non-African evidence, 123-54. In *The Human Revolution: Behavioural and Biological Perspectives on the Origins of Modern Humans*, ed. P. Mellars and C. B. Stringer. Edinburgh: Edinburgh University Press.

Bräuer, G. 1992. Africa's place in the evolution of *Homo sapiens*, 83-98. In *Continuity or Replacement: Controversies in* Homo sapiens *Evolution*, ed. G. Bräuer and F. H. Smith. Rotterdam: A. A. Balkema.

Brown, M. H. 1990. *The Search for Eve*. New York: Harper and Row.

Brown, P. 1990. Osteological definitions of "anatomically modern" *Homo sapiens*: A test using modern and terminal Pleistocene *Homo sapiens*, 51-74. In *Is Our Future Limited by Our Past?* ed. L. Freedman. Nedlands: Proceedings of the Third Conference of the Australasian Society of Human Biology. Centre for Human Biology, University of Western Australia.

Cadien, J. D. 1972. Dental variation in man, 199-222. In *Perspectives on Human Evolution*. Vol. 2, ed. S. L. Washburn and P. Dolhinow. New York: Holt, Rinehart, and Winston.

Cann, R. L. 1987. In search of Eve. *The Sciences* 27:30-37.

Cann, R. L. 1988. DNA and human origins. *Annual Review of Anthropology* 17:127-43.

Cann, R. L., M. Stoneking, and A. C. Wilson. 1987. Mitochondrial DNA and human evolution. *Nature* 325:31-36.

Caspari, R., and M. H. Wolpoff. 1990. The morphological affinities of the Klasies River mouth skeletal remains. Abstract. *American Journal of Physical Anthropology* 81:203.

Chiu Chunglang, Gu Yümin, Zhang Yinyun, and Chang Shenshui. 1973. Newly discovered *Sinanthropus* remains and stone artifacts at Choukoutien. *Vertebrata Palasiatica* 11:109-31.

Coon, C. S. 1962. *The Origin of Races*. New York: Alfred A. Knopf.

Corruccini, R. S. 1992. Metrical reconsideration of the Skhul IV and IX and Border Cave 1 crania in the context of modern human origins. *American Journal of Physical Anthropology* 87:433-45.

Crummett, T. L. 1991. A new handle on the shovel (abstract). *American Journal of Physical Anthropology.* Supplement 12:64.

Day, M. H. 1972. The Omo human skeletal remains, 31-36. In *The Origin of* Homo sapiens, ed. F. Bordes. Paris: Unesco.

Day, M. H., and C. B. Stringer. 1982. A reconsideration of the Omo Kibish remains and the *erectus-sapiens* transition, 814-46. In *L'*Homo erectus *et la Place de l'Homme de Tautavel Parmi les Hominidés Fossiles.* Vol. 2, ed. H. de Lumley. Nice: Centre National de la Recherche Scientifique.

Deacon, H. J., and V. B. Geleijnse. 1988. The stratigraphy and sedimentology of the main site sequence, Klasies River, South Africa. *South African Archaeological Bulletin* 43:5-14.

Deacon, T. D. 1989. The neural circuitry underlying primate calls and human language. *Human Evolution* 4:367-401.

Delson, E. 1988. One source not many. *Nature* 332:206.

Etler, D. A. 1990. A case study of the "*erectus*"–"*sapiens*" transition in Asia: Hominid remains from Hexian and Chaoxian Counties, Anhui Province, China. *Kroeber Anthropological Society Papers* 71-72:1-19.

Excoffier, L., and A. Langaney. 1989. Origin and differentiation of human mitochondrial DNA. *American Journal of Human Genetics* 44:73-85.

Falk, D. 1987. Hominid paleoneurology. *Annual Review of Anthropology* 16:13-30.

Fos, M., M. A. Domínguez, A. Latorre, and M. Moya. 1990. Mitochondrial DNA evolution in experimental populations of *Drosophila subobscura*. *Proceedings of the National Academy of Sciences, USA* 87:4198-4201.

Frayer, D. W. 1992. The persistence of Neanderthal features in post-Neanderthal Europeans, 179-88. In *Continuity or Replacement: Controversies in* Homo sapiens *Evolution*, ed. G. Bräuer and F. H. Smith. Rotterdam: A. A. Balkema.

Frayer, D. W., J. Jelínek, M. Oliva, F. H. Smith, and M. H. Wolpoff. In press. *Upper Pleistocene Human Remains from Mladec Cave, Moravia.* New York: Aldine de Gruyter.

Gould, S. J. 1987. Bushes all the way down. *Natural History* 87:12-19.

Gould, S. J. 1989. Grimm's greatest tale. *Natural History* 89:20-28.

Grün, R., P. B. Beaumont, and C. B. Stringer. 1990. ESR dating evidence for early modern humans at Border Cave in South Africa. *Nature* 344:537-39.

Habgood, P. J. 1989a. The origin of anatomically modern humans in Australasia, 245-73. In *The Human Revolution: Behavioural and Biological Perspectives on the Origins of Modern Humans*, ed. P. Mellars and C. B. Stringer. Edinburgh: Edinburgh University Press.

Habgood, P. J. 1989b. An examination of regional differences on middle and early late Pleistocene sub-Saharan African hominids. *South African Archaeological Bulletin* 44:17-22.

Hedges, S. B., S. Kumar, K. Tamura, and M. Stoneking. 1992. Human origins and analysis of mitochondrial DNA sequences. *Science* 255:737-39.

Holloway, R. L. 1983. Human paleontological evidence relevant to language behavior. *Human Neurobiology* 2:105-14.

Howell, F. C. 1978. Hominidae, 154-248. In *Evolution of African Mammals*, ed. V. J. Maglio and H. B. S. Cooke. Cambridge: Harvard University Press.

Howells, W. W. 1976. Explaining modern man: Evolutionists *versus* migrationists. *Journal of Human Evolution* 5:477-96.

Jacob, T. 1976. Early populations in the Indonesian region, 81-93. In *The Origins of the Australians*, ed. R. L. Kirk and A. G. Thorne. Canberra: Australian Institute of Aboriginal Studies.

Jones, R. 1989. East of Wallace's line: Issues and problems in the colonisation of the Australian continent, 743-82. In *The Human Revolution: Behavioural and Biological Per-*

spectives on the Origins of Modern Humans, ed. P. Mellars and C. B. Stringer. Edinburgh: Edinburgh University Press.
Kidder, J. H., R. L. Jantz, and F. H. Smith. 1992. Defining modern humans: A multivariate approach, 157-77. In *Continuity or Replacement: Controversies in* Homo sapiens *Evolution*, ed. G. Bräuer and F. H. Smith. Rotterdam: A. A. Balkema.
Klein, R. G. 1983. The Stone Age prehistory of southern Africa. *Annual Review of Anthropology* 12:25-48.
Klein, R. G. 1992. The archaeology of modern human origins. *Evolutionary Anthropology* 1:5-14.
Leakey, R. E. F., and R. Lewin. 1977. *Origins*. New York: Dutton.
Lewin, R. 1990. Molecular clocks run out of time. *New Scientist* (2/10/90):38-41.
Lewontin, R. C. 1984. *Human Diversity*. San Francisco: W. H. Freeman.
Li Tianyuan, and D. A. Etler 1992. New Middle Pleistocene hominid crania from Yunxian in China. *Nature* 357:404-7.
Lindly, J. M., and G. A. Clark. 1990a. On the emergence of modern humans. *Current Anthropology* 31:59-63, 64-66.
Lindly, J. M., and G. A. Clark. 1990b. Symbolism and modern human origins. *Current Anthropology* 31:233-61.
Lovejoy, C. O., et al. n.d. *The mtDNA Common Ancestor of Modern Humans Was A Member of* Homo erectus. Unpublished manuscript.
Lü Zun'e. In press. Mosaic evolution of Jinniushan archaic *Homo sapiens*. In *Human Variation in the Pacific Region*, ed. G. Krantz and Ho Chuankun. Seattle: Proceedings of the Circum-Pacific Prehistory Conference.
MacRae, A. F., and W. W. Anderson. 1988. Evidence for non-neutrality of mitochondrial DNA haplotypes in *Drosophila pseudoobscura*. *Genetics* 120:485-94.
Maddison, D. R. 1991. African origin of human mitochondrial DNA reexamined. *Systematic Zoology* 40:355-63.
Maddison, D. R., M. Ruvolo, and D. L. Swofford. 1992. Geographic origins of human mitochondrial DNA: Phylogenetic evidence from control region sequences. *Systematic Zoology* 41:111-24.
Morris, A. G. 1992. *The Skeleton of Contact: A Study of Protohistoric Burials from the Lower Orange River Valley, South Africa*. Johannesburg: Witwatersrand University.
Palca, J. 1990. The other human genome. *Science* 249:1104-5.
Pope, G. G. 1988. Recent advances in Far Eastern paleoanthropology. *Annual Review of Anthropology* 17:43-77.
Pope, G. G. 1991. Evolution of the zygomaticomaxillary region in the genus *Homo*, and its relevance to the origin of modern humans. *Journal of Human Evolution* 21:189-213.
Pope, G. G. 1992. The craniofacial evidence for the emergence of moderns humans in China. *Yearbook of Physical Anthropology* 35:243-98.
Pope, G., Z. An, S. Keates, and D. Bakken. 1990. New discoveries in the Nihewan Basin, Northern China. *The East Asian Tertiary/Quaternary Newsletter* 11:68-73.
Protsch, R. 1975. The absolute dating of Upper Pleistocene sub-Saharan fossil hominids and their place in human evolution. *Journal of Human Evolution* 4:297-322.
Rightmire, G. P. 1979. Implications of the Border Cave skeletal remains for later Pleistocene human evolution. *Current Anthropology* 20:23-35.
Rightmire, G. P. 1986. Africa and the origins of modern humans, 209-20. In *Variation, Culture, and Evolution in African Populations*, ed. R. Singer and J. K. Lundly. Johannesburg: Witwatersrand University Press.
Rightmire, G. P., and H. J. Deacon. 1991. Comparative studies of late Pleistocene human remains from Klasies River Mouth, South Africa. *Journal of Human Evolution* 20:131-56.
Sarich, V. M. 1971. Human variation in an evolutionary perspective, 182-91. In *Background for Man*, ed. P. Dolhinow and V. Sarich. Boston: Little, Brown.
Simek, J. F. 1992. Neanderthal cognition and the Middle to Upper Paleolithic transition, 231-

45. In *Continuity or Replacement: Controversies in* Homo sapiens *Evolution*, ed. G. Bräuer and F. H. Smith. Rotterdam: A. A. Balkema.

Simmons, T., and F. H. Smith. 1991. Late Pleistocene human population relationships in Africa, Europe, and the Circum-Mediterranean. *Current Anthropology* 32:623-27.

Smith, F. H. 1985. Continuity and change in the origin of modern *Homo sapiens. Zeitschrift für Morphologie und Anthropologie* 75:197-222.

Smith, F. H. 1991. The Neandertals: Evolutionary dead ends or ancestors of modern people? *Journal of Anthropological Research* 47:219-38.

Smith, F. H. 1992. The role of continuity in modern human origins, 145-56. In *Continuity or Replacement: Controversies in* Homo sapiens *Evolution*, ed. G. Bräuer and F. H. Smith. Rotterdam: A. A. Balkema.

Smith, F. H., A. B. Falsetti, and S. M. Donnelly. 1989. Modern Human Origins. *Yearbook of Physical Anthropology* 32:35-68.

Soffer, O. 1992. Social transformations at the Middle to Upper Paleolithic transition. the implications of the European record, 247-59. In *Continuity or Replacement: Controversies in* Homo sapiens *Evolution*, ed. G. Bräuer and F. H. Smith. Rotterdam: A. A. Balkema.

Spuhler, J. N. 1988. Evolution of mitochondrial DNA in monkeys, apes,and humans. *Yearbook of Physical Anthropology* 31:15-48.

Stoneking, M., and R. L. Cann. 1989. African origin of human mitochondrial DNA, 17-30. In *The Human Revolution: Behavioural and Biological Perspectives on the Origins of Modern Humans*, ed. P. Mellars and C. B. Stringer. Edinburgh: Edinburgh University Press.

Stringer, C. B. 1988. The dates of Eden. *Nature* 331:565-66.

Stringer, C. B. 1990. The emergence of modern humans. *Scientific American* 263:98-104.

Stringer, C. B. 1992. Replacement, continuity and the origin of *Homo sapiens*, 9-24. In *Continuity or Replacement: Controversies in* Homo sapiens *Evolution*, ed. G. Bräuer and F. H. Smith. Rotterdam: A. A. Balkema.

Stringer, C. B., and P. Andrews. 1988. Genetic and fossil evidence for the origin of modern humans. *Science* 239:1263-68.

Stringer, C. B., and R. Grün. 1991. Time for the last Neanderthals. *Nature* 351:701-2.

Templeton, A. R. 1992. Human origins and analysis of mitochondrial DNA sequences. *Science* 255:737.

Templeton, A. R. In press. The "Eve" hypothesis: A genetic critique and reanalysis. *American Anthropologist*.

Thackeray, A. I. 1989. Changing fashions in the Middle Stone Age: The stone artifact sequence from Klasies River main site, South Africa. *South African Archaeological Review* 7:35-59.

Thorne, A. G. 1980. The longest link: Human evolution in Southeast Asia and the settlement of Australia, 35-43. In *Indonesia: Australian Perspectives*, ed. J. J. Fox, R. G. Garnaut, P. T. McCawley, and J. A. C. Maukie. Canberra: Research School of Pacific Studies.

Thorne, A. G. 1981. The centre and the edge: The significance of Australian hominids to African paleoanthropology, 180-81. In *Proceedings of the 8th Panafrican Congress of Prehistory and Quaternary Studies*, Nairobi, September 1977, ed. R. E. Leakey and B. A. Ogot. Nairobi: TILLMIAP.

Thorne, A. G., and M. H. Wolpoff. 1981. Regional continuity in Australasian Pleistocene hominid evolution. *American Journal of Physical Anthropology* 55:337-49.

Thorne, A. G., and M. H. Wolpoff. 1992. The multiregional evolution of humans. *Scientific American* 266:76-83.

Vandermeersch, B. 1984. Á propos de la découverte du squelette Néandertalien de Saint-Césaire. *Bulletin et Mémoires de la Societe d'Anthropologie de Paris*, série 14, 1:191-96.

Vigilant, L., M. Stoneking, H. Harpending, K. Hawkes, and A. C. Wilson. 1991. African populations and the evolution of human mitochondrial DNA. *Science* 233:1303-7.

Wallace, D. C. 1992. Mitochondrial genetics: A paradigm for aging and degenerative diseases? *Science* 256:628-32.

Wayne, R. K., A. Meyer, N. Lehman, B. Van Valkenburgh, P. W. Kat, T. K. Fuller, D. Girman, and S. J. O'Brien. 1990. Large sequence divergence among mitochondrial DNA genotypes within populations of eastern African black-backed jackals. *Proceedings of the National Academy of Sciences, USA* 87:1772-76.

Weidenreich, F. 1939. Six lectures on *Sinanthropus pekinensis* and related problems. *Bulletin of the Geological Society of China* 19:1-110

Weidenreich, F. 1943. The skull of *Sinanthropus pekinensis*: A comparative study of a primitive hominid skull. *Palaeontologia Sinica*, n.s. D, No. 10 (whole series 127).

Weidenreich, F. 1946. *Apes, Giants, and Man*. Chicago: University of Chicago.

Weidenreich, F. 1951. Morphology of Solo man. *Anthropological Papers of the American Museum of Natural History* 43:205-90.

Wilson, A. C., and R. Cann. 1992. The mother of us all. *Scientific American* 266:68-73.

Wolpoff, M. H. 1980. *Paleoanthropology*. New York: Alfred A. Knopf.

Wolpoff, M. H. 1986. Describing anatomically modern *Homo sapiens*: A distinction without a definable difference. In *Fossil Man. New Facts, New Ideas. Papers in Honor of Jan Jelínek's Life Anniversary*, ed. V. V. Novotný and A. Mizerová. *Anthropos* (Brno) 23:41-53.

Wolpoff, M. H. 1989a. Multiregional evolution: The fossil alternative to Eden, 62-108. In *The Human Revolution: Behavioural and Biological Perspectives on the Origins of Modern Humans*, ed. P. Mellars and C. B. Stringer. Edinburgh: Edinburgh University Press.

Wolpoff, M. H. 1989b. The place of the Neandertals in human evolution, 97-155. In *The Emergence of Modern Humans*, ed. E. Trinkaus. London: Cambridge University Press.

Wolpoff, M. H. 1992. Theories of modern human origins, 25-63. In *Continuity or Replacement: Controversies in* Homo sapiens *Evolution*, ed. G. Bräuer and F. H. Smith. Rotterdam: A. A. Balkema.

Wolpoff, M. H., and R. Caspari. 1990. On Middle Paleolithic/Middle Stone Age hominid taxonomy. *Current Anthropology* 31:394-95.

Wolpoff, M. H., and A. G. Thorne. 1991. The case against Eve. *New Scientist* 22:33-37.

Wolpoff, M. H., Wu Xinzhi, and A. G. Thorne. 1984. Modern *Homo sapiens* origins: A general theory of hominid evolution involving the fossil evidence from east Asia, 411-83. In *The Origins of Modern Humans*, ed. F. H. Smith and F. Spencer. New York: Alan R. Liss.

Wu Rukang and Lin Shenglong. 1983. Peking man. *Scientific American* 248:86-94.

Wu Xinzhi. 1981. A well-preserved cranium of an archaic type of early *Homo sapiens* from Dali, China. *Scientia Sinica* 24:530-41.

Wu Xinzhi. 1987. Relation between Upper Paleolithic men in China and their southern neighbors in Niah and Tabon. *Acta Anthropologica Sinica* 6:180-83.

Wu Xinzhi. 1988. Comparative study of early *Homo sapiens* from China and Europe. *Acta Anthropologica Sinica* 7:287-93.

Wu Xinzhi. 1990. The evolution of humankind in China. *Acta Anthropologica Sinica* 9:312-21.

Xiong, X., W. Li, I. Posner, T. Yamamura, A. Yamamoto, M. Gotto, and L. Chan. 1991. No severe bottlenecking during human evolution: Evidence from two apolipoprotein C-II deficiency alleles. *American Journal of Human Genetics* 48:383-89.

Zhang Yinyun. 1991. An examination of temporal variation in the hominid dental sample from Zhoukoudian Locality 1. *Acta Anthropologica Sinica* 10:85-95.

Chapter **10**

Archaic and Modern Homo sapiens in the Contact Zones: Evolutionary Schematics and Model Predictions

TAL SIMMONS

> The number of different hypotheses erected to explain a given biological phenomenon is inversely proportional to the available knowledge.
> – Edington's Theory, *Murphy's Law Calendar* 1989

Many models have been proposed to explain the emergence of the anatomically modern morphotype in *Homo sapiens.* These models selectively emphasize a variety of evolutionary processes in their attempts to address this phenomenon; included among them are speciation mechanisms, population migrations and replacements, the maintenance of gene flow, etc. In addition, the evolution of modern humans has been related to environmental adaptation (in a phylogenetic biogeographic mode), as well as to cultural or behavioral factors unique to hominid species. The two most prominent models, Out of Africa (*sensu* Stringer and Andrews 1988) and Multiregional Evolution (*sensu* Wolpoff, Wu & Thorne 1984; Smith, Falsetti & Donnelly 1989), adhere to underlying theoretical foundations and perspectives of evolutionary schematics. Differing viewpoints about the very processes of evolution itself are largely responsible for the models currently in vogue, although this bias is seldom fully acknowledged or explored. All inferences and interpretations advanced by paleoanthropologists are inherently biased by prior commitments to other, more basic models of evolutionary theory, regardless of the guise in which they may appear. Clark and Lindly have alluded to this theory dependent bias and state that:

TAL SIMMONS • Department of Anthropology, Western Michigan University, Kalamazoo, MI 49008.
Origins of Anatomically Modern Humans, edited by Matthew H. Nitecki and Doris V. Nitecki. Plenum Press, New York, 1994.

> Put bluntly, there is so little concern with epistemology in palaeoanthropological research designs that it is difficult to find any evidence of an awareness that epistemological issues lie at the heart of the ceaseless debates that characterize the field. (1989a:627)

The fundamental components of the evolutionary theories thus operate as constraints which govern both the structural development of the models and the implications resulting from the application of those theories to the question of modern human origins.

The emergence of anatomically modern humans, the region(s) and date of their origin, and their ubiquitous radiation across the Old World have invited numerous speculations about the process by which *H. sapiens* evolved from *H. erectus* during the Middle to Late Pleistocene. Proponents of the Out-of-Africa theory derive their model from the Punctuated Equilibrium theory of evolution (Eldredge and Gould 1972; Gould and Eldredge 1977) and thus possess a cladogenetic perspective. Both theories posit the existence of static plateaus and definable "speciation events" evinced by any significant morphological change. Essentially, the reductionist structure of the Out-of-Africa model may be presented as a stasis-isolation-speciation-migration-replacement scenario, wherein the origin of the species *H. sapiens* is seen as a monocentric (southern Africa) and monochronic (ca. 200 kya) phenomenon. The Regional Continuity model adheres to a traditional neo-Darwinian perception of the Modern Synthesis and is hence more gradualist in its tenets. This model holds that, regardless of the area(s) of origin of recognizably anatomically modern human form, these humans did not constitute a definable new species. As humans are a highly mobile species, world wide gene flow among populations is assumed to have been significant. The differences between these two models do not necessarily appear to be insurmountable (Smith, Falsetti & Donnelly 1989; Simmons 1990), for as Hoffman (1989: 94) points out, punctuated equilibrium may be regarded either as "merely a wrinkle on the body of the neo-Darwinian paradigm," which does not preclude an allopatric mode of speciation; or "a revolt against the principle of continuity" (Out of Africa tends to portray modern human origins along these lines). Yet because these models are deeply rooted in commitments to antithetical concepts of the processes of speciation, and hence evolution, they are perceived to be irreconcilable.

Both the Out-of-Africa and the Regional Continuity models depend on oft-ignored interpretations of the fossil archaeological records of North Africa and the Levant. Each area represents a corridor, or potential evolutionary contact zone, between Africa and greater Eurasia where populations of archaic and modern *H. sapiens* would have first encountered one another. Unlike the migration of *H. erectus*, which around 1 mya circumscribed areas previously uninhabited by other hominid lineages, the proposed *H. sapiens* radiation included regions where other hominids were established residents and were, subsequently, to be replaced by this new species. The main issue is the contribution of non-African archaic *H. sapiens* hominids, and Neandertals in particular, to the modern human gene pool, and hence, their position in the hominid lineage and their status as populations or separate species (with all that implies). As Trinkaus has stated,

> once there is agreement that at least some of the Neandertals could have contributed to the gene pools of subsequent populations of early anatomically modern humans, it may not be possible to determine from the fossil record to what extent the Neandertals can be included in the ancestry of more recent humans. (1984:259)

Inquiries pertinent to the degree and nature of that contribution, rather than a simple exclusion of Neandertals from the modern human lineage, have recently become the focus of the debate.

The application of new absolute dating techniques – TL and ESR – to fossil localities in the Levant, Africa, and to a lesser extent, Europe, together with the survey of mtDNA variability in extant populations, have cast some doubt on the "humanity" of Neandertals. Dates of great antiquity have been proposed for sites with the skeletal remains of early anatomically modern humans in the Levant; conversely, for nearby sites with Neandertal fossils excavated, the suggested dates are often considerably less ancient. To several researchers these dates imply that the chronological isolation of Neandertals from populations of modern humans precluded their ability to interbreed (Valladas et al. 1987, 1988). Others have extended this argument to infer the inability of Neandertals to interbreed with any group of synchronic anatomically modern humans or, at the very least, to transmit enough of their own genetic material to contribute significantly to the gene pool of such resultant "hybrid" offspring (Stringer and Andrews 1988; Stringer, in press; Bräuer 1992).

SPECIES CONCEPTS AND HYBRIDIZATION ZONES

Inevitably, questions regarding the origin of anatomically modern *H. sapiens* return to definitions of the species *H. sapiens* itself, and to the concept of species in the hominid fossil record in general. (See, e.g., Foley 1991; Kimbel 1991; Delson 1990; Tattersall 1986; Tattersall and Eldredge 1977;

Day and Stringer 1982; Kidder, Jantz & Smith 1992.) As Wiley reasons,

> the concept of species (as taxa) adopted by an investigator will influence his perception of the processes by which species originate . . . the definition of the word species will be built on a species concept and the concept itself will profoundly affect the way in which investigators view the origin of the species they study. (1981:17)

The problem of what is a species has plagued not only paleontologists attempting to classify and organize extinct organisms but taxonomists of extant organisms, whose genetics and interactions can be studied in the living world. Mayr's (1969:26) classic definition of a species as a "reproductively isolated group of interbreeding populations" is clearly difficult to apply to fossil organisms. However, when defining fossil species, paleontologists must be aware of the implications of Mayr's definition. His use of the phrase "group of interbreeding populations" is significant; he is discussing the biology of *groups of populations* of organisms, and he neither emphasizes individual populations nor individual organisms themselves. The applications of this distinction to the Late Pleistocene hominid record require the examination of population distribution, as well as potential migration routes and contact zones in the determination of interbreeding groups. The "biospecies" concept, which defines ecological, genetic, and morphological components in addition to reproductive criteria is equally difficult to apply to the fossil record. Most paleontologists are therefore forced to identify species on morphology alone. These paleospecies (Simpson 1961) show greater morphological similarities to one another than to other groups, and chronological as well as geographic separation assumes importance in their definition. Paleospecies are by nature phyletic species.

Wiley (1981), however, finds the designation of phyletic species to be arbitrary. Therefore, he has introduced a new concept, that of the "evolutionary species," which is

> . . . a single lineage of ancestor-descendant populations which maintains its identity from other such lineages and which has its own evolutionary tendencies and historical fate. (Wiley 1981:25)

He (1981:26) clarifies further the definition, "whereas supraspecific taxa are collections of lineages, species are the lineages themselves." The evolutionary species concept has been criticized by Templeton, who wrote that,

> one fault of the evolutionary species concept is that it provides

> little or no guidance as to which traits are the more important ones in defining a species . . . . (1989:4)

and that

> . . . patterns of continuity/discontinuity vary as a function of the phenotype being measured. (1989:4)

The concept conjoins extremely well, however, with punctuated equilibrium, and the two concepts are integrated by the rejection of phyletic modes of perpetuation (e.g., selection and gene flow). Eldredge and Gould say,

> The coherence of a species, therefore, is not maintained by interactions among its members (gene flow). It emerges, rather, as an historical consequence of a species' origin as a peripherally isolated population that acquired its own powerful homeostatic system. (1972:114)

Another concern relevant to discussions of the recognition of species and, by extension lineages, in the fossil record is the definition of transitional specimens. Whereas phyletic gradualism is inferred from the recognition of transitional fossils, punctuated equilibrium asserts that transitional specimens should be exceedingly rare. Not only should they seldom be encountered, but the location and identification of a "true" transitional fossil must, by the theory's definition, be indicative of the specific area of a speciation event – the peripheral isolate itself. From the phyletic gradualist perspective, transitional specimens must exist because of the mosaic nature of evolution on a wider geographic and genetic scale. Observations on the processes of change are confounded by difficulties in recognizing intragroup variation caused by age, sex, and individual differences, imperfection in chronological definitions and the differential preservation of skeletal remains taphonomically altered. All of these factors prejudice judgments of what constitutes a transitional fossil (Frayer 1992). For example, Jelínek (1985) points to the existence of marked variation within a single hominid population (Shanidar in Iraq) which cannot be explained by chronological differences. He finds that such mosaic changes are characteristic of evolutionary progress and present a suitable model of gradual transition. Kimbel (1991:368), however, cautions that this bias toward a simple and linear progression coupled with "a preoccupation with ever-expanding envelopes of intraspecific variation, will fail to detect taxonomic diversity." McCown and Keith (1939) originally explained hominids at Skhul and Tabun as a population exhibiting great internal synchronic variability, a population in the veritable throes of evolution. Corruccini (1992) finds that the "ignored" Skhul crania, 4 and 9, are considerably less "anatomically modern" in appearance than the oft-cited Skhul 5 cranium. Marked variation exists also within the Qafzeh population (Simmons, Falsetti & Smith 1991; Simmons 1990), where the extremely gracile Qafzeh 9 clusters more closely with Upper Paleolithic hominids from that site than with its contemporaries.

In punctuated equilibrium, a transitional fossil must delineate a speciation event in a peripherally isolated area. Recognition of the mosaic nature of evolution suffers in this model because cladistic arguments cannot adequately categorize features which may be relevant to intragroup variation. With regard to modern human origins, Stringer once wrote:

> The relatively frequent occurrence of "mosaic" fossils in different continents would not be expected if changes from one species to the next was a very restricted event in time and space as claimed by the proponents of punctuated equilibria. The only alternative explanations for these widely distributed mosaics would appear to be less plausible than regarding them as morphological intermediates. They might represent another lineage of hominids living in the same areas as *H. erectus* and *H. sapiens*, displaying mixed characters, or they might be hybrids between the two species, although the extent of temporal or geographic overlap between the species is difficult to establish. (1984:64)

The question of hybridizing species is a very interesting one. In hominid paleontology, Thoma (1965) based on his assessment of several intermediate characteristics postulated that the Skhul Cave hominids were Neandertal/ anatomically modern *H. sapiens* hybrids. Weckler also considered the relationship among these hominid forms, and proposed that Neandertals and modern *H. sapiens*

> . . . after a presumed common origin, evolved independently in almost complete geographic isolation from one another for a long enough time prior to the third interglacial, and that the "progressive" Neanderthals of that time were hybrids between these two forms of man. (1954:1003)

Many recent studies of Middle and Late Pleistocene hominid fossils, particularly those considering possible temporal overlap of Neandertals and anatomically modern humans in some regions, have also speculated about hybridization (see also Simmons, Falsetti & Smith 1991).

Wiley defines a "contact zone" as a region of hybridization or intergradation of populations or species. The recognition of discrete species within these zones is related to the degree of contact (in a geographic context) and the duration of that contact:

> If the hybrid zone is narrow and there is evidence that it is old, then the forms are probably species because they are successfully maintaining their identities in spite of gene flow . . . . If the zone is wide, then the two forms are probably geographic variants of the same evolutionary species. (Wiley 1981:29)

Wiley also makes an interesting statement about speciation events in the context of hybridization which has unique consequences for the application of the evolutionary species concept (Stringer 1992) to the modern human origins debate:

> The concept does not preclude a particular ancestral species from surviving a speciation event. In cases where the mode of speciation is ecological–sympatric, via hybridization–or involves a small isolate, nonextinction of the larger population might be expected . . . . (Wiley 1981:34)

When considering the existence of archaic and modern *H. sapiens* in the "contact zones," Wiley's statement is a clause allowing for the occurrence of "speciation" in which a degree of Neandertal ancestry in modern populations is possible. It also becomes more plausible alternative to a model invoking their complete replacement by a migrating anatomically modern and strictly defined "species." In fact, ancestral species may become extinct during speciation events only "if they are subdivided in such a way that neither daughter species has the same evolutionary tendencies as the ancestral species" Wiley (1981:25). It is doubtful whether such an argument could be made concerning the speciation event of *H. erectus* into *H. sapiens*, particularly in light of the difficulty in delineating members of each species (e.g., Day and Stringer 1982) and the as yet unresolved question of Neandertal inclusion in the modern human lineage.

Furthermore, when considering population speciation modes and hybridization in the contact zones, Wiley (1981:28) cautions that "we must realize that the same populations may show both phenomena *depending on the characters analyzed"* (emphasis added). The ideas of hybridization and the choice of species defining characteristics have particular relevance to the question of Neandertal "humanity" (i.e., inclusion in the lineage leading to modern humans) and the implications of the Out-of-Africa model for modern human origins. The core of this model is dependent on the assumptions of stasis, speciation, migration, replacement, and on subsequent Neandertal extinction without significant contribution to the gene pool of modern human populations.

PUNCTUATED EQUILIBRIUM AND OUT OF AFRICA

The theory of punctuated equilibrium proposes that selection ultimately acts at the level of the species and that speciation events produce morphologies that are "essentially random with respect to the direction of evolutionary change within a clade" (Gould and Eldredge 1977:139). The nondirectional nature of evolution is confirmed by the randomness of that change even though change occurs at a higher taxonomic level. Punctuated equilibrium was proposed to explain the existence of perceived morphological gaps and evolutionary stasis in the fossil record. Morphological gaps are the negative evidence of the paleontological record (traditional explanations regard them as artefacts of various geological processes), but in punctuated equilibrium they assume great significance.

Morphological stasis in this model is regarded as the norm for most species over most of their life span. Visible changes within a species are seen as little more than "minor non-directional fluctuation in form" (Gould and Eldredge 1977:117), or aspects of intrapopulation variation, which have no demonstrable relationship to subsequent speciation events. Hence, the stratigraphy should show an absence of gradual temporal change. Morphological gaps are not really delinquencies of the record but rather breaks indicating the "essentially sudden replacement of ancestors by descendants" (Gould and Eldredge 1977:117). The phyletic gradualism model explains long periods of stasis in the fossil record as plateaus of adaptation and equilibrium wherein a population is not evincing the gradual accumulation of incorporated adaptive changes in its morphology. This is not to imply that those changes did not occur throughout the plateau period, merely that they were not expressed.

Punctuated equilibrium is, therefore, based on the principles of allopatric speciation which implies the existence of a "peripheral isolate," a fragment of a population separated by some natural barrier (or other isolating mechanism) from the larger segment. Because there is reduction in sheer numbers of organisms constituent to this isolated population, genetic changes may accumulate more rapidly on the periphery in this population. Over a relatively short period of time, the changes would be significant enough to isolate reproductively the small peripheral group from the original population. This constitutes a "speciation event." Morphological gaps in the fossil record thus "may record the extinction or emigration of a parent species and the immigration of a successful descendant rapidly evolved elsewhere in a small peripherally isolated population" (Gould and Eldredge 1977:117). In any stratigraphic segment, the fossil record should thus normally indicate a sudden replacement of one species by another. Only in very few and in very rare localities should there be any evidence of "transitional specimens." It is punctuated equilibrium's shift from the emphasis on transitional specimens to the importance of the morphological gap that has transformed the interpretation of the fossil record. Gould's definition of punctuated equilibrium

> . . . proposes that established species generally do not change substantially in phenotype over a lifetime that may encompass many millions of years (stasis), and that most evolutionary change is concentrated in geologically instantaneous events of branching speciation. (1982:383)

Furthermore, punctuated equilibrium redefines macroevolution by associating extreme rapidity of speciation with large scale evolutionary processes and subsuming all variability into macroevolutionary processes, speciation events (Stanley 1976).

In summary, the punctuated equilibrium theory is based on several premises. First, the fossil record is characterized by relatively long periods of stasis followed by relatively rapid bursts of change. Second, any significant change in morphology is regarded, cladistically a true speciation event (and conversely, all speciation events occur by cladogenesis). And third, these episodes of rapid change and speciation occur allopatrically; it is in the peripheral isolates where the mechanisms of evolution acting on the genotype stand a better chance of becoming incorporated and fixed in the population. These alterations in the gene pool will effect substantial change in the isolate whereas in the larger population they would be more easily lost to the effects of random genetic drift. It is primarily these aspects of the punctuated equilibrium theory–the emphasis on rapid change and replacement, peripheral isolation, stasis, and speciation events–which have influenced the Out-of-Africa model of modern human origins. Stringer (1992) explicitly presents the Out-of-Africa model's emphasis on these ideas.

The Out-of-Africa model (Stringer and Andrews 1988; Stringer 1989; Stringer 1992) proposes that (1) genetic and morphological variation is greatest within modern human African populations and that these populations are sharply distinguished in genetic polymorphisms from non-African populations; (2) transitional fossils are not evident in regions external to the African area of modern human origins; (3) population replacement of indigenous non-African archaic humans, resulting from the migration of the original African anatomically modern stock; (4) populations of hominids external to Africa should not display any regional distinctiveness until very late in the Pleistocene; and finally, (5) the earliest modern *H. sapiens* fossils should be found in Africa and those at the peripheries of the African migration should be of less antiquity. In its view of the transition from *H. erectus* to *H. sapiens*, the model also incorporates the idea of stasis preceding speciation. Rightmire (1981, 1984, 1985, 1986) and Wolpoff (1986; Wolpoff, Wu & Thorne 1984) have debated the issue of stasis versus gradual evolution in *H. erectus*.

Stringer (1984:79) applies the idea of stasis to the Out-of-Africa model, stating that in Africa the transition was "gradual and widespread, characterized by mosaicism" whereas external to Africa the appearance of Neandertal and anatomically modern humans was "punctual and localized, leading to easy recognition of taxonomic boundaries" (Stringer 1984:79).

Stringer and Andrews (1988) also cite the molecular data in support of a recent African origin for anatomically modern humans. In particular, indications of low nuclear and mitochondrial DNA genetic variability were a critical facet of the model's formulation. Previously, Stringer had favored a monocentric and monochronic origin for *H. sapiens* and the subsequent replacement of other archaic populations, notably the Eurasian Neandertals (Stringer 1978, 1982; Stringer, Vandermeersch & Hublin 1984). However, "this focus on Africa as the modern human homeland and the emphasis on speciation seemingly coalesced as a result of the mitochondrial DNA analyses" (Smith, Falsetti & Donnelly 1989:38). The mtDNA studies strongly imply that all modern humans evolved from a common African ancestor between 140 ky and 290 ky and that the Eurasian stock diverged later, some 90 ky to 180 ky ago (Cann, Stoneking & Wilson 1987). Stringer and Andrews (1988) certainly seem to share Stoneking and Cann's (1989:28) conclusion that "the rather staggering implication is that the dispersing African population replaced the non-African resident populations without interbreeding."

The Out-of-Africa model may be a more stringent interpretation of previous models, such as those proposed by Bräuer (1984, 1989) and Howells (1974), which emphasized the assimilation of genetic material from archaic humans into anatomically modern human populations via hybridization (see Simmons 1990). As Smith, Falsetti, and Donnelly (1989:37) point out, the significant departure from Bräuer's Afro-European Sapiens Hypothesis is that "no significant admixture would be possible between [anatomically modern humans] and the archaic Eurasians they encountered, because they would have been *different biological species*." Stringer (1992) has responded to this interpretation of the Out-of-Africa model by replying that his use of "*H. sapiens*" is evolutionary, not biological. Regardless of what was meant by Stringer, the reader understands the use of the "*H. sapiens*" nomenclature as a biological species designation. Wiley (1981:29) writes that, following speciation *or* geographic isolation leading to regionalized population differentiation, the resultant species *or* populations could still hybridize and there would be "no reason to reject their being evolutionary species." However, according to Wiley (1978:18), an evolutionary species must "maintain its identity" and have separate "evolutionary tendencies," not only from a sister species, but also from the ancestral species from which it sprung.

Van Valen (1976:233) defines a species as a lineage which "occupies an adaptive zone minimally different from that of any other lineage in its range and which evolves separately from all lineages outside its range." The archaeological evidence from the Levant makes it difficult to reconcile these two morphological types as separate species with regard to their adaptation. Shea (1989a, 1989b) has demonstrated that, in addition to the considerable temporal overlap of archaic and modern hominids in the Levant implied by the new TL and ESR dates, archaic and modern humans in that region made typologically identical tool kits and used them in functionally identical manners. Schoeninger (1982) has demonstrated also that there was no significant difference in the subsistence patterns of the two hominid types. This does not and cannot prove that these hominids were conspecific, but "any paleoanthropological 'solution' to this to this question must involve both biological and cultural remains and adaptations" (Delson 1990:143). There is little evidence to suggest that these two hominid types occupied different economic or ecological niches within this very narrow geographic zone. Intuitively, this casts doubt on the likelihood of two hominid types coexisting for thousands of years without hybridization.

Stringer (1992) has clarified his stance to allow for some hybridization between anatomically modern populations and local archaic groups, but states that the evidence for this has not been established, and that any such nonconspecific interbreeding would not have led to significant contributions to the modern human gene pool by the archaic populations. Both Cann (Cann, Rickards & Lum, this volume; cf. Cann, Stoneking & Wilson 1987) and Stringer (pers. comm.) now discuss hybridization as a distinct possibility, yet they continue to believe that the contributions of archaic populations were insignificant and unlikely to have been sustained for more than a few generations. The difficulties in evaluating this perspective are evident: the question of "significance" is open to interpretation and can only lead to further controversy; and, the question of "sustained" contribution to future generations can only be addressed adequately through the molecular clock in parsimonious computer-projected modeling (as mtDNA offers no direct insight into morphological variability). Clark and Lindly address this issue, stating that:

> genetic data map onto real chronological time more smoothly than do morphological data, with the implication that genetic distance and speciation events are highly correlated (i.e., more highly correlated than morphological distance and speciation events). While the systematics of molecular biological models have an internal logic and a degree of cohesion, the fact that they might be largely independent of morphological divergence will remain an obstacle to wider acceptance in human paleontology so long as there is substantial disagreement about how these models work . . . . (1989b:976)

Again, one is left wondering, first what degree of contribution would be deemed significant – and what is relevant about the *degree* of contribution once the *idea* of Neandertal contribution has been established, and, second, what proof is required to demonstrate that contribution to the satisfaction of Out-of-Africa supporters. Of course, nothing can be determined about the reproductive success of hybrid offspring in the fossil record either. The possibility that morphological contributions (and hence apparent morphological continuity in the fossil record) are not indicative of genetic continuity (in either mitochondrial or nuclear DNA) cannot be ignored.

Apparently, Stringer's views on hybridization do not follow Wiley's (1981), but are rather more consonant with Wiley's "zones of intergradation." Stringer denies that features pointed out by Wolpoff, Smith, Frayer, etc., are representative of continuity and transition but interprets them to be "the result of residual geographic variation left from an ancestral species and either maintained by local selection or chance" (Wiley 1981:29). This is unclear, however, as Stringer has not specifically addressed these proposed regional transitional sequences other than to state that they are not truly transitional, but are rather transitory and not sustained into future generations. The difficulty, of course, is that one cannot identify (either cladistically or phyletically) which characteristics actually reinforce specific identity, and

> In the case of actual intergradation, we must recognize that although a zone of hybridization may be narrow, a zone of intergradation of the same population may be very wide if the particular characters examined are not characters that reinforce species identity. (Wiley 1981:30)

Wiley (1978) does note, however, that when an ancestral species evolves phyletically so that the synapomorphies found in its descendants are fixed very gradually, the problem of separating species on a morphological level will never be resolved. This is because at the time of the initial split (of the ancestor from its own ancestor), the ancestor will presumably have very few of the derived characters which will become fixed in its own populations in the future; and, then left to its descendants when it speciates and itself becomes extinct. In a contact zone situation, the hybridization may either swamp specific identity or reinforce it (Wiley 1978). Assuming for the sake of argument that the origin of anatomically modern *H. sapiens* was a speciation event, in the swamping of one species by another "this usually results in one of the species *incorporating the genes of the other*" (Wiley 1981:42; emphasis added). In either instance, a regional approach should determine which process is occurring and is, therefore, crucial to the identification of transitional fossils.

THE MODERN SYNTHESIS AND REGIONAL CONTINUITY

Gould (1980:119) defines "Modern Synthesis" as a theory consisting of the following components: (1) point mutations are the ultimate source of variation in any population; (2) evolutionary change is the result of a process of gradual change via allelic substitution within populations – "in short, gradualism, continuity and evolutionary change by the transformation of populations"; (3) genetic variation is the only raw material of evolutionary change; (4) evolution is directed by natural selection and hence retained genetic change is adaptive by nature. The Modern Synthesis stresses that the macroevolutionary changes in the fossil population and species are simply accumulated change of microevolutionary proportions (genetic flow and drift as manifested through changes in allelic frequencies over long periods of time). The Modern Synthesis emphasizes selection operating in a Darwinian sense, at the level of the individual organism's phenotype and how that organism's reproductive success ultimately affects the genotypic variance in the population.

Thomson also cautions that

> a true deme or population is more than a chance assortment of individuals. It is defined by breeding systems, patterns of gene flow, balances of polymorphisms, degrees of heterozygosity, and so on. Thus at the population level there are no properties that are more than simply those due to the sum of the constituent parts (individual organisms). This is crucially important because it means that the rules of demes and populations are different from the rules that govern the biology of individual organisms. (1988:11)

Gould (1980:129), on the other hand, argues (contra Simpson 1949, who views natural selection as a creative process as well) that "selection only eliminates the unfit" and that this process does not significantly alter the direction of the population or lead to speciation events. Wiley (1981) does not reject the phyletic character transformation of microevolution, but argues that when applied to the fossil record, the recognition of phyletic criteria is arbitrary (cf. Simpson 1961). While it may be impossible to demonstrate satisfactorily phyletic speciation it proves equally difficult to demonstrate the cladistic mode of speciation events, which are proposed in their stead, in other than a tautological manner. Smith states that each of the regional continuity perspectives

> emphasize that most, if not all, archaic human populations contributed to regional modern human gene pools to some

> extent; and each of them recognizes that gene flow was critical to the emergence of modern humans throughout the Old World. (1992)

Within the context of regional continuity, the multiregional evolutionary model of Wolpoff (1989) attempts to explain several phenomena: (1) the initial contrast between the source population and the peripheral populations; (2) the early appearance of regional features showing evolutionary continuity at the periphery versus the somewhat later appearance of such features at the center of the zone; and (3) the maintenance of the center/periphery contrasts and the persistence of regional continuity into the later Pleistocene. Wolpoff attempts to clear up the confusion of morphological character states generated in a polytypic species such as *H. sapiens*.

In particular, Wolpoff has adamantly defended the regional continuity position against cladistic arguments for the differential speciation of archaic and anatomically modern populations of *H. sapiens*, stating that

> the problem is that the specific pattern of evolutionary change is not the same in each region across the globe, and therefore the potential for confusion between features that uniquely characterize a region (i.e., clade) and features that change over time (grade) is great. (1986:43)

The importance of "population thinking" to models of modern human origins has also been stressed by Mayr (1982). Howells (1974) also urges that a species be defined on the basis of the "total morphological pattern" rather than by individual characteristics. In intra- versus interpopulation variation, the continuity of genes through time and space in human evolution is balanced through the retention and maintenance of enduring clines. The balance between gene flow and selection represents a dynamic process and, in the regional continuity model intermediate (transitional) specimens indicate that human populations were rarely isolated.

> The hominid fossil record and the emergence of modern humans must be seen as a record of change well compatible with neodarwinian theory and one in which it would be, perhaps, more fruitful to look for the conditions which occasionally favor long continued directional change rather than to posit unusually rapid change requiring new mechanisms or even new evolutionary syntheses. (Williams 1987:107)

The larger theoretical context of the regional continuity perspective, first delineated by Wolpoff, Wu, and Thorne (1984) proposes that the fossil record indicates a considerable degree of morphological, and hence genetic, continuity across the archaic/modern *H. sapiens* boundary in many regions of the Old World. According to this model, suites of distinctive morphological features should link archaic and modern humans in specific geographic regions and, hence, transitional fossils may exist in more than one region.

Regional continuity calls neither for orthogenesis nor necessarily for continuity in every region where modern humans appear following the presence of an archaic *H. sapiens* population. Rather the tempo and mode of the emergence (or appearance) of the anatomically modern human morphotype in each region would result from a complex interaction of gene flow, gene drift, and selection unique to each region. The pattern of these evolutionary mechanisms and the extent of their operation would differ with regard to the width of the "hybridization zone" concept described earlier.

Supporters of regional continuity base their beliefs in part on their ability to trace specific regional morphological complexes to "ecological and demographic factors associated with the initial radiation of *H. erectus*" (Smith, Falsetti & Donnelly 1989:39). This model must not be confused with that of Coon (1962), for it does not claim that various modern human lineages evolved independently of one another from *H. erectus*. The regional continuity model asserts that

> gene flow, in the form of both population movement (migration) and genetic exchange across population boundaries would have prevented speciation between the regional lineages and thus maintained human beings as a single, although obviously polytypic, species throughout the Pleistocene. (Smith, Falsetti & Donnelly 1989:39)

The concepts of phyletic gradualism are intrinsic to the regional continuity model, and interpretations of the human paleontological record derived from that model also adhere to this evolutionary approach.

Wolpoff (1989) lists a number of expectations for the Middle to Late Pleistocene hominid fossil record which would be logical consequences of the regional continuity model. First, anatomically modern human forms would differ regionally due to the unique evolutionary history of each area (e.g., various combinations of indigenous and extraneous influences); therefore, no single definition of what constitutes a "modern human" would be globally applicable. Second, there is no reason to expect that anatomically modern humans necessarily appeared initially in the African continent. Third, the earliest anatomically modern peoples of the Eurasian continent should be expected to share some features consistent with previous indigenous archaic populations. Since these hominids are not thought to be solely the result of a mass population influx from Africa, they would lack *distinctively* African regional features. And fourth, these features which are nonadaptive (regional clade features) would be identifiable earliest in the most peripheral zones of the hominid range prior to the appearance of anatomically modern forms and latest in the more central areas.

In conclusion, the regional continuity model affirms that the extent of modern human genetic variability and its morphological manifestations may be best explained by the broad neo-Darwinian paradigm, the Synthetic Theory. This model also asserts that there is no overwhelming evidence to invoke a recent African origin for all anatomically modern humans, nor that the emergence of modern *H. sapiens* constitutes a cladogenetic speciation event with subsequent replacement of other, non-African population of archaic humans.

CONTACT ZONES: NORTH AFRICA AND THE LEVANT

If the origin of anatomically modern *H. sapiens* did indeed constitute a speciation event, then the postulated migration of those hominids out of Africa would have first brought them into contact with indigenous populations of archaic humans in North Africa and the Levant. Evidence of the early appearance of anatomically modern *H. sapiens* has been found in both regions (Valladas et al. 1988; Schwarcz et al. 1988; Stringer et al. 1989; Grün and Stringer 1991) and in the Levant these hominids can no longer be assumed to antedate the Neandertal occupation of this region (Stringer et al. 1989; cf. Valladas et al. 1988). Neither are the dates for the earliest presence of anatomically modern humans in southern Africa (Deacon and Geleijnse 1988; Grün, Shackleton & Deacon 1990) appreciably earlier than for the Levant and North Africa (Jelinek 1992). Yet, again assuming that this migration took place, these regions provide excellent samples for examining the interactions of archaic and modern humans in the contact zones with all the ensuing implications for species recognition outlined above.

Archaeologists Bar-Yosef (1992, this volume) and Shea (1989a) have considered the problem of apparent coexistence of populations of Neandertals and anatomically modern humans in the Levant. Both conclude that there is no definitive evidence for the actual synchrony of cohabitation in this region by the two populations. They suggest, rather, that these hominid populations were probably responding to fluctuating environmental conditions brought about by intermittent glaciation in Europe and desertification in the Sahara; the presence of Neandertals in the Levant was conditional upon the former, while the influx of anatomically modern individuals depended upon the latter. Jelinek (1992) also doubts the coexistence of these populations, but his conclusions are based on perceived discrepancies in the dating techniques. The majority of these arguments are derived from interpretations of new dating techniques. Wobst (1976) predicts that populations similar in size and distribution to those of Late Pleistocene hominids would need to maximize contact with other such populations in order to avoid extinction. The implications for mobility of individual populations and extensive genetic exchange among these contacting populations are clear. Simmons and Smith (1991) and Simmons, Falsetti, and Smith (1991) also examine the question of coexistence from an evolutionary biological perspective, with regard to North Africa and the Levant, respectively.

The relationship of archaic and anatomically modern *H. sapiens* in the contact zones allows for four conceivable scenarios, each entailing a series of predictions for the morphology of the anatomically modern populations involved. First, anatomically modern humans arrived as a new species and replaced, without interbreeding, the indigenous archaic human populations. Here the earliest anatomically modern humans in the region should resemble early anatomically modern Southern African *H. sapiens*, and not bear a particularly close similarity to indigenous archaic hominids – unless one wishes to invoke a great deal of convergence. Any similarity to archaic hominids would be the result of intergradation, not hybridization. Stringer's (this volume) objection to the use of Border Cave as a morphological representative of the anatomically modern human group is perplexing. ESR dates for Border Cave indicate an age from 70-80 ky (Grün, Beaumont & Stringer 1990). Although they (1990:539) state that "genetic data now seem to indicate that regional differentiation of African populations had begun by that time," and that Khoisan origins predate the age inferred for Border Cave, Howells (1989) disputes this on morphological grounds. Corruccini (1992) also demonstrates that this fossil is far from fully "anatomically modern" (with reference to modern Africans). The status of Border Cave is roughly in the same category (magnitude of difference) as the Skhul 9 and 4 hominids are from modern humans, which Corruccini (1992) sees as phenetically intermediate between Neandertals and Upper Paleolithic hominids. Howells (1989:74) states that "Border Cave does not now solidly demonstrate . . . a local ancestry for sub-Saharan moderns, in spite of the conviction of many, myself included, that this is the true case." If Stringer objects to the use of Border Cave, which both he and Bräuer have placed in the early anatomically modern group – the group which has supposedly speciated and is migrating and replacing archaic humans external to southern Africa – which group is migrating and replacing? If Stringer means to imply that it is the African transitional group (Omo 2, Ngaloba, Florisbad, and Djebel Irhoud), one might well argue that no true speciation event has occurred in southern Africa. "Transitional specimens" do not constitute a new species; nor does the area encompassed by the fossils in this group (Ethiopia, Tanzania, Republic of South

Africa, and Morocco, respectively) conform to the degree of isolation necessary to engender a speciation event.

Second, anatomically modern humans of Southern African ancestry arrived as a new species but hybridized with local archaic humans (as both Stringer and Cann suggested recently). The expectation is that the earliest anatomically modern humans in the region should exhibit a mosaic of features, a high degree of intrapopulation variation, and appear to be "transitional" in nature. Support for this scenario on morphological grounds is found in Simmons (1990), Simmons, Falsetti, and Smith (1991), Corruccini (1992), and Simmons and Smith (1991). Yet, both Cann and Stringer (this volume) claim that the morphological and genetic features of the resultant hybrids would not long endure; this certainly implies a lack of viability of those offspring, or the diminished survivorship of successive generations through natural selection.

Third, anatomically modern humans of southern African ancestry arrived as a geographic variant of the same species (or a different subspecies) represented by the local archaic humans, with subsequent interbreeding (or hybridization in the more traditional sense). This scenario is particularly difficult to distinguish on a morphological level from the preceding one; but, the emphasis here is on intergradation (re. the speciation of *H. erectus*) and the interaction of regional morphoclines.

And fourth, anatomically modern humans were derived directly from local archaic ancestors with little extraneous influence from southern African populations. The prediction is that the morphology of the earliest anatomically modern humans should display greater similarity (in having uniquely regional features) to local archaic hominids than to the early anatomically modern southern African morphotype. No regionally specific African characteristics should be present in the earliest anatomically modern hominids. This prediction, too, may be difficult to distinguish morphologically from the second and third scenarios due to the existence of demonstrably non-regionally specific patterns of gracilization and modernization (Smith, Falsetti & Donnelly 1989; Simmons 1990). The important hypothesis to test is whether the first scenario can be rejected and distinguished from the second, third, and fourth, which represent minor variations on the same theme.

Simmons (1990) and Simmons and Smith (1991) have demonstrated a high degree of morphological similarity between the Djebel Irhoud hominids and the Qafzeh population. There is little doubt that Neandertals in Southwest Asia share morphological similarities with the classical Neandertals of Western Europe. However, in the first areas of contact between Neandertals and anatomically modern humans along the proposed routes of migration, there is ample evidence to suggest a relatively close morphological relationship among the anatomically modern humans in these regions and the resident archaic populations. In fact, these individuals resemble one another more closely than either Djebel Irhoud resembles the African transitional or early anatomically modern group, or Qafzeh and Skhul resemble Neandertals. The evidence does not necessarily deny the earlier emergence of anatomically modern human morphotypes in southern Africa and the migration of those hominids into other regions of the Old World, but this theory must be tested by accurate chronometric dates which clearly establish the presence of modern humans in Africa prior to their appearance in the "contact zones." Analyses of archaic and modern *H. sapiens* populations in the contact zones of North Africa and the Levant indicate that indigenous populations of archaic *H. sapiens* did contribute in a qualitatively significant manner to the phenotypic, and hence, genetic, variability of modern human populations. The complexity of population relations in the Late Pleistocene, particularly in the circum-Mediterranean, detracts from a stringent acceptance of the Out-of-Africa model and its implications.

SUMMARY

To understand the role of evolutionary theory and its influence on models of modern humans origin, it is necessary to define explicitly certain perspectives. Much unnecessary confusion (as well as accusations of misinterpretation) has arisen as a result of both lack of an extension of a model's predictions to their wider implications in biological reality and a lack of clarity in the presentation of a model's predictions. "Despite the convictions of some, exactly what the 'biological realities' are is not yet clear" (Smith, Falsetti & Donnelly 1989:41)

An examination of the morphology of related higher primate species and subspecies, such as the baboon is warranted. As the interactions of these populations in overlapping ranges become known (i.e., their capacity and proclivity for interbreeding), perhaps the morphological data from the constituents of those populations and their hybrids would be applicable to the question of modern human origins and the interactions (or isolations, geographic or reproductive) of Late Pleistocene human populations. Even granted that humans are not baboons, nor bioculturally analogous, such an examination of morphologies and genetic data would provide a more comprehensive operational model from which to evaluate the origin of anatomically modern *H. sapiens*.

The existence of a zone of hybridization between anubis and hamadryas baboons in the Ethiopian Awash has been documented since the early 1960s (Nagel 1973). The interactions of the primates within this zone have been observed for the past twenty years, and evidence suggests that it has endured for at least sixty to one hundred years historically (Phillips-Conroy, pers. comm.). Evidence of great phenotypic variability within the hybrid groups themselves, the overlapping of ranges, and the presence of infants have shown these hybrids to be fully fertile (Phillips-Conroy and Jolly 1986). Morphologically (Phillips-Conroy and Jolly 1981) and phenotypically (Phillips-Conroy and Jolly 1986), certain characters appear to be more stable or conservative while other intermediate states are more frequent. It is interesting to compare the Phillips-Conroy and Jolly (1986) model to similar scenarios proposed for Neandertal and anatomically modern human interactions in the Levant (see discussion of Bar-Yosef (1992) and Shea (1989a) above). In the baboon model, however, fluctuating environmental conditions actually lead to increased contact, rather than alternating habitation of the area. Anubis and hamadryas baboons "comprise two distinct functional complexes adapted, respectively, to wetter and more arid habitats," with the hamadryas seen as the more derived form (Phillips-Conroy and Jolly 1986: 347). Developments in the width and persistence of the hybrid zone are related to climatic fluctuations which affect the relevant resource productivity of the habitats. Migrations of either baboon into the other's traditional territory as a result of environmental pressure would lead to a renewal or continuation of hybridization in the contact zones; "the presence of older hybrids reflects a longer-standing and more regular pattern of migration and consequent genetic interchange between coexisting anubis and hamadryas groups" (Phillips-Conroy and Jolly 1986:349). The genetic data for these populations (Shotake 1981; Shotake, Nozawa & Tanabe 1977) also support the existence of a wide and expanding hybrid zone of long duration.

According to Gabow, in allopatric speciation the genetic divergence of populations

> may often involve different kinds of habitat preference such that the two populations become ecologically dissimilar. . . . If geographical barriers between the two populations are then removed, the natural development of allopatric speciation is interrupted: the two populations meet in a zone of sympatry. If reproductive isolating mechanisms have not yet developed, hybridization normally follows in such zones of sympatry. (1975: 701)

A hybridization zone, due to the competitive exclusion principle, will not remain stable; extinction of one species, redevelopment of allopatry, development of reproductive isolating mechanisms, or, the fusion of populations may all eventually result, depending on either restriction or expansion of the contact zone. Interactions in the contact zones between archaic and modern humans are related to the genetic and morphological presence of clines and what they imply about gene flow (Bigelow 1965). The problem remains with the imprecision in determining both the behavioral mechanisms of prehistoric hominids and the duration of contact between them which might aid in explicating the fossil and genetic data. Do the populations maintain their identities after sufficient degree (time depth) of contact to justify considering them separate, reproductively isolated species? Or, instead is there evidence of highly increased variability as a transient stage in the incorporation of "parental" phenotypes into a single population within a few generations?

It is indeed possible that a speciation event, as suggested by expositions of the mtDNA data, did occur in Southern Africa around 200 kya. It is equally plausible that it did not. Many interpretations of the fossil evidence indicate evidence of "transitional" characteristics in regions external to Africa; whether these features are relicts of intergradation, products of hybridization, or result from a true transitional process has not yet been determined. It is doubtful whether the actuality can be realized or known as such. Based on models of interaction in other primate species, it may be possible to speculate on which possibility is more likely to have occurred. However, without a more rigorous examination of the myriad of potentialities of population dynamics in the Late Pleistocene (which is based in implications contingent upon "biological reality") it is prudent at this time not to assume that the course of *H. sapiens* emergence proceeded along any specific pathway.

ACKNOWLEDGMENTS

I would like to thank M. H. Nitecki, F. H. Smith, M. McKinney, B. Shea, R. Walker, and J. Spiegel for their helpful comments on earlier versions of this manuscript. Much of the work leading to the formation of these ideas was supported by NSF grants BNS-8818734 and BNS-8606674.

REFERENCES

Bar-Yosef, O. 1992. Middle Palaeolithic chronology and the transition to the Upper Palaeolithic in Southwest Asia, 261-72. In *Continuity or Replacement: Controversies in* Homo sapiens *Evolution*, ed. G. Bräuer and F. H. Smith. Rotterdam: A. A. Balkema.

Bigelow, R. S. 1965. Hybrid zones and reproductive isolation. *Evolution* 19:449-58.

Bloch, A. 1989. *Murphy's Law Calendar*. Los Angeles: Price, Stern, Sloan, Inc.

Bräuer, G. 1984. A craniological approach to the origin of anatomically modern *Homo sapiens* in Africa and implications for the appearance of modern Europeans, 327-410. In *The Origins of Modern Humans: A World Survey of the Fossil Evidence*, ed. F. H. Smith and F. Spencer. New York: Alan R. Liss.

Bräuer, G. 1989. The evolution of modern humans: A comparison of the African and non-African evidence, 121-54. In *The Human Revolution: Behavioural and Biological Perspectives on the Origins of Modern Humans*, ed. P. Mellars and C. B. Stringer. Edinburgh: Edinburgh University Press.

Bräuer, G. 1992. Africa's place in the evolution of *Homo sapiens*, 83-98. In *Continuity or Replacement: Controversies in* Homo sapiens *Evolution*, ed. G. Bräuer and F. H. Smith. Rotterdam: A. A. Balkema.

Cann, R. L., M. Stoneking, and A. C. Wilson. 1987. Mitochondrial DNA and human evolution. *Nature* 325:31-36.

Clark, G. A., and J. M. Lindly. 1989a. The case for continuity: Observations on the biocultural transition in Europe and Western Asia, 626-76. In *The Human Revolution: Behavioural and Biological Perspectives on the Origins of Modern Humans*, ed. P. Mellars and C. B. Stringer. Edinburgh: Edinburgh University Press.

Clark, G. A., and J. M. Lindly. 1989b. Modern human origins in the Levant and Western Asia: The fossil and archaeological evidence. *American Anthropologist* 91:962-85.

Coon, C. S. 1962. *The Origin of Races*. New York: Alfred A. Knopf.

Corruccini, R. S. 1992. Metrical reconsideration of Skhul IV and IX and Border Cave 1 crania in the context of modern human origins. *American Journal of Physical Anthropology* 87:433-45.

Day, M. H., and C. B. Stringer. 1982. A reconsideration of the Omo Kibish remains and the *erectus-sapiens* transition, 814-46. In *L'*Homo erectus *et la Place de L'Homme de Tautavel Parmi Les Hominidés Fossiles*. Vol. 2, ed. H. de Lumley. Nice: Centre National de la Recherche Scientifique.

Deacon, H. J., and V. B. Geleijnse. 1988. The stratigraphy and sedimentology of the main site sequence, Klasies River, South Africa. *South African Archaeological Bulletin* 43:5-14.

Delson, E. 1990. Species and species concepts in paleoanthropology (Commentary on a paper by David Pilbeam), 141-44. In *Evolutionary Biology at the Crossroads*, ed. M. K. Hecht. Flushing: Queens College Press.

Eldredge, N., and S. J. Gould. 1972. Punctuated equilibria: An alternative to phyletic gradualism, 82-115. In *Models in Paleobiology*, ed. T. J. M. Schopf. San Francisco: Freeman, Cooper & Company.

Foley, R. A. 1991. How many species of hominid should there be? *Journal of Human Evolution* 20:413-27.

Frayer, D. 1992. The persistence of Neanderthal features in post-Neanderthal Europeans, 179-88. In *Continuity or Replacement: Controversies in* Homo sapiens *Evolution*, ed. G. Bräuer and F. H. Smith. Rotterdam: A. A Balkema.

Gabow, S. A. 1975. Behavioral stabilization of a baboon hybrid zone. *The American Naturalist* 109: 701-12.

Gould, S. J. 1980. Is a new theory of evolution emerging? *Paleobiology* 6:119-30.

Gould, S. J. 1982. Darwinism and the expansion of evolutionary theory. *Science* 216:380-86.

Gould, S. J., and N. Eldredge. 1977. Punctuated equilibria: The tempo and mode of evolution reconsidered. *Paleobiology* 3:115-51.

Grün, R., P. B. Beaumont, and C. B. Stringer. 1990. ESR dating evidence for early modern humans at Border Cave in South Africa. *Nature* 344:537-39.

Grün, R., N. J. Shackleton, and H. J. Deacon. 1990. Electron-spin-resonance dating of tooth enamel from Klasies River Mouth Cave. *Current Anthropology* 31:427-32.

Grün, R., and C. B. Stringer. 1991. Electron spin resonance dating and the evolution of modern humans. *Archaeometry* 33:153-99.

Grün, R., C. B. Stringer, and H. P. Schwarcz. 1991. ESR dating of teeth from Garrod's Tabun Cave collection. *Journal of Human Evolution* 20:231-48.

Hoffman, A. 1989. *Arguments on Evolution: A Paleontologist's Perspective*. New York: Oxford University Press.

Howells, W. W. 1974. Neanderthals: Names, hypotheses, and scientific method. *American Anthropologist* 76:24-38.

Howells, W. W. 1989. *Skull Shapes and the Map: Craniometric Analyses in the Dispersion of Modern* Homo. Papers of the Peabody Museum of Archaeology and Ethnology, 79. Cambridge: Harvard University Press.

Jelinek, A. 1992. Problems in the chronology of the Middle Paleolithic and the first appearance of early modern *Homo sapiens* in Southwest Asia. In *The Evolution and Dispersal of Modern Humans in Asia*, ed. T. Akazawa, K. Aoki, and T. Kimura. Tokyo: Hokusen-sha.

Jelínek, J. 1985. The European, Near East and North African finds after *Australopithecus* and the principal consequences for the picture of human evolution, 341-54. In *Human Evolution: Past, Present and Future*, ed. P. V. Tobias. New York: Alan R. Liss.

Kidder, J. H., R. L. Jantz, and F. H. Smith. 1992. Defining modern humans: A multivariate approach, 157-77. In *Continuity or Replacement: Controversies in* Homo sapiens *Evolution*, ed. G. Bräuer and F. H. Smith. Rotterdam: A. A. Balkema.

Kimbel, W. H. 1991. Species, species concepts and hominid evolution. *Journal of Human Evolution* 20:355-71.

Mayr, E. 1969. *Principles of Systematic Zoology*. New York: McGraw-Hill

Mayr, E. 1982. *The Growth of Biological Thought: Diversity, Evolution and Inheritance*. Cambridge, MS; Belknap Press.

McCown, T. D., and Sir A. Keith 1939. *The Stone Age of Mount Carmel: The Fossil Human Remains from the Levalloiso-Mousterian*. Vol. II. Oxford: The Clarendon Press.

Nagel, U. 1973. A comparison of anubis baboons, hamadryas baboons and their hybrids at a species border in Ethiopia. *Folia Primatologica* 19:104-66.

Phillips-Conroy, J. E., and C. J. Jolly. 1981. Sexual dimorphism in two subspecies of Ethiopian baboons (*Papio hamadryas*) and their hybrids. *American Journal of Physical Anthropology* 56:115-29.

Phillips-Conroy, J. E., and C. J. Jolly. 1986. Changes in the structure of the baboon hybrid zone in the Awash National Park, Ethiopia. *American Journal of Physical Anthropology* 71:337-50.

Rightmire, G. P. 1981. Patterns in the evolution of *Homo erectus*. *Paleobiology* 7:241-46.

Rightmire, G. P. 1984. *Homo sapiens* in sub-Saharan Africa, 295-325. In *The Origins of Modern Humans: A World Survey of the Fossil Evidence*, ed. F. H. Smith and F. Spencer. New York: Alan R. Liss.

Rightmire, G. P. 1985. The tempo of change in the evolution of Middle Pleistocene *Homo*, 255-64. In *Ancestors: The Hard Evidence*, ed. E. Delson. New York: Alan R. Liss.

Rightmire, G. P. 1986. Africa and the origins of modern humans, 209-20. In *Variation, Culture, and Evolution in African Populations*, ed. R. Singer and J. K. Lundly. Johannesburg: Witwatersrand University Press.

Schoeninger, M. 1982. Diet and the evolution of modern human form in the Middle East. *American Journal of Physical Anthropology* 58:37-52.

Schwarcz, H. P., W. M. Buhay, R. Grün, H. Valladas, and E. Tchernov. 1988. ESR dates for the hominid burial site of Qafzeh in Israel. *Journal of Human Evolution* 17:733-37.

Shea, J. J. 1989a. A functional study of the lithic industries associated with hominid fossils in the Kebara and Qafzeh caves, Israel, 611-25. In *The Human Revolution: Behavioural and Biological Perspectives on the Origins of Modern Humans*, ed. P. Mellars and C. Stringer. Edinburgh: Edinburgh University Press.

Shea, J. J. 1989b. A preliminary functional analysis of the Kebara Mousterian. Investigations in South Levantine Prehistory. *Prehistoire du Sud-Levant*, ed. O. Bar-Yosef and B. Vandermeersch. Oxford: British Archaeological Reports International Series 497.

Shotake, T. 1981. Population genetical study of natural hybridization between *Papio anubis* and *P. hamadryas*. *Primates* 23:285-308.

Shotake, T., K. Nozawa, and Y. Tanabe. 1977. Blood protein variations in baboons 1. Gene exchange and genetic distance between *Papio anubis, Papio hamadryas* and their hybrid. *Japanese Journal of Genetics* 52:223-37.

Simmons, T. 1990. Comparative Morphometrics of the Frontal Bone in Hominids: Implications for Models of Modern Human Origins. Ph.D. diss. University of Tennessee, Knoxville.

Simmons, T., A. B. Falsetti, and F. H. Smith. 1991. Frontal bone morphometrics of southwest Asian Pleistocene hominids. *Journal of Human Evolution* 20:249-69.

Simmons, T., and F. H. Smith. 1991. Late Pleistocene human population relationships in Africa, Europe, and the circum-Mediterranean. *Current Anthropology* 32:623-27.

Simpson, G. G. 1949. *The Meaning of Evolution*. New Haven: Yale University Press.

Simpson, G. G. 1961. *Principles of Animal Taxonomy*. New York: Columbia University Press.

Smith, F. H. 1992. The role of continuity in modern human origins, 145-56. In *Continuity or Replacement: Controversies in* Homo sapiens *Evolution*, ed. G. Bräuer and F. H. Smith. Rotterdam: A. A. Balkema.

Smith, F. H., A. B. Falsetti, and S. M. Donnelly. 1989. Modern human origins. *Yearbook of Physical Anthropology* 32:35-68.

Stanley, S. M. 1976. Stability of species in geologic time. *Science* 192:267-68.

Stoneking, M., and R. L. Cann. 1989. African origin of human mitochondrial DNA, 17-30. In *The Human Revolution: Behavioural and Biological Perspectives on the Origin of Modern Humans*, ed. P. Mellars and C. B. Stringer. Edinburgh: Edinburgh University Press.

Stringer, C. B. 1978. Some problems in Middle and Upper Pleistocene hominid relationships, 395-418. In *Recent Advances in Primatology*. Vol. 3, *Evolution*, ed. D. J. Chivers and K. Joysey. London: Academic Press.

Stringer, C. B. 1982. Towards a solution to the Neanderthal problem. *Journal of Human Evolution* 11:431-38.

Stringer, C. B. 1984. Human evolution and biological adaptation in the Pleistocene, 55-83. In *Hominid Evolution and Community Ecology: Prehistoric Human Adaptation in Biological Perspective*, ed. R. Foley. London: Academic Press.

Stringer, C. B. 1989. The origin of early modern humans: A comparison of the European and non-European evidence, 232-44. In *The Human Revolution: Behavioural and Biological Perspectives on the Origin of Modern Humans*, ed. P. Mellars and C. B. Stringer. Edinburgh: Edinburgh University Press.

Stringer, C. B. 1992. Replacement, continuity and the origin of Homo sapiens, 9-24. In *Continuity or Replacement: Controversies in* Homo sapiens *Evolution*, ed. G. Bräuer and F. H. Smith. Rotterdam: A. A. Balkema.

Stringer, C. B., and P. Andrews. 1988. Genetic and fossil evidence for the origin of modern humans. *Science* 239:1263-68.

Stringer, C. B., R. Grün, H. P. Schwarcz, and P. Goldberg. 1989. ESR dates for the hominid burial site of Es Skhul in Israel. *Nature* 338:756-58.

Stringer, C. B., J. J. Hublin, and B. Vandermeersch. 1984. The origin of anatomically modern humans in Western Europe, 51-135. In *The Origin of Modern Humans: A World Survey of the Fossil Evidence*, ed. F. H. Smith and F. Spencer. New York: Alan R. Liss.

Tattersall, I. 1986. Species recognition in human paleontology. *Journal of Human Evolution* 15:165-75.

Tattersall, I., and N. Eldredge. 1977. Fact, theory and fantasy in human paleontology. *American Scientist* 63:673-80.

Templeton, A. R. 1989. The meaning of species and speciation: A genetic perspective, 3-27. In *Speciation and its Consequences*, ed. D. Otte and J. A. Endler. Sunderland, MA: Sinauer Associates.

Thoma, A. 1965. Le deploiment evolutif de l'*Homo sapiens. Anthropologia Hungarica* V(1-2):1-179.

Thomson, K. S. 1988. *Morphogenesis and Evolution*. Oxford: Oxford University Press.

Trinkaus, E. 1984. Western Asia, 251-93. In *The Origins of Modern Humans: A World Survey of the Fossil Evidence*, ed. F. H. Smith and F. Spencer. New York: Alan R. Liss.

Valladas, H., J.-L. Joron, G. Valladas, B. Arensburg, O. Bar-Yosef, A. Belfer-Cohen, P. Goldberg, H. Laville, L. Meignen, Y. Rak, E. Tchernov, A.-M. Tillier, and B. Vandermeersch. 1987. Thermoluminescence dates for the Neanderthal burial site at Kebara in Israel. *Nature* 330:159-60.

Valladas, H., J. L. Reyss, J.-L. Joron, G. Valladas, O. Bar-Yosef and B. Vandermeersch. 1988.

Thermoluminescence dating of Mousterian 'Proto-Cro-Magnon' remains from Israel and the origin of modern man. *Nature* 331:614-16.
Van Valen, L. 1976. Ecological species, multispecies, and oaks. *Taxon* 25:233-39.
Weckler, J. C. 1954. The relationship between Neanderthal man and *Homo sapiens*. *American Anthropologist* 56:1003-25.
Wiley, E. O. 1978. The evolutionary species concept reconsidered. *Systematic Zoology* 27:17-26.
Wiley, E. O. 1981. *Phylogenetics: The Theory and Practice of Phylogenetic Systematics*. New York: John Wiley and Sons.
Williams, B. J. 1987. Rates of evolution: Is there a conflict between neo-Darwinian evolutionary theory and the fossil record? *American Journal of Physical Anthropology* 73:99-109.
Wobst, M. H. 1976. Locational relationships in paleolithic society. *Journal of Human Evolution* 5:49-58.
Wolpoff, M. H. 1986. Describing anatomically modern *Homo sapiens*: A distinction without a definable difference. *Anthropos* (Brno) 23:41-53.
Wolpoff, M. H. 1989. Multiregional Evolution: The fossil alternative to Eden, 62-108. In *The Human Revolution: Behavioural and Biological Perspectives on the Origins of Modern Humans*, ed. P. Mellars and C. Stringer. Edinburgh: Edinburgh University Press.
Wolpoff, M. H., Wu Xinzhi, and A. G. Thorne. 1984. Modern *Homo sapiens* origins: A general theory of hominid evolution involving the fossil evidence from East Asia, 411-83. In *The Origins of Modern Humans: A World Survey of the Fossil Evidence*, ed. F. H. Smith and F. Spencer. New York: Alan R. Liss.

Chapter **11**

Samples, Species, and Speculations in the Study of Modern Human Origins

FRED H. SMITH

In many respects, the genesis of renewed interest in the origin of modern humans, which has characterized paleoanthropology in recent years, coincides with a series of ideas published in a single volume eight years ago (Smith and Spencer 1984). In this volume, two basic paleontological viewpoints central to the present debate on the pattern and mode of modern human origins were initially articulated in detail. First, the Afro-European "sapiens hypothesis" (Bräuer 1984) presented a detailed argument for a monocentric origin of modern humans in Africa, based on the pertinent fossil hominid record. Additionally, although long favoring a classical monocentric origin for modern people, Stringer provides his first unequivocal endorsement of Africa as that center of origin (Stringer, Hublin & Vandermeersch 1984). Second, the basis for modern multiregional evolutionary thinking was articulated by Wolpoff, Wu & Thorne (1984); and regional continuity was suggested to characterize various regions of Eurasia, albeit to differing degrees depending on author and region (Smith 1984; Trinkaus 1984)

Since 1984, both the African origins and multiregional models have themselves "evolved" in various ways (see Smith, Falsetti & Donnelly 1989, for a recent review). This evolution, however, has not been due in any significant sense to new discoveries of fossil hominid remains or even to new analytic approaches to previously known paleontological samples and specimens. With very few exceptions, the major fossil players in the game are the same today as in 1984; and a comparison of the articles in Smith and Spencer

FRED H. SMITH • Department of Anthropology, Northern Illinois University, DeKalb, IL 60115.
Origins of Anatomically Modern Humans, edited by Matthew H. Nitecki and Doris V. Nitecki. Plenum Press, New York, 1994.

(1984) with those in Bräuer and Smith (1992) or Trinkaus (1989) reveals use of essentially the same analytical procedures. The major factors influencing evolution of this debate during the past eight years have come from two sources: (1) the application of new chronometric dating techniques to the issue of modern human origins; and (2) various analyses of recent human genetic variability, as revealed in studies of both nuclear and mitochondrial DNA.

CHRONOLOGY, DNA, AND AFRICAN ORIGINS

The impact of thermoluminescence (TL), electron spin resonance (ESR), and, to a lesser extent, accelerator C^{14} dating on recent perspectives concerning post *Homo erectus* human evolution has been tremendous. ESR and TL dates published since 1987 suggest that "modern" human morphology was established in both Africa and western Asia at the sites of Skhul, Qafzeh, Border Cave and Klasies River Mouth on the order of 100 ky ago (Grün and Stringer 1991; Valladas et al. 1988). Additionally, ESR and TL techniques date Neandertals in the Near East to ages ranging from 120 ky BP at Tabun to as recent as 60 ky BP at Kebara (Valladas et al. 1987; Grün and Stringer 1991), or perhaps even 41-50 ky BP on the basis of preliminary ESR dates from Amud (Grün and Stringer 1991). These same techniques indicate that Neandertals persisted in western Europe until < 41 ky BP at Le Moustier (Grün and Stringer 1991) and 36 ky BP at Saint-Césaire (Mercier et al. 1991), and have yet to demonstrate the presence of modern humans in western Europe prior to ca. 30 ky BP. Thus, the argument that modern human morphology appeared roughly contemporaneously throughout the Old World between 40 ky to 30 ky BP, which was eminently defendable in 1984, apparently has been negated by the results of these chronometric techniques (assuming, of course, the accuracy of this chronology). Consequently, models devised to explain such a contemporaneous emergence of modern morphology (e.g., Smith 1985), though theoretically possible, seem unlikely in the light of the present dating framework.

By the early 1980s, studies of polymorphisms reflecting human nuclear DNA variation suggested to some researchers that all modern humans shared a relatively recent common ancestry (e.g., Nei and Roychoudhury 1982). But it was clearly the analyses of human worldwide mitochondrial DNA (mtDNA) variation by Cann, Stoneking, and Wilson (1987) that changed the role of genetic studies in the reconstruction of recent human phylogeny. MtDNA analyses do not require the same assumptions that cast doubt on the reliability of nuclear DNA results (Cann 1988); but, nevertheless, the study by Cann, Stoneking, and Wilson (see also Cann 1988, 1992; Stoneking and Cann 1989) also indicated a recent common ancestry for all modern populations. Furthermore, Africa south of the Sahara was determined to be the ancestral home of modern humans, and the last common ancestor of African and non-African population was estimated to have lived in Africa between 90 ky to 180 ky BP. The African origin of modern humans has been subsequently supported by additional mtDNA studies (Vigilant et al. 1989), Y chromosome markers (Lucotte 1989) and genetic distances derived from nuclear DNA-based polymorphisms (Cavalli-Sforza et al. 1988).

The mainly paleontologically based African origins ideas of the mid-1980s were transformed by these genetic analyses, particularly the mtDNA studies, into a much more influential perspective on modern human origins. It is a perspective which has engendered tremendous interest from a broad range of scholars, the press, and the general public alike. Various problems with both the genetic and paleontological foundations of the African origins model have been noted by both paleontologists (Wolpoff 1989, 1992; Wolpoff and Thorne 1991; Frayer 1992; Smith 1992, in press; Smith, Falsetti & Donnelly 1989; Thorne and Wolpoff 1992) and geneticists (Spuhler 1988; Excoffier 1990; Excoffier and Langaney 1989; Excoffier and Roselli 1990; Templeton 1992). While these criticisms have had reasonable impact within the pertinent specialized disciplines of human paleontology and genetics, they have had little influence on how the issue of modern human origins is portrayed to the public (e.g., Diamond 1989) or in scientific media which have a wide scholarly audience outside of these disciplines, such as *Scientific American*, *Science*, or *Nature* (but see Barinaga 1992, and Thorne and Wolpoff 1992). No matter how much one likes the African origins perspective, and I count myself among those who find aspects of it very attractive, there is a certain "where there's so much smoke, there must be fire" component to this line of thinking. In other words, the unstated implication is that all the indications (paleontological, chronological, and genetic) for a recent African modern human origin cannot possibly be wrong, no matter how problematic each of them individually may be.

My own problems with African origins models are primarily focused on more extreme implications of these models. For example, in their influential *Science* paper, Stringer and Andrews (1988) strongly implied that modern humans represent a separate species from archaic Eurasian hominids, and that the appearance of modern humans outside the African homeland was the result of this new species (*Homo sapiens*) totally replacing indigenous archaic ones (e.g., the Neandertals) throughout Eurasia. Some geneticists have been even more explicit. Commenting on the mtDNA data, Stoneking and Cann (1989:28) state that ". . . the rather staggering implication is that the dispersing African population [of modern humans] replaced the resident populations [of Eurasian archaic humans] without any interbreeding." Finally,

Diamond (1989:59), writing for a popular audience, suggests that it would be wise ". . . to take the negative evidence at face value, to accept that hybridization occurred rarely if ever and, to doubt that any living people carry any Neanderthal genes."

Numerous analyses of the relevant fossil record from Eurasia illustrate that such an extreme speciation/replacement model is incompatible with paleontological reality. Pope (1991), Habgood (1992), and Wolpoff (1989, 1992) have found varying degrees of evidence in East Asia of local continuity from *Homo erectus* to recent humans. Thorne and Wolpoff (1981; see also Wolpoff, Wu Xinzhi & Thorne 1984: Wolpoff and Thorne 1991), Kramer (1991), Habgood (1989), and Wolpoff (1989, 1992) have done the same for Australasia. My own work (Smith 1984, 1991, 1992), as well as that of Frayer (1992), has documented evidence of continuity across the Neandertal/early modern boundary in Europe. Thus, any model which implies a speciation/ total replacement process for the emergence of modern Eurasians is simply not commensurate with the human fossil record.

It should be noted that Bräuer (1984, 1989, 1992) has always accepted Eurasian archaic hominid contributions to early modern populations throughout Europe and Asia. Stringer (1990, 1992; Grün and Stringer 1991) has also recently stated that he never meant to imply that replacement was total nor that a contribution by Eurasian archaics to modern Eurasian populations was impossible. While these assurances that replacement does not have to be total serve to moderate my objections to some extent, it is clear that both Bräuer and Stringer consider the Eurasian archaic human contribution to have been negligible at best. Both view the migration of African-derived populations into Europe and Asia as the major contributor by far, both quantitatively and qualitatively, to the emergence of modern Eurasians. Both argue that evidence for transitional samples and the existence of regional continuity in Europe (and in other areas) is equivocal. Both argue that the presence of modern humans in Africa at or before 100 ky BP is unequivocal. Stringer (1990, 1992) argues that Neandertals virtually have to be recognized as a different evolutionary species than modern people. On each of these points, I disagree.

A TALE OF TWO SAMPLES: THE VINDIJA NEANDERTALS

For several years, I have attempted to demonstrate the existence of a late Neandertal sample in central Europe that exhibits a morphological pattern intermediate between earlier Neandertals and the earliest modern Europeans (Smith 1982, 1984, 1991, 1992; Smith and Ranyard 1980; Smith, Falsetti & Donnelly 1989; Smith, Simek & Harrill 1989; Smith and Trinkaus

Table 1. Maxillary dimensions in mm for Vindija, Western European Neandertals and Altendorf.

	Alveolar height	Nasal breadth	Prosthion to postcanine
Vindija			
225	17.5	28.5	27.3
259	16.3	26.2	24.3
Western European Neandertal			
pooled-sex x (σ)	26.1 (3.0)	33.3 (1.5)	24.9 (1.1)
male x (σ)	28.6 (1.6)	33.2 (1.7)	24.6 (1.4)
female x (σ)	23.6 (1.4)	33.5 (-)	25.3 (0.5)
Altendorf Neolithic			
A222	21.8	15.3	---
Adult x (σ)	23.0 (1.5)	17.1 (1.7)	---

Data for Vindija and Western European Neandertals are summarized from Wolpoff et al. (1981:504). Data for Altendorf were collected by the author.

1991). Indications of the existence of such a sample began to emerge with the description of a maxilla from Kůlna cave in Czechoslovakia by Jelínek (1966, 1969). Jelínek showed that this specimen exhibited a combination of rather typical Neandertal features (e.g., a high alveolar process) and more "progressive" features, particularly the presence of a canine fossa and a relatively narrow nasal aperture. The specimen was associated with Late Mousterian at the site and dated by conventional C^{14} to between 38 ky and 44 ky BP (Smith 1984).

Confirmation of the existence of an intermediate or transitional Neandertal "population" in central Europe came with the excavation of the Vindija cave in Croatia beginning in the early 1970s. A large series of fragmentary Neandertal remains were recovered from a Late Mousterian level of this site and, like Kůlna, revealed an intriguing mosaic of characteristics (Wolpoff et al. 1981; Smith and Ranyard 1980; Smith 1982, 1984). Two partial maxillas (Vi 225, Vi 259) exhibit nasal breadths that fall more than 3 standard deviations below the European Neandertal mean (see table 1). Alveolar heights are also substantially reduced, with values in both specimens falling more than 2 standard deviations below the European Neandertal mean. Furthermore, while both specimens are damaged lateral to the nasal aperture and above the alveolar process, Vi 259 appears to exhibit a shallow canine fossa. Compared to other Neandertals, these features indicate a reduction in facial size and a shift from the facial morphological pattern characteristic of Neandertals (Rak 1986; Smith 1991).

Additional suggestions of facial reduction come from the mandibles. In all four Vindija mandibles with a symphysis, the symphyses are more vertical than other Neandertals, particularly those from the earlier Croatian

Neandertal site of Krapina. In addition, three of the specimens exhibit slight concavities lateral to the midline, just below the alveolar process margin (Wolpoff et al. 1981; Smith, Boyd & Malez 1985). This morphology results in the presence of an incipient mental eminence, and possibly the beginning stages of mental trigone formation in these specimens. This symphyseal configuration, along with the slightly reduced retromolar spaces (Wolpoff et al. 1981), also suggests reduced faces and facial prognathism in the Vindija sample.

The adult supraorbital tori from Vindija exhibit not only an absolute size reduction, but also a shape change involving relatively greater degrees of pinching above each orbit compared to the Krapina and other European Neandertals (Smith 1984). This configuration represents the initial phases in the separation of the torus into the superciliary arch and supraorbital trigone characteristic of the early modern European sample. While African tori also exhibit reduction from archaic to early modern humans, they are not characterized by the pattern of midorbital pinching (Smith 1992). Regression analysis of the Krapina, Vindija, and early modern central European brow ridges also indicate that Vindija occupies an intermediate position relative to the other two samples (Smith, Simek & Harrill 1989). Both the overall reduction and shape change characteristic of the Vindija tori also reflect the facial reduction notable in the maxillae and mandibles.

Finally, the only diagnostic postcranial specimen, a scapular fragment preserving the glenoid fossa and axillary areas (Vi 209), also exhibits a mixture of features. It has a dorsal axillary groove, a feature common in Neandertals but exceedingly rare in early modern or recent human samples (Trinkaus 1977; Frayer 1992). It also exhibits a relatively broad glenoid fossa, which falls out of the range of other Neandertals and approaches the modern European condition (Smith and Trinkaus 1991).

The Vindija specimens, despite their intermediate tendencies, are clearly still Neandertals in a "typological" sense. This is demonstrated by their overall supraorbital torus form, columnar lateral orbital pillars, suprainiac fossae, overall mandibular morphology, axillary scapular border, and other features. Still, they exhibit the same combination of a relatively conservative cranial vault and more reduced modern-shaped face that characterizes specimens like Florisbad and Ngaloba (Smith, Falsetti & Donnelly 1989), which are recognized as unequivocal transitional specimens in the African sequence by supporters of African origins models (Stringer 1988, 1989; Bräuer 1984, 1989). Since a transitional sample should exhibit a mosaic of features, some reflecting the ancestral condition and others the descendant condition, it is difficult *not* to recognize Vindija as a transitional sample between most Neandertals and early modern Europeans.

What explanations can be given for the apparent intermediate or "transitional" nature of the Vindija sample's total morphological pattern? One possibility, often implied by the supporters of African origin models, is that these changes represent parallelism: changes which are unrelated to the emergence of modern Europeans but "mimic" the pattern seen in the "real" African transition. Obviously, this is a possibility, especially considering the fact that parallelisms are probably more common in biological history than is usually recognized (Trinkaus 1992).

Another possibility, suggested by supporters of African origin models, is that the Vindija sample only appears to be transitional because it is made up of small individuals, is predominantly female, and/or contains subadult individuals. Recently, for example, Bräuer has stated the following in reference to Vindija:

> Yet since sex, age, and individual variation in the sample (which consists of only fragmentary pieces) cannot be reliably determined, even Frayer [1992] has stated that it is difficult here to ascertain what is and what's not a transitional fossil. (1992:93)

Regardless of whether Bräuer's paraphrase accurately reflects Frayer's meaning, when taken in proper context (Frayer 1992:180), the implication seems to be that those of us who view Vindija as a transitional sample fail to recognize, or minimize, problems of age, sex, and size differences and the inherent difficulties involved in working with fragmentary remains. Quite to the contrary, we have gone to considerable lengths to evaluate the effects of age and possible differences in size that might provide misleading comparisons, and we fully realize the limitations of small, fragmentary samples (Smith 1984).

In the assessments of the Vindija supraorbital tori, for example, we were careful to exclude specimens that lacked the type of histological remodelling on the external surfaces covering the frontal sinuses and lateral areas of the tori which denote unquestionably adult Neandertals (Smith and Ranyard 1980; Wolpoff et al. 1981). This forced us to exclude specimens like Vi 279 and Vi 224, which are probably adolescents (possibly comparable in age to Le Moustier 1). Inclusion of these specimens would have made our case much stronger both metrically and morphologically, but they were excluded *because* they were not demonstrably adults.

All of the Vindija mandibles are unquestionably adult (Wolpoff et al. 1981), except possibly for the Vi 306 symphysis (Smith, Boyd & Malez 1985). Mandibular corpus dimensions for Vindija are never more than 1 standard deviation smaller, but are sometimes slightly > 1 standard deviation larger than the appropriate Krapina adult means (Wolpoff et al. 1981:508), and the Krapina-Vindija adult means do not differ from each other at a statistically significant level. Mandibular tooth breadths for Vindija also fall within 1 standard deviation of the pertinent Krapina means, except for the canine. Thus, neither age nor systematic size variation appear to be responsible for the systematic morphological differences between Vindija and Krapina mandibles.

Vindija maxilla 259 is also unequivocally adult, in that it preserves a rather worn M^2 and the socket for M^3. It is considered to be a female (Wolpoff et al. 1981), but falls more than 3 standard deviations below the western European Neandertal *female* mean for alveolar height and more than 3 standard deviations below the pooled-sex Neandertal mean for nasal breadth (only one female European Neandertal has a measurable nasal breadth and it falls above the pooled-sex Neandertal mean). Furthermore, the prosthion to postcanine dimension (which measures the anterior breadth of the palate) approximates the female and pooled-sex Neandertal means, indicating that not all facial dimensions decrease (see table 1).

The second maxilla (Vi 225) is slightly larger than Vi 259 in all comparable dimensions. Although I am certain it is an adult, the specimen is broken at the mesial M^2 socket, and M^3 status cannot be evaluated. The specimen had erupted both permanent premolars and M^2, so it would have been *at least* developmentally equivalent to a recent child of approximately 14 years.

I measured nasal breadth and alveolar height on a specimen of comparable dental developmental stage from the early Neolithic stone box grave at Altendorf in southern Germany (Perret 1938) and compared these values with definite adults from this very robust population. The results are presented in table 1 and show that, at this developmental stage, nasal breadth is less than 1 standard deviation and alveolar height only slightly more than 1 standard deviation below the adult means. Thus, even if Vi 225 is subadult (which is *not* likely), this would not totally explain the fact that its nasal breath and alveolar height dimensions are so small relative to other Neandertals.

The final possible explanation for the Vindija morphological pattern is that it results from gene flow into central Europe from more "modern" people outside Europe. If available ESR and TL dates are correct in placing the Skhul and Qafzeh hominids in the Levant at 90 ky BP or earlier, it may be this "population" which provides the source for this gene flow into Europe. If the gene pools of hominids around the Mediterranean were as interconnected during the Late Pleistocene as some studies indicate (Simmons and Smith 1991), it would seem logical for Neandertals to exhibit some evidence of influence from the Skhul/Qafzeh people. However, if this is the proper explanation for the Vindija morphological mosaic, it certainly does not reflect a replacement of Neandertals by a different species of humans but rather the assimilation of some new elements into a basically Neandertal gene pool.

My own preference for this last explanation also is based on the fact that the earliest post-Neandertal Europeans exhibit a number of characteristics that cannot be derived from the Skhul/Qafzeh people or early "modern" Africans. These include occipital bunning, which does not occur in Skhul/Qafzeh, West Asian Neandertals, or archaic and early modern Africans (except for Jebel Irhoud–see Simmons and Smith 1991), but is almost ubiquitous in European Neandertals and early modern Europeans (Smith 1984, 1991, 1992; Smith, Falsetti & Donnelly 1989). Additionally, Frayer (1992) has noted a series of features (mandibular foramen morphology, scapular axillary border morphology, and suprainiac fossa incidence) which also suggests some degree of continuity between Neandertals and early modern Europeans. These patterns, as well as others (Smith 1991), make it extremely difficult to exclude at least some degree of Neandertal contribution to early modern Europeans.

From my perspective, the changes which herald the establishment of modern human morphology in Europe represent a continuation of the process of gene flow into Europe which is also responsible for the morphology of late central European Neandertals. Modern human morphology may certainly be the result of an *increase* in level of gene flow into Europe between 40 and 30 ky BP (Trinkaus and Smith 1985), but it may also reflect the introduction of only a few very important new elements (Smith 1991). Of course, it is also possible that the only significant gene flow occurred with the establishment of modern human morphology in Europe. This is what the African origin supporters argue, and if they are correct, the Vindija morphological pattern is indeed an example of parallelism. Whichever of these explanations one prefers to use as an explanation for the Vindija hominids, one fact is clear: the trends documented in their morphology relative to earlier Neandertal are real, not the result of problematic age or size comparisons. The smallness of the sample sizes could, of course, mean that the Vindija sample is not representative of the Vindija "population," at least in some cases. A sample size of 2 (the size of the Vindija maxilla sample) is far from ideal. But if the Vindija evidence is to be considered "extremely weak" because of a sample of N = 2 (Bräuer 1992:93), it is hard to understand why crucial African samples with N = 1 for key elements do not warrant the same cautious assessment by supporters of the African origin model.

A TALE OF TWO SAMPLES: THE KLASIES RIVER MOUTH HOMINIDS

As I have elaborated elsewhere (Smith 1985, 1992, in press; Smith, Falsetti & Donnelly 1989), the case for a 100 kya age for a modern human presence in Africa rests on skeletal remains from three locations: the Kibish

Formation at Omo (Ethiopia), Border Cave (South Africa), and Klasies River Mouth Caves (South Africa). The crania from Omo-Kibish site KHS (Omo 1) and Border Cave exhibit a robust but essentially "modern" morphology (but see Corruccini 1992), as do the Border Cave 2 and 5 mandibles. However, the dating of Omo 1 is very questionable (Howell 1978; Day and Stringer 1982; Smith 1985; Smith, Falsetti & Donnelly 1989), because the uranium-thorium date of 130 ky BP for member 1 (where the specimen is located) is derived from mollusk shell and thus is open to serious doubt. Also, while the dating of the Border Cave archaeological stratigraphy is more reliable (Grün and Stringer 1991), the stratigraphic contexts of the Border Cave 1 cranium and the Border Cave 2 mandible is uncertain since they were not discovered during controlled excavations. Even the Border Cave 5 mandible, supposedly excavated in situ, may represent an intrusive element into its claimed level of origin, because of ". . . a strong contrast in state of preservation between the human bones and animal bones that occur in the same levels. Unlike the animal bones, which are poorly preserved and fragmented as a result of substantial post-depositional leaching and compaction, the human bones are well-preserved and relatively complete" (Klein 1983:34).

The stratigraphic context of the Klasies River Mouth (KRM) hominid fossils is not at issue. These specimens were recovered during controlled excavations, and the stratigraphic correlation between cave 1 and shelters 1a and 1b now seems clear (Deacon and Geleijnse 1988; Deacon and Schuurman 1992). All of the Middle Stone Age (MSA) associated hominids, except for 2 maxillary fragments, are from the SAS member (Rightmire and Deacon 1991). The SAS member is *maximally* 98 ky to 110 ky BP, based on uranium disequilibrium dating of the top of the immediately underlying LBS member (Deacon 1989). This is supported by linear uptake ESR dates of 94 ky and 88 ky BP for the base of the SAS member (Grün and Stringer 1991). The *minimum* age for the SAS member and KRM hominids is geologically determined to be ca. 60 ky BP (Deacon and Geleijnse 1988).

The hominid fossils from KRM have been described consistently as totally modern in morphology (Singer and Wymer 1982; Bräuer 1984, 1989, 1992; Rightmire 1984, 1989 Rightmire and Deacon 1991) except for the LAS maxillae, which will not be considered here. However, this impression may be misleading, since the KRM specimens have often been lumped together with the more complete and more "modern"-looking Border Cave and Omo-Kibish 1 specimens (Smith, Falsetti & Donnelly 1989). Just how well do the KRM specimens, when considered *alone*, conform to a modern human morphological pattern? Indeed, there have been a few suggestions (Caspari and Wolpoff 1990; Wolpoff and Caspari 1990a, 1990b; Smith 1992, in press; Smith, Falsetti & Donnelly 1989) that the "modernness" of the KRM sample has been overemphasized.

First of all, like the Vindija fossils, the KRM material is quite fragmentary. There is a reasonable sample of mandibles (N = 5), four of which preserve symphyses. Upper facial morphology is revealed in only two specimens: a glabellar fragment of a frontal, also preserving the medial portion of the right supraorbital region (KRM 16425), and a massive left zygomatic (KRM 16651). A few isolated teeth, fragmentary postcranial elements, and cranial vault fragments complete the sample (Singer and Wymer 1982). A complete catalog of the KRM hominids, except for teeth, is given by Rightmire and Deacon (1991:137).

The KRM mandibles differ in several ways from the Vindija and Shanidar Neandertals (Rightmire and Deacon 1991), which is certainly not surprising given the geographic distance between these samples. The KRM mandibles are smaller overall; but when one excludes the aberrantly small KRM 16424 specimen, the size difference between the KRM and the Vindija sample is not impressive. I would argue that the KRM mandibular sample, based on their symphyseal region, is more primitive than Border Cave 2 and 5 or more recent Africans. Of the 4 KRM mandibles preserving the symphysis, only one (KRM 41815) exhibits distinct mental eminence development; and the damage to the alveolar process in the symphyseal region might accentuate this development more than it would have been without the damage. The KRM mandibles 13400 and 21776, as well as 41815, are described as exhibiting moderate to clearly visible mental trigones and eminences. I would argue that, except for 41815, both of these features are at best only slightly more developed than in Vindija mandibles 206 and 231.

I am not convinced that any recent African sample would have 75% of its mandibles exhibiting the relatively weakly developed mental eminences and mental trigones that characterize the KRM sample. Thus, I question whether we should dogmatically refer to the KRM sample as fully anatomically modern. It would indeed be interesting to compare the KRM mandibles to mandibles from the African transitional sample, for example from specimens like Florisbad or Ngaloba. Unfortunately, no mandibles are known for these transitional specimens, which makes assessment of the KRM specimens more difficult. Clearly, the KRM mandibles are not Neandertals. Still, one wonders if the KRM mandibles are really very much more "modern" in an African context than Vindija is in a European one.

The KRM zygomatic (16651) is an extremely interesting specimen. Compared to Neandertals, it exhibits a rather horizontal inferior (zygomatico-alveolar) margin. However, this feature is also characteristic of the African transitional sample (Smith, Falsetti & Donnelly 1989) and thus is not a unique indication of modern Africans. Furthermore, this specimen has a cheek height virtually identical to Kabwe's, larger than Ngaloba's or Florisbad's, and 2.6 standard deviations above a sample of definitely Holocene

Table 2. Horizontal height in mm of the zygomatic [Martin no. 48 (3a)] in Pleistocene and Holocene Africans.

Specimen/Sample	M 48 (3a)
Klasies River Mouth 16651	> 30
Border Cave 1	20.5
Florisbad	24.1
Ngaloba	27.6
Holocene Cape Africans (x,σ)	22.5 (2.87)
Range (N)	15.9-28.2 (33)

These specimens are Cape Africans with secure context, dated to > 4 ky BP. Data courtesy of R. Caspari and M. Wolpoff.

South Africans (table 2). Finally, the frontal process, which forms much of the lateral orbital pillar, is columnarlike (Smith, in press), a condition typical for Neandertals but rare in more recent humans (Smith 1976).

The frontal fragment (KRM 16425) certainly does not exhibit the level of supraorbital projection of earlier African specimens, including Florisbad or Ngaloba. Measured just medial to the supraorbital notch, the Klasies specimen measures 14 mm, compared to 20 mm for Florisbad. However, there is an important question regarding KRM 16425 that has yet to be answered adequately. Is it possible that KRM 16425 is an adolescent? This is difficult to say for certain, but it is important to note that adolescent Neandertals (e.g., Le Moustier 1, Krapina 23, Vindija 224, or Vindija 279) do not exhibit the degree of supraorbital torus development characteristic of adult Neandertals. An adult status for KRM 16425 is supported by the fact that it does appear to have respectable frontal sinus development but adolescent Neandertals also have rather extensive sinus development without extensive development of the supraorbital torus. The point is, that a careful study demonstrating that the KRM frontal is definitely an adult has yet to be published. Even if it is adult, one could argue that KRM 16425 is a small, gracile female, and that a male supraorbital would be considerably more robust.

My impression from these observations on the KRM sample is that the KRM specimens are more primitive than is usually implied by supporters of the African origin models. Perhaps if we had more complete specimens, my impression would prove to be false. But herein lies an important point: it is interesting that the small sample sizes and the possible effects of age and sex variation at KRM seem far less of a problem in the eyes of some African origin supporters than they are for the Vindija sample. In fact, this is rarely mentioned as a potential problem for assessment of the Klasies sample. Perhaps the KRM sample does indeed represent an *early* modern African population, a view that I am not necessarily opposed to (Smith, Falsetti & Donnelly 1989); but the evidence for this is certainly no stronger than the evidence of the transitional morphological pattern in the Vindija hominids.

THE SPECIES CONTROVERSY

Another pervasive issue in recent discussions of modern human origins is the question of the species affiliation of the Neandertals. It is often stated that Neandertals are so distinctive morphologically that science is virtually compelled to recognize them as a separate species from *Homo sapiens* (Tattersall 1986; Stringer and Andrews 1988; Stringer 1992). Tattersall (1986) argues that, using other animals as a model, species are underrecognized in the human fossil record. For example, if one looks at the various species of the genus *Cercopithecus*, which generally represent good biological species, the extent of non-size related skeletal differences between them is minimal (Napier and Napier 1967). Consequently it would be almost impossible to recognize them as species in a paleontological context. Thus, it is argued, when fossil human samples are as distinctive morphologically as Neandertals are, they should be recognized as separate species.

Paleontologists commonly employ some variation of Simpson's (1961) evolutionary species concept rather than a biological one, because of the problems of using the latter concept in the context of significant time depth. Of course, the importance of reproductive barriers to the morphological distinction underlying the evolutionary species concept is clearly recognized (Wiley 1981). Stringer (1992) argues that even if there is some restricted hybridization in contact zones, this would not necessarily demonstrate conspecifity of Neandertals and modern peoples and would not invalidate the recognition of Neandertals as a distinct species. On the other hand, Wiley (1981) points out that if evidence of hybridization is wide spread, it is unlikely that the populations involved represent different species.

There is no question that Neandertals are distinctive from modern humans at some level, so perhaps resurrecting the taxon "*Homo neanderthalensis*" as an evolutionary species has some value in a heuristic sense. Neandertals also have a somewhat of a distinct morphological trajectory, another significant criterion for recognizing evolutionary species (Wiley 1981). However, there is also evidence that late central European Neandertals begin to shift that trajectory in a modern European direction. If one compares La Chapelle-aux-Saints to a recent European, it could perhaps be argued that European Neandertals are distinctive enough to be recognized as a different species. But Neandertals are not as morphologically distinct when one compares late European Neandertals and the earliest post-Neandertals Europeans, and I think this fact makes it difficult to argue unequivocally for species separation. It is also the case that recognizing Neandertals as even a separate evolutionary species seems to lead to the assumption in many circles that such recognition has significant behavioral and reproductive implications. For example, in order to explain why no living people carry "any Neandertal genes," Diamond wrote that:

> If Neanderthal behavior was as rudimentary and their anatomy as distinctive as I suspect, few Cro-Magnons would have wanted to mate with Neanderthals. (1989:52)

There is, however, considerable archaeological evidence that Neandertals and modern humans are not *qualitatively* different behaviorally (Bar-Yosef 1991, 1992; Simek 1992; Hayden and Bonifay, in prep.); and, as I have already argued, Neandertal morphology is not distinctive enough to support such an extreme position.

In the context of arguments about underrepresentation of species in the human fossil record, it is important to bear in mind what *type* of species humans are. Vrba (1980, 1984) has been particularly clear on the differences one should expect in the evolutionary history of generalized, as opposed to specialized, species. Species with a generalized adaptative strategy tend to be geographically widely distributed, polytypic, and relatively long-lived. Furthermore, their lineages tend not to be characterized by much cladogenesis or extreme morphological specialization, and temporal overlap of species is not common. Foley (1991) has also shown that generalized species conform to the pattern outlined above, and that carnivores normally fit this pattern. He then speculates that, once hominids begin to systematically exploit meat (which he equates with appearance of genus *Homo*), they would have become a much more generalized animal (see also Vrba 1988) and less likely to undergo cladogenesis than earlier hominids. Thus, there is a theoretical basis from which to argue that the genus *Homo*, which has been around only slightly more than two million years, should not be characterized by a large number of temporally overlapping species. Viewed in this light, the known morphological variation of Late Pleistocene and recent humans is much more likely to be a reflection of a polytypic single species than multiple species.

Finally, I would argue that other examples from mammals support the assertion that morphological differences on the order of Neandertal/modern human differences do not *necessarily* indicate separate species rather than polytypism. Figure 1 is a comparison of two suids, which are sympatric in the mountains of East Tennessee and other regions. One of them is a domesticated Berkshire and the other a European (Russian) wild boar. Both of the specimens are adult males. I submit that, found in a paleontological context- and applying the same logic some wish to apply to Neandertals and modern humans, these animals would have to be recognized as separate species.

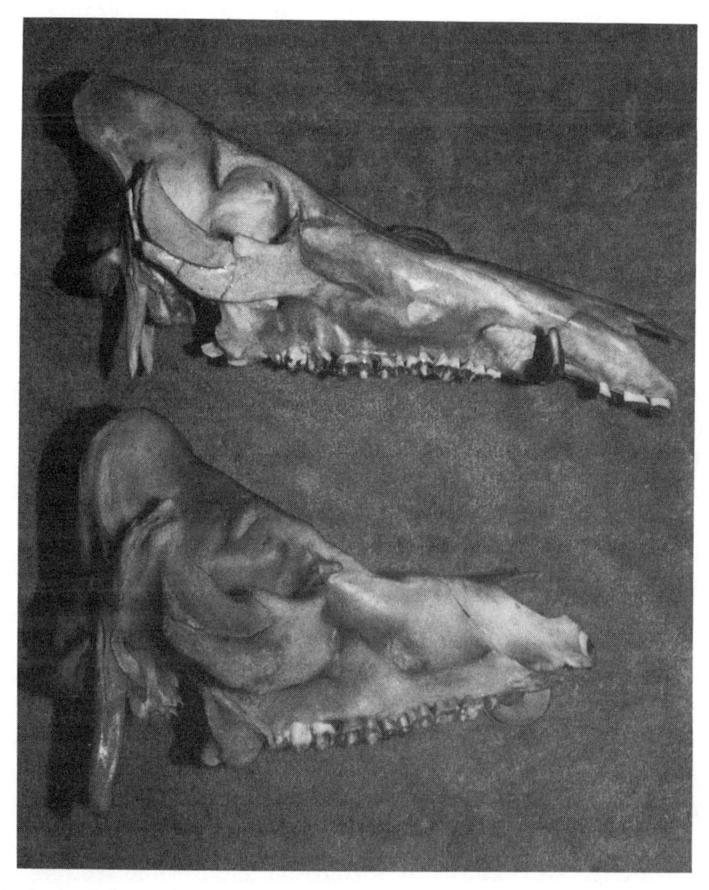

Figure 1. Lateral views of an adult male Berkshire pig and a European (Russian) wild pig. Both are members of the taxon *Sus scrofa*.

However, biologically they are the same animal, regardless of the morphology. Obviously, the role of domestication is a compounding factor; but this example, taken in concert with the other points noted above, indicates that caution should be exercised in asserting that not recognizing Neandertals as a separate species is out of touch with biological reality. Therefore, I would continue to argue that considering Neandertals as a part of a highly polytypic species, which would include recent humans, is quite reasonable (see Smith, Falsetti & Donnelly 1989). This, of course, introduces complexity by extending polytypism back in time; but since there is evidence of a phylogenetic connection between Neandertals and early modern Europeans, I am not convinced that biological reality is better served by the resurrection of the taxon "*Homo neanderthalensis*." This is particularly true because we seem unable to limit use of this taxon to a heuristic construct only.

THE NATURE OF MODERN HUMAN ORIGINS

Recently, Sokal, Oden, and Wilson (1991) have shown that the spread of agriculture into Europe was not simply the result of cultural diffusion, but the spread of new people from western Asia with a revolutionary subsistence technology. The genetic evidence suggests that this process can be described as gradual demic diffusion in which the expanding farmers (moving at the rate of ca. 1 km per year) absorbed or assimilated local genes into their gene pools. A similar process might be an attractive explanation for the origin of modern Europeans and the Neandertals' role therein. In fact, just as the spread of the Neolithic can be traced archaeologically from the Near East (Bar-Yosef 1991, 1992) across Europe (Ammerman and Cavalli-Sforza 1984), there are suggestions that European paleolithic archaeology documents the spread of modern people into Europe and the replacement of Neandertals (Mellars 1989; Klein 1989).

In both western and central Europe, it is generally the case that the earliest modern humans are associated with the Aurignacian (Stringer, Hublin & Vandermeersch 1984; Smith 1984). However, it is important to note a few points, no matter how untidy they make things. First of all, there is some possible evidence of Neandertals associated with early or pre-Aurignacian in central Europe. At Vindija, for example, there are some fragmentary specimens from a possibly early Aurignacian level (Wolpoff et al. 1981) that are morphologically indistinguishable from the Vindija Neandertal sample discussed previously. At Bacho Kiro in Bulgaria, the isolated teeth associated with the "pre-Aurignacian" show stronger affinities with Neandertals than with modern samples (Glen and Kaczanowski 1982). Second, in western Asia, where the early "modern" Skhul/Qafzeh folk are associated with the Mousterian (Trinkaus 1984), the middle/early Upper Paleolithic boundary may be as early as 45-47 ky BP (Bar-Yosef 1991). However, this "transitional" material cannot be classified as Aurignacian. In fact, according to Bar-Yosef (1991), the appearance of the early Levantine Aurignacian at ca. 36 ky BP (Bar Yosef 1992), interrupts the Ahmarian Upper Paleolithic Tradition in the Levant. Thus, while the transition to the Upper Paleolithic may be slightly earlier in the Levant than in central Europe (Allsworth-Jones 1886; Svoboda and Simán 1989), the appearance of the Aurignacian in the Levant seems to postdate the earliest Aurignacian of central/eastern Europe and Spain (Smith 1982; Straus 1989). If the dating is correct, it would be difficult indeed to argue that the Aurignacian develops in the Near East and then spreads into

Europe. It could well be the other way around. Furthermore, there are *no diagnostic fossil human remains* associated with the earliest phases of the Aurignacian in western Europe. Thus, interstratification of Châtelperronian and early Aurignacian levels in the Perigord (Laville, Rigaud & Sackett 1980) of France *does not* directly demonstrate contemporaneity of Neandertals and early modern Europeans. Similar arguments built around the Szeletian and other leaf-point cultures in central Europe (Allsworth-Jones 1986) are equally speculative, because no diagnostic human remains are associated with these complexes (Smith 1984). In the same vein, the diachronic pattern of the spread of Aurignacian into southwestern France, and the concomitant geographic shrinking of the Châtelperronian (Leroyer and Leroi-Gourhan 1983; see also, Simek 1992) does not necessarily reflect the results of an encounter between Neandertals and modern people. Perhaps it will ultimately be proven that early Aurignacian in western Europe was made by people like the evolved Aurignacian-associated Cro-Magnon people; but until that proof is forthcoming, interpretations linking modern humans with the spread of the Aurignacian are only speculation. It may very well be that the origin of the Aurignacian is an indigenous European phenomenon and not related at all to the origin of post-Neandertal Europeans.

Archaeology aside, the presence of the Skhul/Qafzeh people in western Asia at ca. 90 ky BP makes it very difficult to believe that this population did not have a strong influence on the emergence of early modern Europeans. The one point in which I differ with some other purveyors of multiregionalism, is that I think it likely that there was one region where modern human morphology first appeared, perhaps as a shift in growth regulatory mechanisms (Green and Smith 1991; Smith 1991). There is some evidence suggesting this region to be southern Africa, although this interpretation is not problem-free by any means (Smith, Falsetti & Donnelly 1989; Smith, in press; Wolpoff 1989). Regardless of the region of origin, the mechanism by which this morphological complex is established in other regions, including Europe, is far from clear.

Unlike the Neolithic Europe example, "modern" human morphology did not spread from Africa to western Asia, and finally into Europe with a new technology or subsistence strategy. The Skhul/Qafzeh people seem to be identical in both respects to the Mousterian-associated western Asian Neandertals (Bar-Yosef 1989, 1992; Shea 1989). There is also no evidence of a population expansion in southern Africa that would have triggered the spread of modern populations toward the North. Indeed the technological and subsistence innovations that mark the Late Stone Age, compared to the less inspiring Middle Stone Age (Klein 1989), far postdate the appearance of the Skhul/Qafzeh people in the Levant. Thus, no matter how nice a theoretical comparison it might make, there is no convincing evidence for the physical spread of new people with a new technology from their presumably African homeland that is equivalent to the spread of the Neolithic into Europe.

A model of demic diffusion and assimilation is clearly the reasonable way to look at the phenomenon of the origin of modern human morphology, but this does not mean that the major *quantitative* contributors to Eurasian regional gene pools cannot be the indigenous peoples. In some areas, it may well be that the extraneous contributions, while *qualitatively* important, were quantitatively minor. This is particularly feasible if the major changes are in the regulatory portion of the genome. The factors determining relative indigenous versus extraneous contributions in varying regions are far too complex to discuss here (see Wolpoff, Wu & Thorne 1984; Wolpoff 1989, 1992; Smith 1985, 1991; Smith, Falsetti & Donnelly 1989); and, I do not think it is an easy task to sort out the relative contributions of extraneous and indigenous factors in any region. However, such a relatively complex pattern of population dynamics is, in my opinion, more commensurate with the fossil record than biologically more simple models emphasizing replacement, with or without hybridization.

I have attempted to deal with a number of issues in this chapter that are important from my perspective in the ongoing debate regarding modern human origins. I have certainly not dealt with every important issue, and I have likely raised more problems than provided solutions. I hope particularly that I have pointed out some potential difficulties with things which I feel are often not discussed critically enough in the context of the debate. Most of all, I hope I have demonstrated the complexity of a paleoanthropological record that is all too often presented in more simplistic and dogmatic constructs than is advisable at the present time.

We know far more about behavioral and biological factors relating to modern human origins than we know about any other topic within the realm of paleoanthropology. It is often stated that this means we should be able to resolve unequivocally all issues relating to the emergence of modern people, and there seems to be considerable frustration in some quarters that consensus on these issues is not in sight. The reality is that, in historical science, enhanced knowledge allows us to ask progressively more precise questions about the historical record. Consequently, rather than being content with general patterns, debate about modern human origins now centers on more detailed models and more specific issues than is possible for other aspects of paleoanthropology. Viewed in this light, frustration at what we do not yet know should be replaced by an appreciation of the level of precision with which the data and issues pertinent to modern human origins are being investigated. It may still be many years before we can be dogmatic about the exact pattern and mode of modern human origins, but these years promise to be exciting indeed.

ACKNOWLEDGMENTS

I am grateful to the Field Museum of Natural History for sponsorship of the symposium, on which this paper is based. I am also grateful to Ofer Bar-Yosef, Matthew Nitecki, Milford Wolpoff, James C. Ahern, and Louis J. Shirley for their comments on an earlier version of this manuscript.

REFERENCES

Allsworth-Jones, P. 1986. *The Szeletian and the Transition from Middle to Upper Paleolithic in Central Europe*. Oxford: Clarendon Press.

Ammerman, A. J., and L. Cavalli-Sforza. 1984. *The Neolithic Transition and the Genetics of Populations in Europe*. Princeton: Princeton University Press.

Barinaga, M. 1992. "African eve" backers beat a retreat. *Science* 255:686-87.

Bar-Yosef, O. 1989. Geochronology of the Levantine Middle Palaeolithic, 586-610. In *The Human Revolution: Behavioural and Biological Perspectives on the Origins of Modern Humans*, ed. P. Mellars and C. Stringer. Edinburgh: University of Edinburgh Press.

Bar-Yosef, O. 1992. Middle Palaeolithic chronology and the transition to the Upper Palaeolithic in Southwest Asia, 261-72. In *Continuity or Replacement: Controversies in* Homo sapiens *Evolution*, ed. G. Bräuer and F. H. Smith. Rotterdam: A. A. Balkema.

Bräuer, G. 1984. A craniological approach to the origin of anatomically modern *Homo sapiens* in Africa and implications for the appearance of modern Europeans, 327-410. In *The Origins of Modern Humans: A World Survey of the Fossil Evidence*, ed. F. H. Smith and F. Spencer. New York: Alan R. Liss.

Bräuer, G. 1989. The evolution of modern humans: A comparison of the African and non-African evidence, 123-54. In *The Human Revolution: Behavioural and Biological Perspectives on the Origins of Modern Humans,* ed. P. Mellars and C. B. Stringer. Edinburgh: University of Edinburgh Press.

Bräuer, G. 1992. Africa's place in the evolution of *Homo sapiens*, 83-98. In *Continuity or Replacement: Controversies in* Homo Sapiens *Evolution*, ed. G. Bräuer and F. H. Smith. Rotterdam: A. A. Balkema.

Bräuer, G., and F. H. Smith, eds. 1992. *Continuity or Replacement: Controversies in* Homo Sapiens *Evolution*. Rotterdam: A. A. Balkema.

Cann, R. 1988. DNA and human origins. *Annual Review of Anthropology* 17:127-43.

Cann, R. 1992. A mitochondrial DNA perspective on replacement or continuity in human evolution, 65-73. In *Continuity or Replacement: Controversies in* Homo Sapiens *Evolution*, ed. G. Bräuer and F. H. Smith. Rotterdam: A. A. Balkema.

Cann, R., M. Stoneking, and A. C. Wilson. 1987. Mitochondrial DNA and human evolution. *Nature* 325:31-36.

Caspari, R., and M. H. Wolpoff. 1990. The morphological affinities of the Klasies River mouth skeletal remains. Abstract. *American Journal of Physical Anthropology* 81:203.

Cavalli-Sforza, L., A. Piazza, P. Menozzi, and J. Mountain. 1988. Reconstruction of human evolution: Bringing together genetic, archaeological and linguistic data. *Proceedings of the National Academy of Sciences, USA* 85:6002-6.

Corruccini, R. 1992. Metrical reconsideration of the Skhul IV and IX and Border Cave 1 crania in the context of modern human origins. *American Journal of Physical Anthropology* 87:433-45.

Day, M., and C. B. Stringer. 1982. A reconsideration of the Omo-Kibish remains and the erectus-sapiens transition, 814-46. In *L'*Homo erectus *et la Place de l'Homme de Taut-*

avel Parmi les Hominidés Fossiles.Vol. 2, ed. H. de Lumley. Nice: Centre National de la Recherche Scientifique.
Deacon, H. 1989. Late Pleistocene paleoecology and archaeology in the Southern Cape, South Africa, 547-64. In *The Human Revolution: Behavioural and Biological Perspectives on the Origins of Modern Humans,* ed. P. Mellars and C. Stringer. Edinburgh: University of Edinburgh Press.
Deacon, H., and V. Geleijnse. 1988. The stratigraphy and sedimentology of the main site sequence, Klasies River, South Africa. *South African Archaeological Bulletin* 43:5-15.
Deacon, H., and R. Schuurman. 1992. The origins of modern people: The evidence from Klasies River, 121-29. In *Continuity or Replacement: Controversies in* Homo Sapiens *Evolution*, ed. G. Bräuer and F. H. Smith. Rotterdam: A. A. Balkema.
Diamond, J. 1989. The great leap forward. *Discover* 10:50-60.
Excoffier, L. 1990. Evolution of human mitochondrial DNA: Evidence for departure from a pure neutral model of populations at equilibrium. *Journal of Molecular Evolution* 30:125-39.
Excoffier, L., and A. Langaney. 1989. Origin and differentiation of human mitochondrial DNA. *American Journal of Human Genetics* 44:73-85.
Excoffier, L., and D. Roselli. 1990. Origine et évolution de l'ADN mitochondrial humaine Le Paradigme Perdu. *Bulletín et Memoires de la Societe d'Anthropologie de Paris* n.s. 2:25-42.
Foley, R. 1991. How many species of hominid should there be? *Journal of Human Evolution* 20:413-27.
Frayer, D. W. 1992. The persistence of Neanderthal features in post-Neanderthal Europeans, 179-88. In *Continuity or Replacement: Controversies in* Homo Sapiens *Evolution*, ed. G. Bräuer and F. H. Smith. Rotterdam: A. A. Balkema.
Glen, E., and K. Kaczanowski. 1982. Human remains, 75-79. In *Excavation in the Bacho Kiro Cave*, ed. J. Kozlowski. Warsaw: Polish Scientific Publishers.
Green, M., and F. H. Smith. 1991. Heterochrony and Neandertal craniofacial morphogenesis. Abstract. *American Journal of Physical Anthropology*, Supplement 12:82.
Grün, R., and C. B. Stringer. 1991. Electron spin resonance dating and the evolution of modern humans. *Archaeometry* 33:153-99.
Habgood, P. 1989. The origin of anatomically modern humans in Australasia, 245-73. In *The Human Revolution: Behavioural and Biological Perspectives on the Origins of Modern Humans,* ed. P. Mellars and C. B. Stringer. Edinburgh: University of Edinburgh Press.
Habgood, P. 1992. The origin of anatomically modern humans in East Asia, 273-87. In *Continuity or Replacement: Controversies in* Homo Sapiens *Evolution*, ed. G. Bräuer and F. H. Smith. Rotterdam: A. A. Balkema.
Hayden, B., and E. Bonifay. In press. The Neanderthal Nadir. *Journal of Human Evolution.*
Howell, F. C. 1978. Hominidae, 154-248. In *Evolution of Africa Mammals*, ed. V. Maglio and H. Cooke. Cambridge: Harvard University Press.
Jelínek, J. 1966. Jaw of an intermediate type of Neanderthal man from Czechoslovakia. *Nature* 212:701-2.
Jelínek, J. 1969. Neanderthal man and *Homo sapiens* in Central and Eastern Europe. *Current Anthropology* 10:475-503.
Klein, R. G. 1983. The Stone Age prehistory of southern Africa. *Annual Review of Anthropology* 12:25-48.
Klein, R. G. 1989. *The Human Career*. Chicago: University of Chicago Press.
Kramer, A. 1991. Modern human origins in Australasia: Replacement or evolution? *American Journal of Physical Anthropology* 86:455-73.
Laville, H., J.-P. Rigaud, and J. Sackett. 1980. *Rock Shelters of the Perigord*. New York: Academic Press.
Leroyer, C., and A. Leroi-Gourhan. 1983. Problèmes de chronologie: le castelperronien et l'aurignacien. *Bulletin de la Société Préhistorique Française* 80:41-44.
Lucotte, B. 1989. Evidence from a Y-chromosome specific sequence polymorphic DNA probe,

39-46. In *The Human Revolution: Behavioural and Biological Perspectives on the Origins of Modern Humans,* ed. P. Mellars and C. B. Stringer. Edinburgh: University of Edinburgh Press.

Mellars, P. 1989. Major issues in the emergence of modern humans. *Current Anthropology* 30:349-505.

Mercier, N., H. Valladas, J.-L. Joron, J. L. Reyss, F. Lévêque, and B. Vandermeersch. 1991. Thermoluminescence dating of the late Neanderthal remains from Saint-Césaire. *Nature* 351:737-39.

Napier, J., and P. Napier. 1967. *A Handbook of Living Primates*. New York: Academic Press.

Nei, M., and A. Roychoudhury. 1982. Genetic relationship and evolution of human races. *Evolutionary Biology* 14:1-47.

Perret, G. 1938. Cro-Magnon Typen von Neolithikum bis Heute. Ein Beitrag zur Rassengeschichte Niederhessens. *Zeitschrift für Morphologie und Anthropologie* 37:1-101.

Pope, G. 1991. Evolution of the zygomaticomaxillary region in the genus *Homo* and its relevance to the origin of modern humans. *Journal of Human Evolution* 21:189-213.

Rak, Y. 1986. The Neanderthal: A new look at an old face. *Journal of Human Evolution* 15:151-64.

Rightmire, G. P. 1984. *Homo sapiens* in sub-Saharan Africa, 295-325. In *The Origins of Modern Humans: A World Survey of the Fossil Evidence*, ed. F. H. Smith and F. Spencer. New York: Alan R. Liss.

Rightmire, G. P. 1989. Middle Stone Age humans from eastern and southern Africa, 109-22. In *The Human Revolution: Behavioural and Biological Perspectives on the Origins of Modern Humans,* ed. P. Mellars and C. B. Stringer. Edinburgh: Edinburgh University Press.

Rightmire, G. P., and H. Deacon. 1991. Comparative studies of late Pleistocene human remains from Klasies River Mouth, South Africa. *Journal of Human Evolution* 20:131-56.

Shea, J. 1989. A functional study of the lithic industries associated with hominid fossils in the Kebara and Qafzeh caves, Israel, 611-25. In *The Human Revolution: Behavioural and Biological Perspectives on the Origins of Modern Humans,* ed. P. Mellars and C. Stringer. Edinburgh: University of Edinburgh Press.

Simek, J. 1992. Neanderthal cognition and the Middle to Upper Paleolithic transition, 231-45. In *Continuity or Replacement: Controversies in* Homo Sapiens *Evolution*, ed. G. Bräuer and F. H. Smith. Rotterdam: A. A. Balkema.

Simmons, T., and F. H. Smith. 1991. Late Pleistocene human population relationships in Africa, Europe and the Circum-Mediterranean. *Current Anthropology* 32:623-27.

Simpson, G. G. 1951. The species concept. *Evolution* 5:285-98.

Singer, R., and J. Wymer. 1982. *The Middle Stone Age at Klasies River Mouth in South Africa*. Chicago: University of Chicago Press.

Smith, F. H. 1976. The Neandertal Remains from Krapina: A Descriptive and Comparative Study. *University of Tennessee Department of Anthropology Reports of Investigations* 15:1-376.

Smith, F. H. 1982. Upper Pleistocene hominid evolution in south-central Europe: A review of the evidence and analysis of trends. *Current Anthropology* 23:667-86.

Smith, F. H. 1984. Fossil hominids from the Upper Pleistocene of central Europe and the origin of modern Europeans, 137-209. In *The Origins of Modern Humans: A World Survey of the Fossil Evidence*, ed. F. H. Smith and F. Spencer. New York: Alan R. Liss.

Smith, F. H. 1985. Continuity and change in the origin of modern *Homo sapiens. Zeitschrift für Morphologie and Anthropologie* 75:197-222.

Smith, F. H. 1991. The Neandertals: Evolutionary dead ends or ancestors of modern people? *Journal of Anthropological Research* 47:219-38.

Smith, F. H. 1992. The role of continuity in modern human origins, 145-56. In *Continuity or Replacement: Controversies in* Homo Sapiens *Evolution*, ed. G. Bräuer and F. H. Smith. Rotterdam: A. A. Balkema.

Smith, F. H. In press. Models and realities in modern human origins: The African evidence. *Transactions of the Royal Society* (London).

Smith, F. H., D. C. Boyd, and M. Malez. 1985. Additional Upper Pleistocene human remains from Vindija cave, Croatia, Yugoslavia. *American Journal of Physical Anthropology* 68:375-83.

Smith, F. H., A. B. Falsetti, and S. Donnelly. 1989. Modern human origins. *Yearbook of Physical Anthropology* 32:35-68.

Smith, F. H., and G. C. Ranyard. 1980. Evolution of the supraorbital region in Upper Pleistocene hominids from south-central Europe. *American Journal of Physical Anthropology* 53:589-610.

Smith, F. H., J. Simek, and M. Harrill. 1989. Geographic variation in supraorbital torus reduction during the later Pleistocene(c. 80 000-15 000 BP), 172-93. In *The Human Revolution: Behavioural and Biological Perspectives on the Origins of Modern Humans,* ed. P. Mellars and C. B. Stringer. Edinburgh: University of Edinburgh Press.

Smith, F. H., and F. Spencer. 1984. *The Origins of Modern Humans: A World Survey of the Fossil Evidence*. New York: Alan R. Liss.

Smith, F. H., and E. Trinkaus. 1991. Les orígínes de l'homme moderne en Europe centrale: un cas de contínuité, 251-90. In *Aux Origines de L'* Homo sapiens, ed. J.-J. Hublin and A. M. Tillier. Paris: Presse Universitaires de France.

Sokal, R. R., N. Oden, and C. Wilson. 1991. Genetic evidence for the spread of agriculture in Europe by demic diffusion. *Nature* 351:143-45.

Spuhler, J. 1988. Evolution of mitochondrial DNA in monkeys, apes and humans. *Yearbook of Physical Anthropology* 31:15-48.

Stoneking, M., and R. Cann. 1989. African origin of human mitochondrial DNA, 17-30. In *The Human Revolution: Behavioural and Biological Perspectives on the Origins of Modern Humans,* ed. P. Mellars and C. B. Stringer. Edinburgh: University of Edinburgh Press.

Straus, L. 1989. Age of the modern Europeans. *Nature* 342:476-77.

Stringer, C. B. 1988. The dates of Eden. *Nature* 331:565-66.

Stringer, C. B. 1990. The emergence of modern humans. *Scientific American* 263:98-104.

Stringer, C. B. 1992. Replacement, continuity and the origin of *Homo sapiens*, 9-24. In *Continuity or Replacement: Controversies in* Homo Sapiens *Evolution*, ed. G. Bräuer and F. H. Smith. Rotterdam: A. A. Balkema.

Stringer, C. B., and P. Andrews. 1988. Genetic and fossil evidence for the origin of modern humans. *Science* 239:1263-68.

Stringer, C. B., J.-J. Hublin, and B. Vandermeersch. 1984. The origin of anatomically modern humans in western Europe, 51-135. In *The Origins of Modern Humans: A World Survey of the Fossil Evidence*, ed. F. H. Smith and F. Spencer. New York: Alan R. Liss.

Svoboda, J., and K. Simán. 1989. The Middle-Upper Paleolithic transition in southeastern central Europe (Czechoslovakia and Hungary). *Journal of World Prehistory* 3:283-322.

Tattersall, I. 1986. Species recognition in human paleontology. *Journal of Human Evolution* 15:165-75.

Templeton, A. 1992. Human origins and analysis of mitochondrial DNA sequences. *Science* 255:737.

Thorne, A., and M. H. Wolpoff. 1981. Regional continuity in Australasian Pleistocene hominid evolution. *American Journal of Physical Anthropology* 55:337-49.

Thorne, A., and M. Wolpoff. 1992. The multiregional evolution of humans. *Scientific American* 266:76-83.

Trinkaus, E. 1977. A functional interpretation of the axillary border of the Neandertal scapula. *Journal of Human Evolution* 6:231-34.

Trinkaus, E. 1984. Western Asia, 251-93. In *The Origins of Modern Humans: A World Survey of the Fossil Evidence*, ed. F. H. Smith and F. Spencer. New York: Alan R. Liss.

Trinkaus, E. 1992. Cladistics and later Pleistocene human evolution, 1-7. In *Continuity or Replacement: Controversies in* Homo Sapiens *Evolution*, ed. G. Bräuer and F. H. Smith. Rotterdam: A. A. Balkema.

Trinkaus, E., ed. 1989. *The Emergence of Modern Humans: Biocultural Adaptation in the Later Pleistocene*. Cambridge: University of Cambridge Press.
Trinkaus, E., and F. H. Smith. 1985. The fate of the Neandertals, 325-33. In *Ancestors: The Hard Evidence*, ed. E. Delson. New York: Alan R. Liss.
Valladas, H., J.-L. Joron, G. Valladas, B. Arensburg, O. Bar-Yosef, A. Belfer-Cohen, P. Goldberg, H. Laville, L. Meignen, Y. Rak, E. Tchernov, A.-M. Tillier, and B. Vandermeersch. 1987. Thermoluminescence dates for the Neanderthal burial site at Kebara in Israel. *Nature* 330:159-60.
Valladas H., J. Reyss, J. Joron, G. Valladas, O. Bar-Yosef and B. Vandermeersch. 1988. Thermoluminescence dating of Mousterian 'Proto-Cro-Magnon' remains from Israel and the origin of modern man. *Nature* 331:614-16.
Vigilant, L., R. Pennington, H. Harpending, T. Kocher, and A. Wilson. 1989. Mitochondrial DNA sequences in single hairs from a southern African population. *Proceedings of the National Academy of Sciences, USA* 86:9350-54.
Vrba, E. 1980. Evolution, species and fossils: How does life evolve? *South African Journal of Science* 76:61-84.
Vrba, E. 1984. Evolutionary pattern and process in the sister-group Alcelaphini-Aepycerotini (Mammalia: Bovidae), 62-79. In *Living Fossils*, ed. N. Eldredge and S. M. Stanley. New York: Springer-Verlag.
Vrba, E. 1988. Late Pliocene climatic events and hominid evolution, 405-26. In *Evolutionary History of the "Robust" Australopithecines,* ed. F. E. Grine. New York: Aldine de Gruyter.
Wiley, E. O. 1981. *Phylogenetics: The Theory and Practice of Phylogenetic Systematics*. New York: John Wiley and Sons.
Wolpoff, M. H. 1989. Multiregional evolution: The fossil alternative to Eden, 62-108. In *The Human Revolution: Behavioural and Biological Perspectives on the Origins of Modern Humans,* ed. P. Mellars and C. B. Stringer. Edinburgh: University of Edinburgh Press.
Wolpoff, M. H. 1992. Theories of modern human origins, 25-63. In *Continuity or Replacement: Controversies in* Homo Sapiens *Evolution*, ed. G. Bräuer and F. H. Smith. Rotterdam: A. A. Balkema.
Wolpoff, M. H., and R. Caspari. 1990a. Metric analysis of the material from Klasies River Mouth, Republic of South Africa. Abstract. *American Journal of Physical Anthropology* 81:319.
Wolpoff, M. H., and R. Caspari. 1990b. On Middle Paleolithic/Middle Stone Age hominid taxonomy. *Current Anthropology* 31:394-95.
Wolpoff, M. H., F. H. Smith, M. Malez, J. Radovicíc, and D. Rukavina. 1981. Upper Pleistocene hominid remains from Vindija cave, Croatia, Yugoslavia. *American Journal of Physical Anthropology* 54:499-545.
Wolpoff, M. H., and A. Thorne. 1991. The case against Eve. *New Scientist* 22:37-41.
Wolpoff, M. H., X. Wu, and A. Thorne. 1984. Modern *Homo sapiens* origins: A general theory of hominid evolution involving the fossil evidence from East Asia, 411-84. In *The Origins of Modern Humans: A World Survey of the Fossil Evidence*, ed. F. H. Smith and F. Spencer. New York: Alan R. Liss.

Part V

Synopsis and Prospectus

Why is our book important? The origins and histories (we use the plural advisedly) of humans are of paramount importance to all. We want to know who we are, how we arrived here, and how we became what we are and by what stages. This fundamental biological question of who we are is based in great measure on the question of human origin, and the answer, or an assumed answer, colors all our knowledge, and all our interpretations, of the cosmos. We ask questions only when we believe that we have alternative answers or solutions, when we assume the range of answers into which the correct solutions fit.

Clark Howell has not been, up to this point, actively involved in the problems of the origins of modern humans. In his chapter he provides his valuable views and conclusions on the current debate. As with all other of Clark Howell's work, this is an edifying middle-of-the-road view. It will undoubtedly become a great source of information for all premises on paleoclimatology, geomorphology, and vertebrate paleontology. So what do we have, know, or think that we know? We believe that Clark Howell in his chapter provides the best answer, and, therefore, we will let him speak.

Chapter **12**

A Chronostratigraphic and Taxonomic Framework of the Origins of Modern Humans

F. CLARK HOWELL

BACKGROUND AND ROOTS OF *HOMO*

Earlier (Pliocene) hominid evolutionary history was evidently wholly confined to the African continent, so far as all available evidence demonstrates. The origins of *Australopithecus*, both phylogenetically and temporally, still remain obscure, but the complexity of its differentiation is manifest in the now very substantial fossil record in parts of eastern and southern sub-Saharan Africa. Several spatially and temporally delimited speciation events are appropriately recognized, particularly, but not only, on the basis of craniodental morphology.

The emergence of *Homo* is seemingly also an African and, now undoubtedly, Pliocene event. Thus, the roots ultimately of the source of *Homo sapiens* are fundamentally African. Over the past two decades an ever-increasing and convincing body of evidence has come to suggest both a more ancient origin and an earlier diversification within *Homo* than had once been envisaged. That evidence can now be reasonably subsumed within a polyspecific framework, encompassing both successive and sympatric-contemporaneous taxa–*Homo rudolfensis, Homo habilis* and *Homo ergaster* (ex. *H.* cf./aff. *erectus*)–as recently formalized by Wood (1991, 1992; also in Howell 1991). The respective age ranges and principal localities are: 1.9-1.85 (1.7) ma (Turkana Basin, and Malawi Basin of presumably equivalent age); 1.9-1.85 ma (Turkana Basin), and ~1.85-1.4 ma (Olduvai Basin), and of un-

F. CLARK HOWELL • Department of Anthropology, University of California, Berkeley, CA 94720.
Origins of Anatomically Modern Humans, edited by Matthew H. Nitecki and Doris V. Nitecki. Plenum Press, New York, 1994.

certain age in Sterkfontein, Member 5; and 1.85-1.5 ma (Turkana Basin, and southern Ethiopian Rift; an age within this span is also probable for its occurrence at Swartkrans, Member 1). The former taxa are largely Pliocene, and the last is largely earliest Pleistocene in age.[1] The first taxon may extend to 2.4 ma (Chemeron, Baringo Basin) (Hill et al. 1992); and elsewhere, in the lower Omo Basin, dental remains (Members E, F, Shungura Formation) attributed to aff. *Homo* sp. indet. have an age of > 2.34 ma (Howell, Haesaerts & de Heinzelin 1987).

Archeological occurrences in immediate association, or in lithostratigraphically contiguous or proximate circumstances, occur with all these taxa, as well as with two derived "robust" australopithecine (*Paranthropus*) taxa at particular localities, and in fact as early as ~2.4 ma. The attribution, therefore, of culturally patterned behavior to one such taxon, to the exclusion of another, must now be considered from another and quite different perspective. Furthermore, the enhanced sample of artifactual occurrences in the upper Pliocene, now becoming known from localities in rift basin settings in both northern Kenya and in several parts of Ethiopia, afford another, and indeed, requisite basis for assessment of the Oldowan Industrial Complex as manifest in the lower reaches of Olduvai during the eponymous (N) subchron.

RANGE EXTENSION AND DISPERSAL INTO EURASIA

As hominid remains are unknown in the Eurasian Pliocene (termination ~1.8 ma), from which there are numerous localities that yield occurrences of very substantial and diverse vertebrate fossil assemblages, the African hominid record affords the essential perspective for ascertaining the overall morphological pattern, or grade of organization, characteristic for the subsequent hominid dispersal into those continents. That dispersal, or substantive range extension, occurred at least significantly early in the Lower Pleistocene to judge from the extensive and successive lithic assemblages from littoral/limnic situations at 'Ubeidiya (Israel) (Bar-Yosef and Goren-Inbar 1993; also Goren 1981), and which largely mirror, on techno-typological criteria, the Developed Oldowan/Early Acheulean of Olduvai. The best age estimate for this occurrence, largely on biostratigraphic grounds, is 1.3-1.4 ma (Tchernov 1986, 1987, 1988; also 1992a, 1992b). The several hominid fragments (cranial vault and I_2 and M^3), most of which were found in situ in excavations, are unfortunately scarcely very informative as to possible phylogenetic affinities, other than exclusionary attribution to *Homo* sp. indet. (Tobias 1966).

In this regard the recent recovery (Dzaparidze et al. 1989 [1992]) of hominid remains, with archeological and substantial faunal associations, at

Dmanisi (southern Georgia) is particularly welcome and of major paleoanthropological significance. The underlying lava falls within the Olduvai (N) event on the basis of polarity and radiometric age measurement. The vertebrate fauna is of upper Villafranchian aspect. In contrast to 'Ubeidiya, in which some of the same taxa also occur but in conjunction with demonstrably Ethiopian provincial species, this is a predominantly palearctic fauna. An age probably in excess of that attributed to 'Ubeidiya is thus most likely, on magnetostratigraphic and geochronological grounds. This occurrence now ranks as the oldest, best documented hominid occurrence in Eurasia. The mandible, with permanent dentition, appears referable to *H. ergaster*. The associated artifactual assemblage (with which is associated broken, incised and worked long bones and antler) bears similarities both to the Developed Oldowan industry and, in some respects as well, to the Karari industry of eastern Africa.

For now I remain rather skeptical of very ancient, indeed Pliocene, hominid occupation elsewhere in Europe in spite of some strong assertions to the contrary (E. Bonifay 1989, 1991; Bonifay, Consigny & Liabent 1989). However, there is increasingly strong evidence from the Massif Central, with upper Villafranchian faunal associations, particularly from localities in Haute-Loire including Ceyssaguet (~1.2 ma) (M.-F. Bonifay 1986, 1991) and Nolhac-Biard (~1.3-1.4 ma) (Bracco 1991), for a post-Olduvai (N) subchron hominid presence (cf. Rolland 1992). And, it cannot now be excluded that such evidence will be forthcoming, if not already in hand, from southern Spain (Aguirre 1991) and, perhaps, as well in northern and central Italy (cf. Peretto 1991, 1992). The hominids themselves have until now still eluded discovery.

MALAYSIAN REGION

In Asia no hominid occurrences as yet demonstrate surely such an early Lower Pleistocene antiquity. There does appear to be relatively close congruence in age between ancient hominid occurrences in both continental eastern Asia and in peninsular southeastern Asia on Sundaland, one of the world's most extensive continental shelves, (generally shallower than 100 m, and often only 40-60 m). Substantially enhanced understanding of late Cenozoic stratigraphy, tephrochronology, magnetostratigraphy, and geochronology on Java is a consequence of major international, interdisciplinary efforts by a joint Indonesian-Japanese research program (in Watanabe and Kadar 1985), coupled with important revisions of biostratigraphy by Sondaar (1984), de Vos et al. (1982), and associates. Three principal efforts (by Sémah 1986, by Shimuzu et al., and by Yokoyama and Koizumi 1989) have been made to unravel the paleomagnetic stratigraphy at various Javan localities and to effect correlation with the geomagnetic reversal time scale (GRTS, table 1). These efforts have not afforded fully commensurate results; but, whereas there are sometimes differences in interpretation and attribution to GRTS events, there is a substantial measure of congruence in overall polarity results. Marine conditions persisted over much of the area through the Pliocene and, in some parts of south central and eastern Java, into the (Lower) Pleistocene. The (upper) marine units, including the Puren (Kalibeng) Formation, record the Olduvai (N) subchron, as apparently also does the (unconformably overlying) Pucangan Formation at Sangiran. There is a single, imprecise K/Ar age measurement (2.0 ± 0.6 ma) on an andesitic pyroclastic flow (lahar) in the latter formation, neither confirmatory nor incompatible with the age of that event (although almost exactly concordant with an age estimate based on marine diatoms T.5 and 6) have confirmatory fission track (F/T) ages of 1.49 and 1.51 ma in a reversed sedimentary sequence. Much of that formation comprises dark fresh-water (palustral) clays and silts with eleven successive interbedded tephra, as well as evidence of one lower and a higher triplet-cluster of marine (lagoonal) incursions (with marine fossils and diatoms) (Yokoyama and Koizumi 1989). Above the Lower Lahar the formation has largely R polarity until the uppermost two tuffs (T.10 and 11) in which a R-N-R sequence occurs. This corresponds to the Jaramillo (N) subchron (T.11 has an F/T age of 1.16 ± 0.24 ma) and the subsequent end-Matuyama reversed interval.

The Brunhes-Matuyama boundary is recorded, at some still ill-fixed horizon, within the lower extent of the succedent Babang (Kabuh) Formation, and certainly above the well-defined Grenzbank Zone (a calcareous conglomeratic sandstone). The boundary apparently occurs above the Lower Tuff, it and the underlying fluviatile deposits exhibiting R NMR. F/T ages of the Middle Tuff (0.78 ma) and another above it (0.71 ma) are congruent, as are tektite occurrences, with an early Brunhes (N) chron age, although both such tuffs are sometimes reversed, but in other sections are of normal polarity throughout.

The vertebrate fauna is typified by that of Kedung Brubus (which locality yielded the first early hominid mandible fragment KB-1 and a partial femur). It is quite diverse (at least 21 spp.), balanced in composition, and reflects a major influx of mainland-derived species via a substantial continental connection. A diversity of hominid cranial remains (S. 3, 10, 11, 12, 13, 14, 15, 17, 18, 19, 20, 21, 23, 34, etc.) are known from within the Babang (Kabuh) Formation at Sangiran and from several horizons, all attributable to *H. erectus*, and about which there is reasonable consensus among those familiar with the sample. Lithic artifacts are now definitely documented, in stratigraphic context, at Ngebung (NW Sangiran). They occur in the basal unit of

Table 1. Chrons and subchrons of GRTS since the Gauss (N) chron; (subchrons after Champion and Lanphere 1988).

CHRON	APPROXIMATE AGE (ma)
BRUNHES (N)	
Laschamp sc	0.042
Blake sc	0.128 ± 33
Jamaica sc	0.182 - 0.170 ± 33
Levantine sc	(0.310 - 0.280)
Biwa sc	(0.400 - 0.380)
Emperor sc	(0.460 - 0.450)
Big Lost sc	0.565 ± 14
Delta sc	(0.640 - 0.630)
MATUYAMA (R)	*boundary* = ~ 0.78
Kamikatsura sc	~ 0.850
Jaramillo sc	~ 1.02 - 0.91
Cobb Mountain sc	~ 1.14 - 1.10
Olduvai sc	~ 2.0 - 1.75
Réunion 1	2.04 - 2.01
Réunion 2	2.14 - 2.12
GAUSS (N)	*boundary* = ~ 2.6

a series of fluvial and tuffaceous sediments which overlie a thick deposit of heavily weathered clayey tuff above the Grenzbank (Sémah et al. 1992).

Two pre-Jaramillo vertebrate biozones, the earlier Satir (only 3 species) and the later Ci Saat (more diversified, 7 species), as best known at Bumiayu (lower Kali Glagah Formation), are each absolutely or relatively impoverished and thus unbalanced faunas, and markedly to somewhat insular in aspect. Hominids are still unknown in these or comparable-aged deposits elsewhere (see below). However, hominids (e.g., S.*22* mandible) are definitely known in the uppermost Pucangan Formation (below T.11), about 1.1 ma, and thus bear witness to more enhanced continental connection around that time. The succedent Trinil H.K. fauna, now equated broadly with the Sangiran Grenzbank horizon, is substantially more balanced in composition (at least 18 spp.), but contains endemics suggestive of some isolation effects. Hominids are known in substantial numbers from this interval, ~1.0 ma, both at Trinil (type locality, T.1, calvaria, etc.) and especially at Sangiran (e.g., S, *1*$_b$, 2, 4, *5, 6, 8 ,9* to mention only the more securely provenanced examples).

It is well known that several hominid taxa have been proposed, often inconsistently, by various investigators for the pre-Kabuh Formation hominids, and also that this taxonomy is unduly and unwarrantedly complex. LeGros Clark (1964) was among the first to stress that "the whole problem of *H. erectus* in Java has unfortunately been much confused by the multiplicity of the taxonomic terms which have been applied to the various remains." He considered that it "may readily be admitted that the available fossil material from Java is not yet adequate to decide finally whether there was more than one genus, or more than one species, of hominid living in Java during the Early Pleistocene," and concluded "there is at present no really convincing morphological basis for the recognition of more than one species, *H. erectus*," although he recognized "that there may well have been more than one sub-species or 'race'." Subsequently, investigators who have followed this perspective and sought to "lump" all such Javan hominids in a single taxon (= *H. erectus)* include Lovejoy (1970), Campbell (1973), Howells (1980), Kramer (1984, 1989, in press; Kramer and Konigsberg in press), Pope and Cronin (1984), Wolpoff (1975, 1980), and Zhang (1984, 1985). A contrary, more "splitting" position, but with varied attributions, has been espoused especially by von Koenigswald (1950, 1953, 1954a, 1954b, 1957a, 1957b, 1968, 1973a, 1973b, 1980), and by Franzen (1985a, 1985b), Howell (1978), Jacob (1981), Krantz (1975), Orban-Segenbarth and Procureur (1983), Sartono (1976, 1982), Tobias and von Koenigswald (1964), and Tyler (1991, 1992).

I believe, as before, that the former position is, in fact, wrong. I consider that the available evidence, such as it is, merits reevaluation from another perspective, and one which includes consideration of demonstrably earlier hominid taxa now well known in sub-Saharan Africa. Sometimes claims have been made for the potential affinity of some Javan specimens (e.g., S.6, 8 = "*Meganthropus*"; and S.5, 9 = "*P. dubius*") to the African genus *Australopithecus.* However, neither known cranial vault morphology (e.g., S.4 and S.27, 31 and unnumbered fragments), gnathic morphology, nor that of the dentition favor such an attribution (see also Kramer 1989, in press, in this regard). Thus, it is within the genus *Homo* that comparisons ought to be effected and phylogenetic affinities sought.

There are unquestionably several distinct morphotypes within this sample; they differ not merely in respect to size, but in respect to certain proportions and especially structure. One, represented by S.22 (ex-mandible F, a well-preserved specimen, and its dentition), is in size and morphology distinctly comparable to *H. erectus* examples (of younger geologic age). This is among the oldest, adequately provenienced hominid specimens from Sangiran, and thus demonstrates an early appearance of this morphotype (and species) in Sundaland, at a time perhaps even broadly equivalent to the emplacement of Ci Saat faunal elements (apparently *after* the Olduvai (N) subchron).[2] This taxon, of course, persists into the Grenzbank at Sangiran (S.1b, 4) and equivalently at Trinil (Tr.1, etc.) and is subsequently known, in quantity, from the Babang (Kabuh) Formation at Sangiran (see above).

Another (or other) morphotype(s) appears discernible among specimens recovered from the Grenzbank conglomerate and, apparently, also the underlying, uppermost Pucangan palustral clays. These include mandible (S.5, 6, 8, 9) and cranial elements (S.27, 31, and perhaps 26) as well as, probably, some of a variety of largely unprovenienced isolated teeth, and to which species taxa of *palaeojavanicus* and *dubius* have been employed. The mandibular dentition differs in size and cheek tooth morphology from *H. erectus* (*s.str.*), and bears some general, and even particular, resemblances to *H. rudolfensis* and to *H. habilis* of eastern Africa. Especially distinctive is exceptional mandibular corpus robusticity, including that of the symphysis and which may effect the basal (S.6) as well as the subalveolar portion (superior transverse torus); the genioglossal fossa is variably expressed. The mental foramen is large and relatively highly situated. The canine may be large, or more reduced (S.9, alveolus of S.5), and M_3 may be relatively reduced within the large cheek tooth series. Mandibular tooth rows are rather convergent (S.9). Elements of the dentition recovered from Chinese apothecary shops (and attributed to *Hemanthropus peii* von Koenigswald), and maybe several others from a cave in Jianshi district, Hupei (and associated there with *Gigantopithecus blacki*), perhaps represent a comparable continental taxon. A derived morphology is also apparent in the still ill-known calvarium (esp. S.31), said to derive from the marine Kalibeng Formation, in which there is enhanced vault bone robusticity, thick mastoid, massive angular torus, and extensive confluent, markedly elevated and thickened occipital torus with marked supratoral depression. Another, a crushed cranium (S.27), including dentition, apparently confirms this distinctive morphology. The substantially derived morphological pattern might be hypothesized in part as a consequence of isolation effects in an insular/peninsular situation in this southeast Asian setting; it is not known to persist after ~1 ma when *H. erectus* is the sole known hominid species.

Finally, it should be noted that there are several problems, both in respect to age (both absolute and relative) and species attribution which attend the infant hominid calvarium (Perning 1) recovered (1936) at Sumbertengah near Perning in eastern Java. The specimen was given provisional (vorläufig) attribution to *H. modjokertensis* von Koenigswald, without either adequate differential diagnosis or differentiation from *H. erectus*, comparable-aged calvarial elements of which are still unknown. Thus, this would appear to constitute a *nomen nudum* under the rules (ICZN 1985), and the attribution of that nomen also to Sangiran specimens S. 4 (and attributed S.1b) would similarly appear both ill-considered and inappropriate. The provenance of the specimen and of a substantial (14-16 spp.) vertebrate assemblage is reportedly not in doubt, being derived from a fluviatile sandstone clearly interstratified between molluscan horizons (II and III) in a substantial fluviatile succession of interbedded sandstones, conglomeratic and tuffaceous deposits (and which overlie marine dark clays) (Cosijn 1931, 1932). This succession has been attributed, originally, to the Pucangan Formation (its volcanic facies), although this view has been subsequently questioned and attribution to the Kabuh Formation proposed (Sartono et al. 1981). Two varying interpretations have been afforded paleomagnetic results, including recognition (or not) of the Olduvai (N) subchron (in the upper marine sediments), and attribution of normal NRM within the overlying fluviatile series to either the Jaramillo (N) subchron (Hyodo, Sunata & Susanto 1992) or to the Brunhes/Matuyama boundary (Sémah 1983, 1986). The vertebrate assemblage, with substantial mainland influence, shares resemblances with both the Trinil H.K. and Kedung Brubus local faunas, but it is not wholly adequate to resolve the issue. However, neither attribution would necessarily support a reported age by K/Ar of 1.9 ± 0.4 ma (pumices some 50 m *below* the fossil horizon) (Jacob and Curtis 1971), or even that recently obtained (by SCLF, ~1.75 ma), again for a pumice tuff underlying the hominid/faunal level (G. H. Curtis and C. C. Swisher III, pers. comm.). Such ages would support the hypothesis of land mammal dispersal, including hominids, into Indonesia at a time substantially earlier than is currently envisioned.

A continental source has long been recognized for the past and now-insular vertebrate faunas (including Hominidae) of Sundaland. At one time more-or-less separate, and temporally successive, dispersals from Siva-Malayan (south Asian) and Sino-Malayan (east Asian) sources were proposed. With a substantially enhanced late Cenozoic faunal record in southern China and important refinements in the geochronological placement of south Asian (Siwaliks) faunas, this issue is now afforded a new perspective. Moreover, the continental and Javan occurrences can be linked through a combination of GRTS placement and isotopic age determination. Unfortunately, the small mammal component is very poorly known in the Javan record, and this important data base is thus largely unavailable.

The oldest Javan occurrences (Satir, Ci Saat) are of later Matuyama, post-Olduvai (N) subchron age. They are thus broadly coeval, in part, with late Upper Siwaliks "Pinjor" provincial faunas which span the ~2.5-1.5 ma interval. These are known from both historical and more recent, adequately provenanced collections from Pakistan (Potwar Plateau, including Campbellpore Basin and Pabbi Hills) (de Terra and Teilhard de Chardin 1936) and northwestern India (Haritalyangar and Chandigarh districts). The Javan occurrences lack primates (including hominids), rodents (except perhaps *Rhizomys)*, pholidots, lagomorphs, and perissodactyls (except the chalicothere *Nestoritherium*, a rare "Pinjor" species, also present in Yunnan). Only three carnivores are certainly or probably represented: a sabertooth (*Megantereon*) shared with both Upper Siwaliks and Yunnan; a tiger, possibly only shared with Yunnan; and an otter with southeast Asian affinities and no definite continental documentation. There are three proboscideans: "*Mastodon*," a *Stegodon*, both of which have southeast Asian counterparts, and *Elephas (P.) planifrons* of Pliocene age from the (upper) Siwaliks, appearing ~3 ma and which persists to perhaps 1.5 ma, the last age possibly comparable to its earliest documentation in Java.

Artiodactyls are a more diverse order in these older faunas. An anthrocothere (*Merycopotamus*) has a LAD in the upper Siwaliks in the late Gauss chron. A single suid (*Sus*) may be represented with some counterpart in the "Pinjor" faunal complex. One or several hexaprotodont hippo taxa in the Javan localities occur in both Upper Siwaliks faunal complexes, and *Hippopotamus sivalensis* persists there into the late Gauss chron. As many as five cervid species, of both *Axis* and *Rusa* affinities, are reported in Java and several are definitely parallel in both the "Pinjor" faunal complex and at Yuanmou (Yunnan); and at the latter, the muntjac (*Muntiacus*) is also represented. There appears to be only a single bovid, *Duboisia*, perhaps with two species, and which is a derived boselaphine, exhibiting divergences from both extant (*Boselaphus, Tetracerus*) and extinct (*Selenoportax, Pachyportax*) forms known from the Upper Siwaliks series.

About a million years ago or so and broadly coincident with the Jaramillo (N) subchron, there is evidence of substantial change in the Javan mammal fauna. The most important and diverse assemblages are those from Sangiran-Grenzbank (N = 30) and from Trinil H.K. (N = 20); the latter, in addition to mammals, also includes four species of birds, all migrants from southern Asia or farther north. Endemism is apparent in *Stegodon*, hexaprotodont hippo, most cervids, and a boselaphine antelope. Otherwise, the fauna is substantially more balanced, with new additions of primates (langur, macaque, probably orang, and fairly abundant hominid(s); rodents (porcupine, murids); rabbit(s); canids (extinct cuonines), an ursid, giant hyaenid (*Pachycrocuta*), an additional sabertooth (*Homotherium*), and several felines (bengal cat, clouded leopard) in addition to tiger among the carnivores. Ungulates are more diverse and include two one-horned rhinos, a tapir, large and small *Sus* species, a tragulid, and three bovines (Groeneveldt's wild ox, Asiatic buffalo, and banteng). The absence of a well-documented sub-Himalayan fossil record of this time range hinders comparisons with continental southern Asia; the Pabbi Hills (northern Pakistan) fluvial molasse sedimentation (+1000 m thickness) evidently samples the post-Olduvai/pre-Jaramillo interval (Keller et al. 1977; Jenkinson et al. 1989), and affords a substantial, fairly diverse vertebrate fauna, unfortunately as yet unstudied or unpublished. A number of fossiliferous karstic infillings in southern China (Yunnan, Kwangsi, Sichuan) afford diverse faunas of presumed Lower to Middle Pleistocene age, and in stratigraphic contexts often difficult to decipher and whose age relationships are largely estimated (seriated) through comparative biostratigraphy. And, there is still a real dearth of useful, well-determined paleontological collections from well-provenienced late Cenozoic occurrences in peninsular southeast Asia. Thus, it is often difficult to ascertain and differentiate past biogeographic affinities within the general Malaysian Faunal Region. There are, however, some specific links with the (upper) Siwaliks as there are also with southern China.

Further slight changes in faunal composition are evidenced around the time of the Lower/Middle Pleistocene boundary, that is at Matuyama/ Brunhes (R-N) boundary, just less than 0.8 ma. The major vertebrate localities are those of Kedung Brubus (Kendeng hills) (N = 19) and Sangiran (Babang Formation) (N = 22). Noteworthy additions are a very large pangolin (scaly anteater), true *Cuon*, spotted hyena (*Crocuta crocuta)*, mostly suggestive of southeast Asia/China derivation, and an advanced proboscidean (*Elephas hysudrindicus*) with Indian affinities. Various other ungulate taxa, including a boselaphine and bovines persist as do (rare) cervids, suids, hippo, *Stegodon*, both rhinos, tapir, and some earlier known carnivores.

The hominid population of this interval is *Homo erectus erectus*, best known from Sangiran, and with substantial overall resemblances or affinities with *H. e. pekinensis* of continental eastern Asia. The succedent hominid populations of late Middle/earliest Upper Pleistocene age, known from Ngandong (Solo River), and perhaps (perhaps not) including the Sambungmachan (Solo River) cranium, reveal consistent, overall morphometric affinities with their antecedents, and are appropriately attributed to *H. e. soloensis*. There are no early representatives of *H. sapiens* and this species must be considered an immigrant, along with a substantial diversity of other mammal taxa in the late Pleistocene.

It is worthy of note that *H. e. soloensis* is unknown elsewhere in the Malaysian faunal region. Thus, it may best be envisaged as a terminal, presumably insularly derived product of the *H. erectus* clade. As many as 18, and as few as 11, mammalian taxa have been reported from Ngandong, with as many as 4 others from other localities (Kuwukulon, Watualang, Pandejan) of the Notopoero (or Pohjajar) Formation attributed to this faunal zone. However evaluated, these taxa are almost exclusively continuations of a persistent, formerly immigrant fauna, with only a single new species of canid (*Cuon*), and two differentiated subspecies (cf. tiger and *Sus*). Thus, there is no evidence to indicate any substantial mainland connection and enhanced faunal exchange. The Ngandong hominid sample, and an often well-preserved large mammal fauna, derived from the basal sediments of a High Terrace (ca. 20 m) remnant west of an outbend of the Solo and adjacent to a small affluent stream. This drainage system postdates the mid-Pleistocene sedimentary regime, and includes the Notopoero fluvio-volcanic lahars from the developing Lawu volcano to the south (Bartstra 1977, 1987; Bartstra, Soegondho & van der Wijk 1988; also Sartono 1976). Pumices, in fluvial context, from the latter have an age (F/T) of 0.250 ± 0.7 ma, indicating a late mid-Pleistocene age. Solo High Terrace accumulations are demonstrably younger as they may overlie, in erosionally unconformable position, such older formations; at Ngandong the former rests on a Neogene marine marl. Efforts to date terrace bone radiometrically (U-series) afford ages wholly within the middle to earlier Upper Pleistocene. Thus, the most appropriate age estimate for the Ngandong fauna (and hominid sample) is probably earlier Upper Pleistocene (ca. 85-115 ka).

In a comparable, and perhaps also slightly older, Pleistocene context on mainland southern Asia are evidences of another distinctive hominid. Specimens derive from the isolated Shizi Hill, a karst tower fissure fill near Maba, Qujiang county (Guangdong) (24°50'N, 113°30'E), and from near Hathnora, near Hoshangabad, in the middle reaches of the Narmada Valley (Madhya Pradesh, India) (22°52'N, 77°53'E) from a basal conglomerate (apparently Conglomerate-1 of Agrawal, Kotlia, and Kusumgar 1988; see also Badam 1989) overlying channel sands of an ancient terrace fill. This is a part of the Lower Group of the Older Alluvium (of De Terra and Paterson). The Maba occurrence, found during fertilizer exploitation by local farmers, is associated with an advanced *Stegodon-Ailuropoda* fauna (N = 25), but only several lithic artifacts. A single U-series age determination (129 +11/−10 ka, on teeth) suggests an age near the Middle/Upper Pleistocene boundary. The Hathnora occurrence is in a geomorphological context of inferred comparable Pleistocene antiquity (Badam 1989; also H. de Lumley and A. Sonakia 1985), and one which yields (as elsewhere in the basin) a diverse large mammal fauna and lithic artifacts of developed Acheulean aspect.

The Maba specimen comprises the frontoparietal portion and right orbital area of a calvarial calotte. The Hathnora specimen comprises most of the right half, including the cranial base and orbit, and partial left frontoparietal area of a calvarium. Initial (Wu and Peng 1959) and subsequent (Wu and Wu 1985) descriptions of the Maba individual noted significant deviations from the *H. erectus* morphotype, and in this regard derived features indicative of an "advanced" grade (often termed "archaic *H. sapiens*" in more recent parlance). Although attributed to *H. erectus* by M.-A. de Lumley and Sonakia (1985), their description of the Hathnora calvarium, in fact, indicated the presence of features not congruent with such attribution. Recent morphometric comparative analysis (Kennedy et al. 1991; Kennedy and Chiment 1991) sets out more explicitly and fully the melange of primitive, derived, and unique features exhibited by Hathnora, the fundamental divergences from "*erectus*" condition, and the derived characters which approximate the "archaic *sapiens*" (*s.l.*) condition. Strangely, no direct analysis or comparison has been effected *between* the Maba and Hathnora calvaria, which are fundamentally similar in total morphological pattern. This pattern diverges from both *H. erectus* (including its several nominate subspecies) and from *H. sapiens (s.str.*, and specifically the Liujiang [Guangxi] cranium of uncertain, but surely Late Pleistocene age), and as such indicates requisite phylogenetic reevaluation and taxonomic reassessment is surely warranted.

EASTERN (ORIENTAL) PALEARCTIC REGION

Hominids are not definitely known from the Eastern (Oriental) Palearctic Region prior to the Lower Pleistocene. There is no documentation of hominids, on either paleontological or archeological evidence, in the region within the Olduvai (N) subchron of the Matuyama (R) chron (GRTS). In fact the evidence remains scant even within the post-Olduvai interval of that chron. Fortunately, through extensive studies of the lithostratigraphy and loess/paleosoil sequence of the extensive Loess Plateau (some 320,000 km^2) of central/northern China, coupled with attendant paleomagnetic analyses of those and other critical sections in certain basins (e.g., Yushê of western Shaanxi, Nihewan of northern Hebei) having substantial vertebrate paleontological resources, there is now the basis for both correlation with marine isotopic events (Kukla et al. 1990; Kukla 1987; Kukla and An 1989) and a temporally defined, rather than wholly relative biostratigraphy (Tedford et al. 1991; Qiu 1990; and Li, Wu & Qiu 1984).

In northern Yunnan (Yuanmou county), the Yuanmou Basin (25°36'N, 102°E) in the southern headwaters of the Yangtze River preserves a remarkably long succession of late Cenozoic sedimentation, often rich in vertebrates of both Pliocene (Xiophe, Longchuan formations) and Pleistocene (Yuanmou Formation) age. Various and successive efforts there, by a diversity of investigators, have advanced understanding of many aspects of the tectonic setting, geological succession, paleomagnetics, paleoenvironmental conditions, biostratigraphy and aspects of paleoanthropological interest over nearly seventy years. Various frameworks, hypotheses, and solutions have been proposed, however, further studies are evidently still required.

The Yuanmou Formation, in excess of 670 m thick, comprises a succession of lacustrine, fluvio-lacustrine and largely fluviatile sediments, commonly subdivided into 4 members and 28 beds (numbered upwards) (Member

A is termed the Shangnabang Formation by some investigators, Qian et al. 1991; Zhou and Zhang 1984). This formation and its sedimentation was disrupted by Qinghai-Tibetan tectonism, and subsequent Middle/Upper Pleistocene sedimentation is largely reflected in repetitive terrace formation (Longchuan River). A palynological record and substantial and diverse vertebrate fauna (N = 46) occurs in all members of the formation. Hominid evidence occurs first in Member 4 (N = 37 mammal taxa) and is largely restricted to its bed 25, a fluviatile accumulation, particularly around Shangnabang. Artifacts are few, but have been collected in surface situations indicative of derivation from this or adjacent units; they include cores, flakes, and a few retouched pieces, and the artifactual nature of some is not wholly convincing. Two complete hominid I^1 have been recovered as well as a tibial diaphysis, the latter perhaps from a higher horizon. Much of the incisor crown and root morphology can be duplicated in ZKD-1 sample of *H. erectus*, including pronounced marginal ridges; a median lingual crest is very strongly expressed. This formation surely samples the Gauss (N) chron (Members 1 and 2) and the Matuyama (R) chron (Members 3, 4), but there is no consensus among several investigators (Li et al. 1977; Cheng, Li & Lin 1977; Liu and Ding 1983; Qing 1985) and other commentators of the overall validity, meaning and correlation (with GMPS) of the upper members' reversal chronology. Bed 25 is evidently between two N-intervals in R-imprinted sediments. It is not impossible that the lower (N) represents the Olduvai (N) subchron; on the other hand both might as well represent the Cobb Mountain (N) and Jaramillo (N) subchrons, and a subsequent N-interval (Bed 28) might even represent a final Matuyama N-subchron. The matter cannot now be satisfactorily resolved, and clearly more intensive paleomagnetic study is called for.

Gongwangling (GWL no. 63706) locality (Lantian district, southern Shaanxi; 34°11'N, 109°29'E) affords the oldest well-provenanced hominid fossil occurrence in the region. Much of a calvarium, eroded and somewhat distorted, and maxillae from a level in one (no. 3) of five paleosoil zones in a substantial loess/paleosoil sequence, represent part of a Pliocene to Upper Pleistocene sedimentary succession in this area of the Baho drainage at the northern foot of the Tsinling mountain range (Gu et al. 1986; Zhang et al. 1978). Paleomagnetics (An and Ho 1989) document the Brunhes (N) chron and the Jaramillo (N) subchron of the Matuyama (R) chron, some meters below which the remains occur in a reversed sequence (estimated age > 1.2 ma, considering the proposed revised age of this event). Mammals (N = 20) in association reveal both southern affinities (6 spp.) and resemblances to some degree with both Nihewan (~40%) and even more so with Zhoukoudian 1 (~60%). This fauna has a more subtropical aspect than is reflected in the (limited) pollen spectra of more temperate character. A limited number of lithic artifacts (some 20, especially cores, some flakes and scrapers) are associated in the same fossiliferous horizon, others are from adjacent or broadly comparable horizon(s) in the environs. This calvarium (Wu 1964, 1966) has been attributed to *H. erectus lantianensis,* of which the type is a mandible from Chenjiawo, some 10 km distant, previously suspected and now known to be geologically younger (= early Brunhes (N) chron, > 0.65 ma) than Gongwangling proper. The calvarium, in some features of vault morphology and thickness and in relatively low cranial capacity, resembles some Javan hominids from the Grenzbank horizon at Sangiran, perhaps especially Sangiran H.4, and thus diverges from the Zhoukoudian *H. erectus* sample. Similarly, some features and proportion of the mandibular corpus of Chenjiawo are also less readily matched in the latter sample. It is as yet unclear, due to dearth of specimens in the sample, whether any such distinctions should be afforded taxonomic (and, thus, phylogenetic) distinction, but rather reflect only aspects of populational variation within a polytypic species. Donghecun Cave (Luonan County), in the southern foothills of the Qinling Mountains, yields a fauna with some southern elements, and is considered to be perhaps as old as GWL. It also affords a single hominid right M_1 of *erectus*-like morphology.

The Nihewan Basin, some 200 km^2, is a tectonically-controlled subsidence depression with over 600 meters (documented by cores) of fluvio-lacustrine infilling. The succession encompasses much of the later Cenozoic from the Pliocene through the Quaternary, the paleomagnetic record documenting the Gauss (N), Matuyama (R) and Brunhes (N) chrons and most well-known subchrons. The fossiliferous Nihewan Beds disconformably overlie earlier/mid-Pliocene *Hipparion* clays and gravels, and have been variously subdivided in times past, and are currently termed (upwards) the Nihewan Formation and Xiandukuo Formation, disconformably separated and disconformably overlain by lacustrine and loessic deposits of Middle-Upper Pleistocene age. The Nihewanian local fauna, as traditionally known since its study by Teilhard de Chardin and Piveteau (1930), derives at least from both those formations, and more recent collections from well-provenanced localities have led to refinement in vertebrate biostratigraphy (Chen Maonan et al. 1988).[3] Artifactual occurrences with associated faunal residues are known both from the largely mid-Pleistocene Xiandukuo Formation (at Qingsiyao; also at Maliang and Chenjiawo, the former in higher energy fluviatile sands, and the latter in low energy silts or clays and thus with greater contextual integrity), and in the older Nihewan Formation proper (at Xiaochangliang, the first discovered *in situ* locality, and at Banshan and Donggutuo). The latter locality (40°13'N, 114°40'E), which is in a *pre*-Jaramillo (N) subchron reversed sedimentary context, has been extensively investigated, preserves rich and diverse but quite simple lithic assemblages made on local chert and more rare volcanics. It exhibits some resemblances overall to the African Developed Oldowan, and occurs in scarcely disturbed perifluviatile circumstances, with faunal residues indicative of both hominid (bone breakage, cut marks, etc.) and carnivore activities (Schick et al. 1991).

Homo erectus is consistently present throughout most of the Middle Pleistocene at least in central and northern China. Zhoukoudian Loc. 1 (Hebei) (39°40'N, 115°15'E), in the limestone western hills (Hsishan) southwest of Beijing, constitutes the classic locality for this interval in respect to both paleoanthropological and paleontological documentation (Wu et al. 1985; Jia and Huang 1990). Over 40 meters of infilling, comprising 17 sedimentary beds, span the final Matuyama (R) chron (17-14) and much of the subsequent Brunhes (N) chron (13-1) prior to the Upper Pleistocene (represented by other localities – 3, 4, 15, etc. in the environs). A combination of U-series, ESR, TL, and F/T age assessments (Zhou and Ho 1990) afford a fairly consistent temporal framework for the successive strata, their paleontological (mammals – 97 spp., birds – 62 spp.) remains and palynological records, and the evidences of repeated hominid occurrences, represented by skeletal remains, lithic artifacts and associated residues.

This succession affords evidence (sedimentology, minerals, weathering, brecciation, and travertines) that enables tentative and partial correlation with the well-defined loess/paleosoil sequence of the central China Loess Plateau, and thus to some extent with oxygen isotope events documented in deep-sea cores (perhaps including stages 16 though 6). Fluctuations in pollen frequencies, among which grasses and herbaceous species may vary from 20 to 50%, reflect variations in coniferous and deciduous forest cover, and the presence and varied expansion of grassland over a series of warm to cool temperate, and even colder intervals. More than some 45 hominid individuals are represented by 6 calvaria, 12 calvarial elements/fragments, 15 mandible parts, 157 teeth (73 isolated), and some 14 fragmentary postcrania; at least some 20% are immature individuals. Hominids derive from 7 successive beds, most numerously in 10 and 8 (ca. 460-420 ka) and in 4 and 3 (~300 ka), but uncommonly in beds 11, 7, and 5 (Wu and Lin 1983). Lithic artifacts (a few) occurred from bed 11 upwards, with major densities in 8-9 (1,336), 6 (1,045, lacking hominid skeletal parts), and especially 4-5 (6,651) and 3 (3,484, including a few from 2 and 1). Major circumscribed evidences of fire (ash, carbon, charcoal, charred and calcined bone, and crackled stones) occurred principally in (lower) 3, 4, between 8-9 and in 10. A fair estimate of the overall duration of occupation is in excess of 300 thousand years,[4] between ~575 and ~250 ka. Occupation evidently occurred both on an intermittent and sometimes repetitive and protracted basis, with both natural agencies and the activities of carnivores (hyaenids, canids, felids), particularly (but not only), as in the lowermost infilling and in bed 5, accounting for some faunal accumulations in addition to human activities involving animal resource procurement and processing. There are almost 15,000 provenienced artifacts in the total ZKD-1 assemblage of which about 20% are retouched and utilized pieces (Pei and Zhang 1985). Various changes are evidenced through the successive lithic assemblages in respect to the exploitation and utilization of primary raw material resources, in the differential employment of primary reduction and flaking techniques, in the increased production of flake blanks for further modification and retouch, in the relative frequency of large (core-based) pieces vs. increasingly flake-based tools, in size reduction, in elaboration of retouch on particular flake tool categories, and in the introduction and technological refinement of distinct tool classes and specific categories. Nonetheless, the lithic industry maintains its distinctive nature and individual integrity, with an aspect of "informality" broadly comparable with that evidenced in the African Oldowan/Developed Oldowan Complex, and from which tradition the "Choukoutienian," to use Movius's original terminology, evidently once derived.

Hominid skeletal elements attributed to *H. erectus* have been reported from at least seven other provinces of China, and include more than a dozen localities.

In Shaanxi both cave and open-air situations afford remains. At Chenjiawo (34°14'N, 109°15'E), along a tributary of the Bahe River, a mandible with dentition (Wu Rukang 1964), together with a savanna/wooded steppe faunal assemblage (N = 11, without "southern" elements), came from a loess-paleosoil succession (S6 paleosoil) just above the Brunhes/Matuyama geomagnetic boundary (An and Ho 1989). This is as old (~0.70 ma) or older than any continental representative of the species within the Brunhes (N) chron. Artifacts are known from adjacent and, presumably, penecontemporaneous fluviatile deposits of higher terraces and upland paleosoil situations; these are not unlike other "Choukoutienian" assemblages, except for the (few) additional bifacial and unifacial (tetrahedral) picklike elements.

In central Shandong, Qizian Hill Cave (Yiyuan county) in the Tumen karst area yields fauna in infillings at several localities, and in the Xiaya Cave group, situated some 5-15 m above the valley floor, human remains are also known. These include much of a calvarium, apparently parts of another, and several teeth (Loc. 1), and seven upper teeth and unspecified postcranial parts (Loc. 3). All are reported to have distinctive *erectus* morphological features (Lü et al. 1989).

In Henan, teeth attributed to *H. erectus* are known from two localities. At Xinghuashan (Nanzhao county) (32°25'N, 42°41'E) a substantial fauna (N = +20, with "southern" and northern elements) derive from a Jihe River terrace, and a single hominid P_4 is also known. Elsewhere (Xichuan County) some dozen teeth with reported *erectus* features are known from apothecary collections, but of otherwise undetermined provenience.

In Hubei at least three karstic caves have afforded teeth attributed to *H. erectus*. These include Longgushan (32°58'N, 110°57'E) and Builongdong (32°58'N, 110°45'E) caves, both in Yunxian county, from which 4 and 7 hominid teeth respectively, and substantial faunas (N = +20), but limited artifacts, are known (Wu and Wu 1982). Longgudong Cave (Jianshi county) has yielded three lower molars (another was of apothecary shop origin) in association with a fauna having "southern" elements (Gao 1975; Liu 1985; Gu et al. 1986).

In Anhui, Longtandong Cave (31°45'N, 118°20'E), in Hexian County from hills north of the Yangtze River, yielded a substantial fauna (N = 40, both micro- and large mammals) including a number of "southern" elements, and a largely complete human calvarium, mandible fragment, frontoparietal part of another individual and 10 teeth; as many as four individuals might be present, as there are four left M_2, including that in the mandible fragment (Wu and Dong 1982, 1985; Huang, Fang & Ye 1982). U-series ages between ~145, 180 and 194 ka have been obtained (6 samples) on associated mammal teeth, all suggestive of an age late in the Middle Pleistocene (Chen et al. 1987). This compares with results by the same investigators (Chen and Yuan 1988) for the upper beds (3-1) at ZKD-1 of ~290-220 ka. Although fundamentally consistent with the known *H. erectus* morphological pattern, including endocranial volume (Hexian = 1025 cc; ZKD-1, no. V = 1140 cc, and not the largest in that sample), attention has been directed to some attenuated features shared with Calvaria V, ZKD-1, including lesser frontal constriction, more slender supraorbital torus, thinner vault bones, elevated and superiorly arched temporal squama, reduced occipital torus, and closer approximation of internal and external occipital protuberances. It is, however, unclear as to what extent this constitutes a real and meaningful (mean) shift of phyletic significance of *erectus* morphology rather than an aspect of the expected normal range of individual variation.

In Guizhou the Yanhuidong karstic cave (28°15'N, 106°45'E) (Tongzi County) has afforded a substantial mammal fauna (N + 25, of "southern" aspect), some artifacts, and seven isolated human teeth, all of the upper dentition (Wu Maolin et al. 1983). These have a characteristic *erectus*-like morphology. The artifacts, as far as known, are relatively simple in both technology and limited typological characteristics, and bears some overall resemblances to the now well known multilevel (7 layers) locality of Guanyindong Cave (Li and Wen 1986) in Qianxi County to the southeast, and which has come to represent a comparative base against which other South China archeological assemblages may be evaluated. It also yielded a substantial fauna (N = 23) and over 3000 artifacts of a core and flake industry (eponymously named), technologically simple in respect to flake production (by direct percussion), and having abundant cores (polyhedral, single platform), rather amorphously retouched flakes with steep edges, a small variety of flake tool classes (mostly scrapers, becs, denticulates, etc.) and types, and some core-choppers. Although considered to be of substantial mid-Pleistocene age Guanyindong has unexpectedly young (and inconsistent) U-series ages, between 110-170 ka (l. 5) and 109-122 ka (l. 8). The same investigators (Chen and Yuan 1988) obtained ages of 113 (± 11) and 181 (+ 11/9) ka (for level 4) at Yanhuidong; newer determinations (Shen and Li 1991) afford older ages, greater than ~200 ka (upper capping travertine) and ~240 ka for flowstones related to layers 3 and 4. Such ages are consistent with evidence for *H. erectus* elsewhere in northern/central China in later mid-Pleistocene times.

Homo erectus is a well-defined, clearly delineated Asian hominid species. Its roots evidently lie in the Lower Pleistocene, perhaps even close to the Plio-Pleistocene boundary. There are marked phenetic differences, as noted already by various authors, between mid-Pleistocene human populations distributed in eastern Eurasia and those occurring in western Eurasia. Such distinctions, which are also reflected to a degree in the archeological record, cannot probably be considered as merely clinal, but appear to indicate a discontinuous distribution and attendant weakness, even absence, of gene flow in an east-west latitudinal direction. Paleogeographic and paleoenvironmental circumstances may well have played a forceful role in effecting such disjunctions.

There is some evidence to suggest substantial and, at least, intermittent isolation of eastern Asia from western Eurasia, particularly Europe (west of the Ural range), during the last eight major glacial cycles. This was a consequence of glacial effects in the northern reaches of western central Asia. The evidences and attendant consequences are briefly as follows (see especially Velichko 1984; also Gerasimov and Velichko 1982; Frenzel, Pécsi & Velichko 1992):

Arctic glaciation repeatedly developed over the Barents-Kara Seas, forming an ice sheet of at least 2500 km diameter, and extending over the north Siberian mountains and uplands to about 65°N. Lobes and their moraines of this sheet dammed major northward-flowing Siberian rivers (Irtysh, Ob, Yenisei-Angara), producing the formation of very extensive proglacial lakes (Pur-Mansi, and Yenisei farther east) across western Siberia over at least 1.5 million km^2, between 65° and nearly 50°N (Grosswald 1980). In southern and western central Asia, respectively, the Aral Sea was substantially enlarged (inflow via the Turgai drainage from northern lakes), and the Caspian attained at least 2.5 times its present size, and elevations of some 75 m above modern level. The development of ice-wedge polygons and associated structures testify to permafrost formation as far south as 48°-49°N in the Last (Sartan) Glacial alone, and are indicative of air temperatures some

10°C below present, and glacial ground temperatures of −20° (8°-10°C lower than present). (Even in the early Holocene relict permafrost had a latitudinal breadth of 1200 km in western Siberia, indicating persistence of such ground conditions into an early interglacial-equivalent environment). Glacial vegetation patterns had no close modern analogues and also apparently had a hyperzonal character compared with interglacial (postglacial) latitudinal zonation (and which can be deduced from a well-documented Holocene record). Open periglacial forests (birch, larch, pine) and a complex of tundra and steppic herbaceous species were distributed from eastern Siberia to the margin of the Scandinavian ice sheet, and reaching to 60°N in western Siberia (and 60°-55°N to the west of the Urals); this association was succeeded southward by drier forest-steppe, extending to 56°-57°N in central/western Siberia. Southward, the vast stretches of central Asia between 60°-85°E, including Kazakhstan in large part, and to below 50°N, comprised a periglacial steppe dominated by herbaceous species. Grass/*Artemisia* dry steppe and semidesert predominated, between 47°-40°N, including Uzbekistan and environs.

Although there is increasingly substantial documentation of the Mousterian (often of Levallois facies) industrial complex over central Asia (Tadjikistan, Uzbekistan – where Teshik Tash demonstrates Neandertal presence – parts of Kazakhstan, and parts of southern Siberia, including the Altay) there is an *almost* utter absence of demonstrable hominid presence earlier. In particular, Xinjiang, southern Siberia, Kazakhstan and Uzbekistan are devoid of such traces. It has been suggested that human occupation failed to extend north of 48°N in Early and Middle Pleistocene times, but in fact even 40°N may have been a closer approximation to the limit. There are exceptions and they are briefly discussed because of their singular nature.

Loessic deposits are widespread, discontinuously, across western/central Asia, between 40°-60°N between the Urals and as far as Lake Baikal. Major thick accumulations occur, some 5° on either side of the 40° parallel in Uzbek-Tadjikistan, 65°-80°E. In southern Tadjikistan a loess-paleosoil sequence, some 200 m thick, encompasses essentially the entire span of Pleistocene (Dodonov 1986; Dodonov, Melamed & Nikiforova 1977), with substantial vertebrate fossil occurrences, as at Kuruksay and Lakhuti (Nikiforova and Dodonov 1980; Nikiforova and Vangengheim 1988). There are 9 pedocomplexes (pc) post-dating the Brunhes/Matuyama boundary, with human occupation documented by archeological evidence in pc 5 (N = 6), pc 6 (11), pc 7 (3), pc 8 (1), pc 9 (1), and pc 11/12 (1), and thus indicative of recurrent or persistent habitation through the Middle Pleistocene. Well-documented localities are particularly those of Lakhuti-I, in pc 5, and attributed to $^{18}0$ stage 6 (TL age, 130-150 ka), with almost 1200 artifacts (pebble/flake industry, including Levallois technique), and Karatau-I, in pc 6, and attributed to 18O stage 7 (TL, ~200 ka), with over 500 artifacts, less concentrated in distribution than the former, but equally lacking in other associated organic, etc., residues. The oldest, until now, occurrence is that of Kul'dara, in the Khovaling region (~37°N, 68°E), situated in pc 11/12 in end-Matuyama (R) sediments, and above the documented Jaramillo (N) subchron. The estimated age is about 800 ka. Only 40 artifacts are known, from limited excavation, and distinguished by small size, lack of refined retouch or consistent original blank form derived from cobble reduction (Ranov, Davis & Dodonov, in press).

In Kirgizia, the multichambered cave of Sel'Ungur is situated in the Obisir intermontane valley (1900 m), north of the Amu Dar'ya, and south of Fergana, at 40°N (Islamov 1990; Islamov, Zubov & Kharitonov 1988). It contains 6 meters of infilling, including 5 archeological horizons (in 12 distinguished beds) with diverse macrofauna of ungulates (N = 10), carnivores (N = 6), and rather limited (N = 7 spp.) microfauna. The mammal species all suggest a late-Middle Pleistocene age: U-series dating of capping travertines afford an age of 126 ± 5 ka (early in 18O stage 5), which would suggest occupation occurred over a single climatic cycle earlier, in stage(s) 6/7 most probably. The largely flake-based lithic assemblage (N = 1417, including cores, about half waste, and some 300 retouched pieces) rather resembles overall the comparably-aged Karatau-I assemblage. Human remains are 10 (worn) teeth (of several individuals) and much of a (distal) humerus, largely distinguished by its thick cortical walls and relatively reduced medullary cavity. The available material is hardly adequate for comparative treatment or taxonomic assessment.

A series of localities in both central and northern China afford skeletal evidence of non-*erectus* hominid(s) in upper mid-Pleistocene to initial Upper Pleistocene contexts. Such remains have increasingly been placed in an informal category "archaic *Homo sapiens*." This categorization is taxonomically both invalid and otherwise inappropriate, since its basis is essentially an effort to distinguish a collective grade and thus, in my view, obscures rather than clarifies phylogenetic (including cladistic) relationships. The most extensive descriptive and evaluative treatment of this body of material is that recently published by Pope (1992; also 1991); useful earlier overviews are afforded by Wu Xinzhi (1988, 1989) and Wu Xinzhi and Wu Maolin (1985). Thus, only a very abbreviated overview is offered here. There are reportedly 11 such localities (in 7 provinces), 7 of which are caves (including fissure fills and rock shelters), and 4 of which are open sites (peririverine, or loess-paleosoil situations). The available human sample is not large, comprising 4 crania, all of which require some reconstruction (and two substantially distorted), 3 small maxillary fragments (and some dentition), 1 small mandible fragment, 19 isolated/associated upper (permanent) teeth (and a partial dentition in a cranium), four lower permanent teeth, and some postcranials (with its cranium). Moreover, much of this material (except for Xujiayao) has received only limited and preliminary description and analysis, and, accordingly, as yet lacks any rigorous comparative treatment as well.

Some characteristics, as recorded so far, of cranial morphology of these specimens are set out in tables 2 and 3. There are clear gaps, and important ones to be sure in the available documentation, and these, as well as the often subjective nature of observations, make this a very provisional and indeed quite imperfect assessment. With such caveats in mind even such a limited data base is informative. Most importantly, there is a substantial range of variation in a variety of cranial features such that this sample should in no way be considered as constituting either a single population or even morphological "type" (grade). Attribution to "archaic" *Homo sapiens* is both presumptive and misleading as the latter specific morphological pattern is not replicated or, in many respects, approximated. This suggests, not surprisingly, that several and distinctive populations are sampled, both over a substantial geographic area (still greater if the variability evidenced in the Maba and Hathnora calvaria is included into consideration), and as well over a not insubstantial temporal span. The available, admittedly incomplete, provisional, and sometimes unconfirmed results of isotopic age determination would suggest a span on the order of 200 x 10^3 years. In order of *decreasing* age the most important specimens may be considered as (?) Quyuanhekou (Yunxian); Tianshuigou (Dali) and Walongdong (Changyang); Yanshan (Chaoxian) and Dingcun; (?) Jinniushan (Yingkou); and Xujiayao (Datong). The morphological structure evidenced overall is divergent from that exhibited by (Asian) *H. erectus*, such that distinction from that species is mandatory and valid (and attribution to *sapiens*, correspondingly, is neither). However, it is not to be too readily rejected that these constitute samples of successive and phylogenetically closely related populations. There is consistent divergence from penecontemporaneous European human populations, so that attribution to *Homo neanderthalensis* is likewise unwarranted. Resemblances to, and potential affinities with, other western Asian human samples are briefly examined in due course.

These specimens evince (? progressive) increase in endocranial volumes, and a set of transformations in overall cranial vault morphology is reflected in altered proportions and conformations of its individual components. Among the latter are the length and reduced curvature of the parietal, the elevated and curved temporal squama, the tendency toward more superiorly broadened transverse (para-coronal) vault contour, the generally enhanced occipital curvature (reduced angulation), altered proportions of its upper and lowered scales, and both transverse and some vertical increase in frontal size and its squamal disposition. There is variable, but consistent,

Table 2. Some non-*Homo erectus* (archaic) hominid occurrences in China.

Locality	Site type	Isotopic Dating	Faunal; biostratigraphy	Artifactual residues	Hominid remains
ZHEZIANG					
Wuguidong, Jiande co. (29°20'N, 119°05'E)	karstic cave-upper of 2 levels	110-90 ka. [1] U/s	N=11 *Stegodon /Ailuropoda*, 'late'	none	rt. C
HUBEI					
Walongdong, Changyang co. (30°15'N, 110°05'E)	rock-shelter	215-175 ka. [1] U/s	N = 18 'southern spp.'	none	left maxilla with P3, M1, isolated P2.
Guojiujan, Changyang co.	rockshelter - lower of 2 levels	?	some, no details	?	calvaria, 2 mand. ff, some pc's
Quyuanhekon, Yunxian co. (32°50'N, 110°35'E)	Han R., up. terrace; l.3, sandy calc. clay	none (? >300ka) U/s	N = 10+ 'southern spp.'	present (44), ?Dingcun affinity	crania, 2 - crushed, distorted
ANHUI					
Yanshan, Chaoxian co. (31°33'N, 117°52'E)	karstic cave l.2 of 5 beds	200-160 ka. U/s [2]	N = 15	? none	occipital ,inc.; rt. maxilla, P3 - M1, l. maxilla f.; l. P4, M1, M2, isol.
SHANXI					
Dingcun, Xiangfen co. (35°50'N, 111°25'E)	Loc.54:100 Fen R. ter. 3, fluv. sands	210-160 ka. U/s [1]	N = 15	Abundant Dingcun industry	isol. r.I1,-I2; r.M2 (juv.ind.); parietal (infant)
Xujiayao, Datong basin, Yangguo co. (40°06'N, 113°59'E)	Liyigou R., Ter.2. fluv/lac,silts/ clays	125-100 ka. 88 ± 5,114±17[1] ka. U/s	N = 20	Very abundant = >14,000 Xujiayao industry	ca. 10 inds. = 12 pariet., 2 occip., 1 temp.; l. max, I1,C,M1,M2; mand. fr., l. M 1/2, l. M2, M.
SHAANXI					
Tianshuigou, Dali co. (34°52'N, 109°43'E)	Loess Plateau, Luo R., Ter. 3, l. 3 gravel	230-180 ka. [1] U/s	N = 12	present (181 directly assoc.)	Cranium, largely complete, face displaced, c/o dentition.
HEBEI					
Xindong (New Cave), Zhoukoudian (39°40'N, 115°53'E)	cave/fissure = Loc. 4 + inner chamber	ca. 100 ka.[1] l.6/7=175-135 ka. U/s base=257±36[4] T/L upper trav. = ~ 88 ± 3 [3] low. trav. = [3] 179±13/11 ka 125 ± 7 ka 139± 5 ka U/s	N = 40 (cf. correlative with Locs. 3, 15	present, few flakes	isolated l. P3, worn
LIAONING					
Miaohoushan, Benxi co. (40°15'N, 124°08'E)	Cave. Loc. A; l. 5,6	l.4= 420-280 ka l.5= 330-200 ka l.6e= 280-180 ka l.6u= 155-130 ka U/s [5]	N = 19	present, 88 (l.6) Miaohoushan industry	lev. 5 = C; lev. 6 = r. M1, juv. femur diaph.
Jinniushan, Yingkou co. (40°34'N, 122°26'E)	karst fissure fill, Loc. A, levels 1-7, down	l.5 = 260-210 ka [1] l.6 = 300-230 ka between 134-135 and 257 ka [6] l.4 = 240 ka [7] l.5/6 = 270 ka l.7 = 280 ka all U/s	N = up to 26 (various levels)	? None	l. 6/7 = partial cranium, dentition, vertebrae (5/6), ribs (2), ulna, l. innominate, patella, manus & pes elements

Footnotes: [1] Chen and Yuan 1988; [2] Chen et al. 1987; [3] Zhao and Liu 1991; [4] Pei and Zhang 1985; [5] Yuan et al. 1986; [6] van der Plicht, van der Wijk, and Bartstra 1989; [7] Lü 1985, 1990.

reduction of various cranial reenforcement superstructures, including occipital torus (its robusticity and lateral extent), angular torus, supramastoidal crest(s), and frontal and parietal keels (and associated depressions). Cranial thickness is uncommonly variable, and may be substantially thinned (Jinniushan, Yanshan), or even exceptionally thick (Xujiayao, Dali). A distinct continuous suprafacial (supraorbital) torus is present, with glabellar linkage, although that area tends to be somewhat depressed, and with variable expression of medial/central (superciliary) component, and variable but overall robusticity reduction laterally (supraorbital component), the latter associated with some posterolateral inclination. The posttoral sulcus is moderately to shallowly expressed, even as an ophryonic depression. Frontal sinus may be large. The infratemporal exhibits usually broadened, shallow mandibular (glenoid) fossae, moderate to rather thinned tympanic plates, some coronal orientation of tympanic axes, mastoid (tympanomastoid) fissure, and variable, but sometimes well-expressed, inferiorly directed mastoid processes. Other important characters of the middle and anterior cranial base are unstudied (and perhaps only adequately preserved in the Dali individual). In several instances in which the middle meningeal vascular pattern has been elucidated, it has enhanced complex ramification, and a persistent well-developed inferior temporal branch.

The facial skeleton, known in four crania, all of which require some to extensive reconstruction, reveals both variability and some distinctive divergences from regionally antecedent hominids and from penecontemporaneous European samples. The face is overall broad and relatively long. The orbits are rectangular or subrectangular and may exhibit a measure of medial obliquity. Interorbital breadth is consistently wide, the nasal root is elevated, the nasal profile angle large, and the frontonasal and frontomaxillary sutures are aligned. There is substantial midfacial flattening. The zygomatic root has a relatively elevated origin, the inferior zygomaticomaxillary margin (IZM) is generally elevated in height and set quite horizontally, with a shallow or more substantial malar notch (incisura malaris). There is moderate to more notable infraorbital depression, and pneumatization may be restricted, particularly into the superior maxillary area, but instances of invasion into the premolar area do occur. The nasal aperture is consistently broad, the inferior margin (superior alveolar clivus) well delimited, and an anterior nasal spine is well expressed. Subnasal alveolar prognathism is moderate to minimal. The palatal/dental arcade is generally U-shaped.

The upper dentition, still often poorly described, is tolerably well represented within crania (Yunxian, Jinniushan), in maxillary fragments (Walongdong, Yanshan-Chaoxian, Xujiayao) and as isolated teeth from some half dozen localities. The lower dentition is almost unknown, there are no mandibles, and only four isolated cheek teeth in all. No elements of the deciduous

Table 3. Some aspects of cranial morphology of some East Asian Later Middle Pleistocene hominids.

CHARACTER STATE	TIANSHUIGOU (DALI)	QUYUANHEKAN (YUNXIAN)	JINNIUSHAN	XUJIAYAO	YANSHAN (CHAOXIAN)	WALONGDONG (CHANGYANG)
Cranial Vault						
Cranial capacity	1120 cc	(>1350 ?) cc	<1300 cc ?	---	---	
Overall shape	long, rel. low	long, low	long, low	---	---	
Transverse conformation	more gabled; max. br. post.-sup. temp.	rather bell-shaped; very wide bi-supramastoid.	more gabled; max.br.post.-sup. temp.	---	---	
Cranial thickness	very thick	?	relatively thin	very thick	relatively thin	
Frontal flatness	rel. flat, receding	flat, receding	rel. elevated	---	---	
Frontal squama inclination	moderately low	moderately low	moderately low	---	---	
Parietal length & curvature	short, flat	short, flat	?	elongated, flattened	---	
Temporal squama ht/curvature	elevated, curved	elevated, curved	elevated, curved	elevated, curved	---	
Occip. proportions (squama/nuchal)		squama<nuchal	squama>nuchal	squama>nuchal	---	
Occip angulation	substantial	substantial	substantial	more rounded	moderate	
Occip torus, develop., extent	moderate, reduc. lat.; shallow supratoral sulcus	strong, continuous	slight, reduc.lat.; ?supratoral sulcus	narrow, reduc. lat.; no/trace supratoral sulcus	slight, reduc. lat.; small supratoral fossa	
End-inion/inion relationship	approximated (11 mm)	?	?	approximated, 10-15 mm.	less approximated, 22 mm.	

Angular torus	present, reduced	absent		absent	---	
Frontal/parietal keels	mid-frontal; parietal	none	frontal; ant. parietal	--; parietal none	---	
Bregmatic eminence	present	minimal	?	---	---	
Basilar-temporal region						
Glenoid fossa depth	broadened, rel. shallow	broadened, shallow	broadened, shallow	broadened, shallow		
Tympanic thick. (a-p)	moderate	thick, long	moderate	moderate		
Medial (entoglenoid) recess	present - ?	absent - ?	?	?		
Mastoid (tympano-mastoid) fissure	present, narrow	present, broad	present	present, broad		
Supratubarius process	blunt					
Tympanic orientation (re coronal plane)	in coronal plane	in coronal plane	?	in coronal plane		
Mastoid/supramastoid crests	not united	not united, weak		? not united		
Mastoid process	moderate	massive, inf. projecting; medially oriented	relatively small	small, very thick		
Meningeal vessel pattern	complex, ramified; pronounced inf. temp. br.	?	?	complex, ramified; pronounced inf. temp. br.		

Table 3. (continued)

CHARACTER STATE	TIANSHUIGOU (DALI)	QUYUANHEKAN (YUNXIAN)	JINNIUSHAN	XUJIAYAO	YANSHAN (CHAOXIAN)	WALONGDONG (CHANGYANG)
Facial Region						
Supraorbital torus, development, shape	glabella/superciliary linked medianly , massive; glabella depr.; s-o thick, curved latero-posteriorly	glabella/superciliary linked; slight glabella dep., s-o thick, curved latero-posteriorly	glabella/superciliary linked; s-o thick, curved latero-posteriorly			
Posttoral sulcus	shallow	ophryonic depression	quite shallow, broad			
Postorbital constriction	mod. reduced	mod. reduced	mod. reduced			
Frontal sinus	?	?	large			
Interorbital breadth	wide	wide	wide			
Nasal root, x-sectional shape	elevated	elevated	elevated			
Orbital shape	rectangular, rounded inf.	rectangular	sub-rectangular, rounded inf.			
IZM * height	low	elevated	elevated			
IZM * inclination	more horizontal	horizontal	inclined			
Incisura malaris (malar notch)	present, shallow	present, substantial (as in ZKD-1)	present, shallow			
Upper/mid-facial flattening	quite flat	quite flat	quite flat			
Zygomatic process root	high origin	high origin	lower origin			

Facial proportions	shorter, broad	long, broad	longer, broad			
Mid-facial pneumatization	superior maxillary;infra-orbital depression	absent; infraorbital depression	superior maxillary; slight infraorb. depr.	juv. in P area	extends ant. to P^3	extends to P^3
Fronto-sphenoidal malar proc.,orientation	anteriorly	? rather anteriorly	antero-obliquely			
Nasal aperture	very broad	very broad	broad	?	? rel. broad	prob. rel. broad
Inferior nasal margin	sharp; nasal spine	sharp; nasal spine	sharp; nasal spine	juv.- nasal spine	--; weak nasal spine	sharp; nasal spine prob. substantial
Subnasal alveolar prognathism	minimal	moderate	moderate	juv. - reduced	slight	slight
Lateral alveolar prognathism	none	none		none		
Palatal/dental arcade shape	divergent U-shaped	U-shaped	U-shaped	---		

* IZM = inferior zygomaticomaxillary margin
Data after Weidenreich 1943; Pope 1991, 1992; Li and Etler 1992; and other authors.

dentition are known. Upper incisors are usually shovel-shaped, with substantial marginal ridges, some development of lingual basal tubercle, and attendant enamel ridges; apparently none of these features, all of which are features of ZKD-1 sample, are expressed in the Yunxian individual(s). Of the few (3-5) canines known and adequately described, that from Xujiayao approaches the ZKD-1 condition in robusticity, cingulum, vertical grooves, basal tubercle, and associated enamel ridge development; those from Yunxian and Wuguidong (Jiande) largely do not. The few premolars are reportedly like ZKD-1 (e.g., Walongdong and Yanshan-Chaoxian), or in a single isolated instance (P3, from Xindong, New Cave, ZKD) diverge in size, occlusal morphology and accessory structures, and root structure. The molar series is generally inadequately described, but some at least (Yanshan-Chaoxian, Xujiayao) approach in morphology and, sometimes in size, ZKD-1 homologues. Overall, there may be substantive and substantial approximations to the *H. erectus* morphotype well documented at ZKD-1.

The postcranial skeleton is represented only, in part, at Jinniushan. Ultimately, it will afford useful comparative information, but only an ulna (Lü 1989) has been discussed until now, and such postcrania of other hominids remain largely or wholly unknown elsewhere in eastern/southern Asia.

WESTERN ASIA (AND LEVANT)

In the Transcaucasus the large karstic cave system of Azykh (39°32'N, 47°E) is situated at 900 m on the southern slope of the Karabakh upland, above the left bank of the Karuchai River (Nagorni Karabakh, southern Azerbaidzhan). The karstification is considered to have occurred during the Akchagylian (Gauss/early Matuyama) and earlier Apsheronian (Olduvai sc. and later) transgressive intervals. There are 14 meters (10 beds) of infilling with successive human occupations in the lower 6 units (10-5), and a subsequent Mousterian (denticulate, with bifaces, of Levallois facies) occupation (bed 3) after a sterile deposit. The artifacts of the earliest occupations (Beds 10-7), termed Kuruchai industry by Guseinov (1981, 1985), comprise marginally and terminally flaked/trimmed cobbles, including chopping tools/cores, and irregular, often cortical flakes, variably retouched. These assemblages do show some very general resemblances to the Developed Oldowan. It is associated with rare, fragmentary, and apparently indeterminate fauna. Although considered of upper Apsheronian-equivalent age (as much as 1.2 ma), and, if so, should have a Tamanian faunal composition, there is apparently little evidence to support such an antiquity. The Acheulean industrial complex is documented in bed 6 and 5, subdivided into Early (Bed 6) and Middle (Bed 5) on techno-typological bases. It is also based on cobbles, and both bifaces (knifelike in the latter manifestation), choppers/cores, and flakes (with a diversity of retouched margins) may often preserve cortical surfaces. The very substantial vertebrate fauna from these occupations (Baryshnikov 1989) comprise micromammals (bats, lagomorphs, and some 13 rodent species) (Markova 1982), many carnivores (15 spp.), perissodactyls (7 spp.), and artiodactyls (11 spp.), including a suid, cervids (5), bison, and caprines (3). The lower occurrence (Bed 6) certainly shares forms characteristic of the Tiraspolian faunal complex of early Brunhes chron age (Baku stage equivalent), whereas the upper (Bed 5) is reminiscent of the Singilian complex (= lower Khazar = Gyurgyan stage) attributed to $^{18}0$ stage(s) 14-12. Substantial variations in winter temperature (9°-14°C) and precipitation (500-1000 mm) have been inferred for the lower set of occupations, with subsequent reduction of precipitation and temperature during the Middle Acheulean (Velichko et al. 1980). That occupation (Bed 5) yielded a human right mandible fragment with M_3 (and roots of the other molars), largely distinguished by the robusticity of the corpus and the elongate, unreduced M_3.

The Acheulean is known in two other caves in Georgia: in Koudaro-1 and -3 caves, at 1600 m above sea level (and 260 m above the Djorjori River valley), and in Tsona Cave, 5 km to the south, both in northeast Colchis, western Georgia. In each locality there are subsequent Mousterian (typical of Levallois facies) occupations as well. (At least 20 Mousterian occupation occurrences, almost all caves except Ilskaya, and three with Neandertal teeth/maxillary parts, are well documented in Georgia [Lubine 1977]). Kudaro-1 (Lubine et al. 1985; Lubine and Baryshnikov 1984; Baryshnikov 1989) is of particular interest because: (1) its basal layer records the Matuyama/Brunhes reversal; (2) there are Acheulean occupation horizons (5B, 5b, 5a), the lowest of which (5B) contains a Tiraspolian-type mammal fauna (19 spp.) analogous to that at Azykh (level 6), and diverse earlier species which are lacking in the two successive occupations; and, (3) there is a substantial pollen spectrum throughout, which records, differentially, three warm climatic optima within the Acheulean occupation span, and which are separated by cold/humid to temperate intervals,clearly indicative of mid-Pleistocene interglacial or interstadial to glacial events.

The Acheulean industrial complex, in a diversity of situations and in a succession of geological contexts, is now well represented over greater southwestern Asia, particularly in the Levant. Among the older, broadly "middle," Acheulean occurrences (f = faunal association) are those of Latamne (f, Orontes River) and the Nahr El Kebir (Syria); Joubb Jannine-II (southern Bekaa) and Wad Aabet, at +50-52 m above sea level (Lebanon), and Evron-Quarry (f, coastal plain) and Gesher Benot Ya'acov (f, Jordan Valley), (Israel).

Only one locality of substantial Middle Pleistocene antiquity, has afforded hominids from such an archeological context–at Gesher Benot Ya'acov in the upper Jordan Valley (Israel). The eponymous formation (BYF), named after a bridge across the Jordan, outcrops over a distance of several kilometers, initially some 2 km south of former Lake Hula, and west of the basaltic plateau forming the Golan Heights, about 70 m above sea level (Goren-Inbar and Belitzky 1989; Picard 1963, 1965; Schulman 1978). The paleontological and archeological potential was recognized early in the 1930s, and surface collection and excavation pursued between the mid-1930s to early 1950s by M. Stekelis (1960), briefly by D. Gilead in 1967 and 1968, and most recently (since 1989), by N. Goren-Inbar and colleagues in newly revealed outcrops about 0.67 km farther south. The formation, forming a partial anticline, is in unconformable contact with the underlying Yarda Basalt, most recently dated (K/Ar) between 900 and 800-880 ka (Heimann, Mor & Steinitz 1987; Goren-Inbar et al. 1992a). This sector of the Jordan rift is fault-bounded (NNE-trends) of course, and subject to strike slip displacement, and the formation has been either slightly or strongly tilted, even back-tilted, depending on the outcrop area, subsequent to its emplacement. The BYF, truncated by an erosional terrace of the modern Jordan River, has a known thickness of 20-25 m, representing (upwards) low-energy (marls, peats) to increasingly high-energy fluviatile (basaltic and rarer limestone cobbles, boulder conglomerates) deposits with interbedded finer sediments. The age and polarity of the Yarda Basalt is broadly consistent with emplacement during the Jaramillo (N) sc., and normal polarity of sampled BYF indicates an (earlier) Brunhes (N) chron age, i.e., less than 0.78 ma. At least four horizons, in both more limnic, paludic, and fluviatile sedimentary regimes, within the BYF afford Acheulean lithic artifacts, and also freshwater mollusca, (some of which are known to be extinct by ~240 ka), vertebrate fossils, remains of woody plants, some with apparent human workmanship, and fruits and seeds. Some of the latter suggest both riparian and probably more open, even upland settings, as well as now-distant plant distributions (*Pistacia*). The previously known fauna comprised a limited number (N = 10) of ungulate and proboscidean (2) taxa, and will doubtless be enhanced by newer investigations. The Acheulean at BYF is in aspect certainly more "Middle" than "Early" or "Late"; it comprises abundant bifaces (largely in basalt, also limestone and flint), including both handaxes and cleavers, produced especially by the block-on-block (centripetal) and Kombewa flake production techniques, and a diversity of flakes and flake tools, with extensive evidence of the Levallois prepared core technique (Goren-Inbar 1992; Goren-Inbar, Zohar & Ben-Ami 1991; Goren-Inbar et al. 1992b).

The hominids were retrieved from older, unstudied faunal collections, and thus their geographic and stratigraphic provenance within the BYF unknown. These are two femoral diaphyses, the larger (more robust) comprising about three-fourths of the shaft, and the smaller, right diaphysis more nearly complete, broken just below the neck and just above the distal condyles. Probably, a complete femur would have been rather short (Geraads and Tchernov 1983). There is a limited amount of useful morphology: muscle insertion areas are not well defined (that of *v. medialis* slightly rugose, *g. maximus* slightly better defined, no definitive pectineal line); an elongate narrow, hypotrochanteric fossa; low, divided linea aspera, elevated only along middle third of shaft; elongate, somewhat concave popliteal surface; very platymeric shaft, both proximally and midshaft; and thickened cortex. Overall, there is a distinct absence of Neandertal autapomorphies, and so far as ascertainable, preservation of some distinctly plesiomorphic features.

Occurrences attributed to the "late" Acheulean are both more numerous (over 200 records overall; Bar-Yosef 1984) and rather more widespread into now steppic and subdesert situations. These include significant localities in Iraq (Barda Balka); Jordan (Azraq, Fjaje); Syria (Jerf 'Ajla Cave, Yellow-2; F, Yabrud-I shelter, l. 23, 18, 17, 12, 10, 8; Nahr-el-Kebir; Palmyra and El Kowm basins; and Jrabiyate and Hama-Latamne, Orontes River); Lebanon (Ras Beyrouth-1; Ard es Saouda); and Israel, in open sites, on the coastal plain (f, Holon and Evron-Zinat; Kefar Menachem; Kissufim, Negev), and in inland upland situations (Ramat Yiron; Berekhat Ram, Golan Heights, sealed by a basalt dated [K/Ar] 233 ± 3 ka; and Ma'ayan Baruch, northern Hula Basin), and in coastal (f, Tabun-F) and inland (f, Oumm Qatafa-E and D, Dead Sea, with a possible U-series age of 115 ± 19 ka on l. E) caves. None of these or other comparably aged situations have, unfortunately, afforded hominid skeletal parts, although some have preserved associated faunal remains.

The "late" Acheulean in coastal Lebanon (Sanlaville 1977) can be shown, in part, to *predate* the Enfean I (+15 m) transgression attributed to isotopic stage 5e, and to *postdate* the second or Jboailian (II) transgression (+40-45 m, as at Ras Beyrouth), presumably correlative with stage 9, and whose antecedent transgression (I) is considered to represent isotopic stage 11 (Holsteinian). Thus, this industry apparently could span upwards of some 225 ka, approximately < 350-125 ka. As will be seen below, an age of as much as, or, indeed greater, than 200 ka for such occurrences is, in fact, likely in view of both U-series and ESR, and (some) TL age determinations on industrial/stratigraphic complexes which are, in some situations, known to succeed the "late" Levantine Acheulean. Diverse facies of "late" Acheulean have been recognized in Syria (Samoukian facies at Nahr el Kebir, Defaian facies in the Orontes, both of Levallois flake production technique) and in southern Levant (Evron-Kissufim group, based on both coastal and inland localities, also with Levallois technique, and Koussin-Yiron group, in inland situations, and also employing Levallois technique).

In several situations there are stratified successions of lithic industrial complexes of the Levantine (late) Lower and Middle Paleolithic. These comprise "late" Acheulean and the Acheuleo-Yabrudian (A-Y) complex–sometimes subsumed under the appellation Mugharan Tradition (MT) (of Jelinek 1981, 1982a, 1982b; Jelinek et al. 1973)–and the Mousterian *s.str.* The MT (A-Y) is recognized generally to comprise 3 to 4 expressions or facies: *Acheulean facies* (as at Yabrud-I, l. 23, 18, 17, 12, 10, 8), with Levallois flake technology, relatively few (15-20%) retouched tools, and substantial (to 15%) bifaces, small and pointed/oval in form; *Yabrudian facies* (as at Yabrud-I, l. 25, 22, 16, 14), overall relatively uncommon, with scant or no employment of Levallois flake technology, a dominance of retouched flake tools, particularly canted and transverse sidescrapers on thick flake blanks, few denticulates, and rare Upper Paleolithic tool types; *Acheuleo-Yabrudian facies* (as at Yabrud-I, l. 24, 11), with characteristics of both the foregoing; and, *Amudian* (so-called Pre-Aurignacian) *facies* (as at Yabrud-I, l. 15, 13, and Tabun-Ea, = Unit XI-Beds 75 I1-2, 75s), exhibiting some Levallois flake technology, rare bifaces and denticulates, and having quantities of Upper Paleolithic tool types (blades, knives, endscrapers, burins). The A-Y (MT), known from less than 20 localities in all, is restricted to the central and northern Levant, being unknown south of the Carmel range, and extending inland into now xeric locales (such as Yabrud, Anti-Lebanon, and El Kowm oasis, Syria). In recorded instances of industrial superposition it *always* precedes Mousterian, (as in Yabrud-I, l. 5, 6, 7, 9; Tabun-B-C-D = Units I-IX; Zuttiyeh, etc.) and succeeds and/or has intercalated "late" Acheulean manifestations (Yabrud-I, Tabun, which have very deep fillings, 11 m and 20 m, respectively).

The A-Y is known from oasis/springside localities in xeric settings, at Ain El-Assad (Azraq) in eastern Jordan (Rollefson 1980), and in the El Kowm Basin, north central Syria (Besançon et al. 1981; Copeland and Hours 1983). In the latter instance, the northernmost locale known (+34°N), there are 11 such sites, and almost all contain a few bifaces. Three localities have U-series ages (Hennig and Hours 1982) on spring travertines (tufas) associated with the Yabrudian, with ages of 156 (138-179) ka (Hummal), 139 (131-150) ka, an underlying travertine 245 (222-278) ka (Oumm el Tell), and 99 (95-110) ka (Aarida). A TL age (160 ± 22 ka) at Hummal is comparable to the U-series results. These determinations, taken as given, all suggest ages largely within isotopic stage 6, and extending into (earlier) stage 5. The several coastal sites in Lebanon are in relation to former higher marine levels. At El Masloukh, on the northernmost coast, the occurrence (lacking bifaces) overlies a +45 m beach remnant (Skinner 1970). At Adlun (south of Beirut) (Roe 1983) the occurrence in Bezez Cave of Acheuleo-Yabrudian directly overlies a +15 m beach, attributed to the Enfean-I (= stage 5e, ca. 125 ka); above an intervening sterile zone there is also Mousterian (stage 3 or Tabun-B type) occupation. At Abri Zumoffen, 100 km to the north, a 12-13 m beach attributed to the Enfean-II transgression (= stage 5c, ca. 112 ka) has repeated (8 in all) overlying Amudian or Yabrudeo-Amudian facies (including one with bifaces) occupations.

At the shelter of Yabrud-I, as aforementioned, there is an evident interstratification of MT facies(late) Acheulean (l. 23, 18, 17, 12, 10, 8), Acheuleo-Yabrudian (24, 11), Yabrudian (25, 22, 16, 14) and two levels (15, 13) of Amudian ("Pre-Aurignacian"). Isotopic dates from l.18 (Acheulean), one of two such levels which succeed four basal levels (25-22), two (25, 22) of which are typically Yabrudian, and one (24) Acheuleo-Yabrudian, afford substantially older ages largely within stage 6 (Farrand 1993). A TL determination (Oxford) on flint gave 195 ± 15 ka, and ESR determinations (McMaster University) on teeth gave 177 ± 20 ka (EU) and 231 ± 19 ka (LU), the first falling between the ESR results.

Isotopic dating results from Mount Carmel caves and the caves of Qafzeh and Zuttiyeh (Wadi Amud, Kinneret Lake) are set out in table 4. There are some concordances and many discrepancies between methods and with reference to cultural (industrial)-stratigraphic (C-S) entities, (see also Jelinek 1992), and also in respect to comparative stratigraphy (Farrand 1993). Concordant results obtain at Qafzeh and at Kebara between ESR (EU) and TL. The results at Skhul are concordant between U-series and ESR (EU), but not with TL. These are concordant in terms of C-S units, with Mousterian Ph. 3 (Kebara) younger than Ph. 2 (Skhul, Qafzeh); however, the Tabun-C (upper) Ph. 2 Mousterian falls within the age range (ESR-EU) of Ph. 2 at those localities, although it is appropriately younger (~86 ka) than Ph. 2 (~102 ka) of (lower) Tabun-C. Mousterian Ph. 1 (Tabun-D) falls within isotopic stage 5-e, and the ESR (EU) results are again consistent. The ESR (EU) results for Tabun-E (Mugharan Tradition) are largely consistent, and indicate ages equivalent to isotopic stage 6. At Tabun the basal infilling (Layer G of Garrod, Unit XIV of Jelinek), a somewhat thin loamy sediment, dips downward into an underlying karstic cavity ("swallow hole"). The unconformably overlying and also dipping sediments, comprising Layer F (up to 3.6 m) and basal E (Ed of Garrod), of comparable or even greater thickness, are strongly sandy and in composition comparable to coastal (cemented) eolianites, or *kurkar* formations, which form several littoral cordons along the present coastal plain (Goldberg 1973); these are generally considered to be of isotopic stage 5 age (Givirtzman et al. 1985). Tabun Cave is now some 15 m above the plain, and over 50 m above sea level Subsequent sediments (upper deposits of layer E, and layer D) are substantially less sandy, and instead strongly silty and loesslike, indicative of a major change in sedimentary source and regime. Apparently there is a continuous sedimentary sequence through E (Unit XIII) and D (Unit VII) layers, and Farrand (1993)

Table 4. Radiometric datings (in K/Ar) of some hominid-bearing localities in the Levant.

Locality/Horizon	Industry	T/L	ESR EU	ESR LU	U/Th	Ref
AMUD B *	Mousterian-Levallois facies (B-type)		41/42 ± 3	49/50 ± 4		2
ZUTTIYEH travertine	< Acheuleo-Yabrudian				148 ± 6	5
travertine	> Yabrudian				97 ± 13	5
Travertine, above basal hominid breccia					95 ± 10 (u) 164 ± 21 (l)	5
QAFZEH	Mousterian-Levallois facies (C-type)					
XXIII-XVII *		92 ± 5 (85.4 ± 6.9 - 109.9 ± 9.9)				10
XXI-XV *			100 ± 10	120 ± 8		6, 2
KEBARA	Mousterian-Levallois facies (B-type					
VI		48.3 ± 3.5				9
VII		51.9 ± 3.5				9
VIII		57.3 ± 4.0				9
IX		58.4 ± 4.0				9
X		61.6 ± 3.6	60 ± 6	64 ± 6		7, 9
XI		60.0 ± 3.5				9
XII *		59.9 ± 3.5				9
SKHUL B *	Mousterian-Levallois facies (C-type)	120 ± 9 119 ± 18	81 ± 15	101 ± 12		3, 4 8

continues to regard this as "accumulated within a single glacial-interglacial cycle." He considers the F/basal Ed sequence to be correlative with isotopic stage 5e, whereas ESR (EU) dates suggest attribution to isotopic stage 7 (see also Bar-Yosef 1992). The TL results from Tabun give consistently much higher ages, often by a factor of two, in comparison with ESR (EU) results, and are, not uncommonly, out of stratigraphic order as well (Mercier 1992). At Yabrud-I the MT facies, on the basis of both TL and ESR results on level 18, is as old or older than its counterpart in the Tabun-E succession, and perhaps equates with isotopic stage 7.

Table 4. (continued)

TABUN B *	Moust.-Lev. facies (B-type)		86 ± 11	103 ± 16		1
C (Unit I, Beds 1-6) *	Moust.-Lev. facies (C-type)	184 ± 29 (161 ± 24) +	102 ± 17	119 ± 11		1, 4
D (Units II-IX) *?	Moust. Lev facies (D-type)		122 ± 20	166 ± 20		1
II, 27-32		263 ± 25 (242 ± 24)+				4
V, 41 - 42		307 ± 30 (281 ± 27) +				4
IX, 62 - 69		297 ± 57 (272 ± 52)+				4
Ea *	Mugharan Tradition		154 ± 34	188 ± 31		1
X, 70-72		270 ± 23 (249 ± 21) +				4
xi, 73-77		296 ± 18 (273 ± 17) +				4
Eb *	Mugharan Tradition		151 ± 21	168 ± 15		1
XII, 78-80		350 ± 37 (319 ± 34) +				4
Ec	Mugharan Tradition		176 ± 10	199 ± 7		1
XIII, 81-85		331 ± 30 (302 ± 27) +				4
Ed	Mugharan Tradition		182 ± 61	213 ± 6		1

Explanations: * = human remains; + = adjusted ages, reduced by 15% for moisture.
References: 1. Grün, Stringer and Schwarcz 1991; 2. Grün and Stringer 1991; 3. Mercier et al. 1993; 4. Mercier 1992; 5. Schwarcz, Goldberg, and Blackwell 1980; 6. Schwarcz et al. 1988; 7. Schwarcz et al. 1989; 8. Stringer et al. 1989; 9. Valladas et al. 1987; 10. Valladas et al. 1988.

Two isolated finds of human remains were recovered in association with the MT in Tabun Cave, Mt. Carmel (32°40'N, 35°05'E); these occurred in the upper part of the 4 m thick infilling of Member E (stratigraphy of D. A. E. Garrod), specifically subdivisions Ea and Eb, and which comprise (stratigraphy of A. J. Jelinek and W. R. Farrand) Units X-XI (Beds 70-72, 73-77, respectively) and Unit XII (beds 78-80). In the former instance (lower Ea = beds 73-77 equivalent) both Garrod (1956; also Garrod and Bate 1937) and Jelinek (1990, also 1982b) recorded concentrations of Upper Paleolithic-type lithics. Garrod's (1956) lowest samples comprised additionally 7-8% bifaces and some 20-25% Yabrudian-type sidescrapers (60%, if all scrapers are included). In Jelinek's finely sampled Units XI-X, with 16 beds (horizons), there are (upwards), recognizably, an initial Yabrudian-facies horizon, five horizons of Acheulean facies, three horizons of Amudian facies (Upper Paleolithic-like with 35-50% backed knives in proportion to bifaces and sidescrapers), and 6 overlying horizons of Yabrudian-facies. One human specimen, a right femoral diaphysis (specimen E_1, at a depth of −5.6 m), occurred within this lower sequence, but whether in a Yabrudian-facies or Amudian-facies context is not now clear. Another specimen, an isolated right $M_{1/2}$, occurred (depth of −6.5 m) in the underlying Eb subdivision (= Unit XII, beds 78-80), in which there was (following Garrod) a concentration of Upper Paleolithic-like lithics in a zone above the mid-depth of the deposit; however, Jelinek (1982b) attributes all of this equivalent deposit to an Acheulean facies. Unfortunately, neither human specimen is very informative (McCown and Keith 1939). The lower molar is substantially worn, small, and apparently had a +5 crown pattern. The (superior) femoral diaphysis exhibits substantial flattening (platymeria) and very weakly expressed linea aspera, features generally considered plesiomorphic among mid-Pleistocene hominids.

The U-series results from Zuttiyeh (Schwarcz, Goldberg & Blackwell 1980), which derive from travertines between (95-97 ± 10-13 ka) Mousterian and Acheuleo-Yabrudian, and from below (148-164 ± 6-21 ka) the latter, are consistent with the Tabun-E results (ESR-EU). Zuttiyeh Cave (32°51'N, 35°30'E) is situated shortly before the outlet of the Wadi Amud into the northwest margin of Lake Kinneret (Sea of Galilee). The very large chamber (20 x 12-18 m) had an extensive infilling (+5 m thick), extensively removed in two (1925-26) field seasons by F. Turville-Petre (1927), when Mousterian horizons with limited fauna (N = 7) were found to occur beneath a late (Bronze Age) occupation and an underlying major rockfall. Some unusual aspects of the lithic assemblage were noted by the excavator and confirmed by Garrod (1962), who noted resemblances to the Yabrudian/Amudian facies (or MT). Subsequently, limited soundings (1973) clarified the otherwise unduly simplified stratigraphy, afforded microfauna (undescribed to my knowledge), and revealed Mousterian (of Levallois facies) occupation underlain by multiple levels of Yabrudian at the entrance (or Acheuleo-Yabrudian, within the chamber), and including Upper Paleolithic-like (Amudian) lithic components (Gisis and Bar-Yosef 1974). A human craniofacial fragment, recovered in the excavations of the 1920s from below such pebbly-sandy infilling, in a shallow fossa in the basal sterile clayey sands, evidently was antecedent to those occupations. Thus, its "cultural" (industrial) associations might expectably lie with the MT, and correspondingly afford some insight into an otherwise almost unknown human population within the Levantine late mid-Pleistocene.

Some aspects of the known craniofacial morphology of the Zuttiyeh specimen are set out in table 5, along with comparable observations on the

ZKD-1 (*H. erectus*) sample, and on the most complete (no. 1, adult cranium) of the Jebel Irhoud (Morocco) human crania.

Zuttiyeh is comparable in frontal length (chord, arc), angle and curvature with ZKD-1 examples. The relative position of maximum frontal curvature and frontal shape (flatness) is also within that range, at the upper end. The Hexian specimen has a shorter frontal, and the curvature (chord/arc) index (82.5) is substantially lower. The relative position of maximum frontal curvature in Zuttiyeh is elevated, in the upper ZKD-1 range, and the frontal shape (flatness) also at the top of that range. Maximum frontal breadth is barely greater, and minimum frontal breadth notably greater in Zuttiyeh. The Hexian specimen is only slightly larger in the former and smaller in the latter dimension, although frontal constriction is relatively reduced. Frontal height is elevated in Zuttiyeh relative to ZKD-1, the frontal is notably thinner, and the frontal boss only faintly expressed. The Hexian specimen resembles ZKD-1 in these aspects. Zuttiyeh has a well-developed supraorbital torus, attenuated laterally, delimited behind by a substantial supratoral sulcus (as also in ZKD-1), but also with a large postglabellar depression (unlike ZKD-1, whereas Hexian has some development of this condition). In Zuttiyeh each torus element is continuous, is thick medially, thins in the middle and laterally, and has a flattened (s-i) configuration; this form differs from ZKD-1 and from Hexian, the latter being thickened medially (18-19 mm), much thickened centrally (16-17 mm) and also laterally (12-13 mm). The Zuttiyeh toral structure is more arched, with accentuated medial to lateral curvatures, the glabella is less elevated, rather flat in respect to the toral elements, and thus unlike ZKD-1 (or Hexian). There is some depression between the medial (sc) and lateral (so) toral components at the orbital margin, and a tuber (proc. supraorbitalis) is moderately expressed; and, there is a weak notch/depression of the superior orbital margin. These conditions differ from ZKD-1 and Hexian. The frontal sinus is enlarged, though centrally situated, unlike ZKD-1 (in Hexian the sinus is somewhat enlarged).

The nasal region, so far as preserved, exhibits a series of deviations from ZKD-1. The nasal root is shallow, the frontonasal suture convex (superiorly), and there is slight nasion projection relative to glabella. Both posterior interorbital and superior nasal breadths are reduced. There is slight projection of the nasals relative to the frontal process of maxilla. Orbital dimensions are scarcely different from ZKD-1, the shape rather subrectangular and more mesoconch in Zuttiyeh. However, in Zuttiyeh the orbital pillar is more or less straight in alignment, and there is a substantially larger projection of the lateral orbital border relative to the sphenoidal greater wing. The superior orbital margin is salient rather than having the rounded configuration, due to torus development, of ZKD-1 or Hexian.

Unfortunately the midfacial region of the Zuttiyeh specimen is only

Table 5. Some morphological aspects of Zuttiyeh, Jebel Irhoud-1, and ZKD-1 hominids.

CHARACTER STATE	ZKD-1 *Homo erectus* (N = 5)	ZUTTIYEH	JEBEL IRHOUD - 1
FRONTAL			
Length, na-br chord (M-29)	109.8 (102-115)	112.	109.
Arc, na-br (M-26)	122.6 (115-126)	125.	127.
Angle (M-32(5))(FRA)	141.7° (134°-147°)	140.6°	133.
Curvature (m-29/m-26)	89.9% (86.9-91.8)	89.6%	85.8%
Relative position (re na-br chord) Max. frontal curvature	39.4 (32.6-48.1)	47.8	40.4
Shape/flatness (st/M-29)	16.4 (14.8-18.6)	18.3	20.2
Maximun breadth (M-10)	106.7 (101.5-108.)	109.	120.
Minimum breadth (M-9)	85.9 (81.5-91.)	97.	106.
'Maximum' (eu-eu) breadth (M-8)	137. - 143.	(138.)	149.5
Height (br above FH)	96.8 (93.-104.)	116.	109.
Thickness	moderate/thinner	thick	moderate
Boss	none/faint	distinct boss/keel	short, modest upper keel
Supraorbital torus -	medial middle lateral	medial middle lateral	medial middle later
thickness	14.9 14.9 12.3 (13.-19.6) (12.1-17.4) (11.2-14.)	15.8 10.6 10.9	(14.5) 10.5 (11.
mid-line projection	19.2 (17-22)	23.	20.
general form	very broad, massive	broad, attenuated laterally	broad, slightly biarched.
horizontality	more horizontal & only somewhat curved.	more arched, with stronger medial to lateral curvature.	sub-horizontal, with inferior curvature laterally.
glabellar area	elevated, continuous /confluent with supraorbital components.	less elevated, flat, supraorbital components more individualized.	elevated, and essentially co-p with supraciliary segment.
supraorbital element	s-c and s-o continuous, with latter still strongly thickened down to f-z proc. *may be* (Sk.XII) oblique, elongate, shallow depression* (s-l to i-m) delineating lateral (s-o) from s-c portions. (* = homologous to the *incisura supraorbitalis* in mod. man). Maybe rather strong tubercle (= *processus supraorbitalis)* bounding the lateral/and orbital termination of this depression (Sk. X, XII).	s-c and s-o continuous; latter (=s-o) much thinner (s-i) & thus rather flattened; there is an inf. depression between med. (s-c) and lat. (s-o) parts at the orbital margin, but it is *not* elongated, etc.; there is a mod. tubercle fm., stronger on left than rt.; overall there is greater semblance of angulation between medial & lateral segments of the torus cf. to ZKD. Overall rather like Dj. Ir.-1 but latter with rather more hor. overall torus, weaker s-c seg., glabella more elevated, & overall less broad s-t sulcus.	s-c and s-o continuous, form linked by raised glabella, latt thickened/projected ant. to s with gentle infero-lat. curvat fronto-malar process; scarcel perceptible depression at orb border between s-o and s-c, distinct tubercle (proc. supra
'foramen' supraorbitals	none developed	none developed; an indentation/'notch' on the left side, & weak depression on left.	none developed, slight inden both sup. orbital margins.
Supratoral sulcus	well-developed; small/limited postglabellar depression	substantial; lg. postglabellar depression	moderate, shallow grades int squama; stronger laterally.
Frontal sinus	small	large, central	large, central
NASAL REGION			
Root	slightly elevated	shallow	depressed
Fronto-nasal suture	horizontal, straight	upwardly convex	slightly upwardly convex
Superior internasal suture	may be patent	fused	fused
Position of nasion re glabella	moderate projection	slight (3.0) projection	retracted (-5.)
Post-interorbital (bi-lachrymal) br. (M-49)	30.	~ 23.	(~ 27.)
Superior nasal br. (M57 (2))	17. - 17.3	14.	(16.5)
Projection of nasals (re frontal process of maxilla)	substantial	slight	slight
ORBITS			
Orbital pillar alignment	rather bent; ant.-lat. oriented	straight; ant.-lat. oriented	straight; oriented ant.
Projection lateral orbital border re sphenoid gt. wing	(~ 30), shorter	39. (substantial)	substantial (35)
Superior orbital contour	rounded inf. s-o torus	distinctly salient margin	distinctly salient margin
Orbital shape	chamaeconch (81.8)	mesoconch (84.1)	chamaeconch (75.-78.)
Orbital breadth (M-51)	44.	(44.)	48.
Orbital height (M-52)	36.	37. (large)	36 - 37.5
MID-FACIAL REGION			
Infraorbital plate orientation	slightly latero-anteriorly; well-behind s-o torus overhang	latero-anteriorly; slightly behind s-o torus overhang	antero-inferiorly; rather behi overhang

Table 5. (continued)

Nasio-frontal angle fmo-na-fmo (M-77)	142 ° (no.12)	high (159 °)	146°
Zygomatic body re orbital pillar	rel. large; robust	rel. small; gracile	rel. large; pillar
Inferior zygomatic border			horizontal, incurved
Zygoproptosis	substantial	slight	substantial
Upper, facial breadth, fmt-fmt (M-43)	121.	119.	(128.)
TEMPORAL/SPHENOIDAL AREA			
Temporal fossae, and internal wall	short; quite curved	short; quite vertical	short; quite vertical
Sphenoid cerebral facies	small, low	expanded transversely, vertically	?
Sphenoidal sinus	none	enlarged	? (present in Irh.- 2)
Lateral pterygoid-sphenoid hafting	?	slight	?
CRANIAL CAPACITY (cc)	1059 (915 - 1225) N = 6 1054 (incl. Hexian)	?	1305 (RLH)

Data after Weidenreich 1943; Keith 1927; Hublin 1991; Simmons 1990; Simmons, Falsetti, and Smith 1991; and own observations.

preserved laterally. Nonetheless, there are a set of divergences from the ZKD-1 condition. The former has more lateroanterior orientation of the infraorbital plate. The nasiofrontal angle (fmo-na-fmo) is notably higher than in ZKD-1, Ngandong (#11 = 146°) several African mid-Pleistocene hominids (Kabwe = 133°, Bodo = 139°), some west Asian Neandertals (Amud, Tabun-1, Shanidar-5 = 128°, 145°?, 138°, respectively), European Neandertals (130°-143° in 6 individuals), or in their European antecedents (Arago = 146°, Petralona = 137°). The zygomatic body relative to the orbital pillar is gracile and relatively small. Zygoproptosis (*sensu* Keith) is slight. Upper facial breadth is less than or comparable to ZKD-1. In the temporal region the temporal fossae are relatively short with quite vertical walls. The cerebral facies of the sphenoid are expanded, vertically and transversely. The sphenoidal sinus is enlarged.

The Zuttiyeh specimen has often, indeed usually, been considered to represent some variant within the Neandertal group, however the latter might be defined and a particular hypodigm constituted. This was the original opinion of Keith (1927, 1931) and of Hrdlicka (1930, and who did allude to potential ZKD-1 resemblances, and the latter he initially thought had Neandertal affinities). This was also the view of McCown and Keith (1939), who considered there to be various resemblances/affinities with the Tabun/Skhul samples, and by Weidenreich (1943), who repeatedly made some comparisons with the specimen in his treatment of the ZKD-1 sample and ultimately considered it allied most nearly to the Skhul sample. More recently Trinkaus (1983, 1984) considered it to have affinities with some (earlier) Shanidar Neandertals or, subsequently (1989), as an appropriate antecedent to all west Asian hominids of late Pleistocene age. Smith (1985) also followed this perspective. More recently Simmons, Falsetti and Smith (1991; also, Simmons 1990), in one of the more inclusive, multivariate efforts focused on frontal bone morphometrics in west Asian hominids (employing 8 major parameters and 12 dimensional measures), concluded "that Zuttiyeh does not share any clearly derived traits with the Skhul/Qafzeh sample . . ." and that "Zuttiyeh is primitive compared to both Levantine Neandertals and the Skhul/Qafzeh hominids. . . ." This perspective may be compared (contrasted) with that of Vandermeersch (1989; also 1981) who, in a way followed Weidenreich's interpretation, but extended it (cladistically) to place Zuttiyeh in an ancestral position in a modern-*sapiens* lineage, due to a few, but, for him, definitive characters shared with the Skhul/Qafzeh human sample. This, in its way, extended Hublin's (1976) conclusion, in a comparative treatment that included the ZKD-1 sample, European and Levantine Neandertals (and European antecedents), and the Kabwe (Zambia) individual, that Zuttiyeh was more archaic (plesiomorphic) than (any) Neandertals or modern *H. sapiens*, lacked Neandertal autapomorphies, and in "the majority of its characters could constitute a morphological intermediate between *Homo erectus* and the first modern men of which the oldest are appropriately known in Palestine."

In their recent reexamination, in a comparative perspective, of the Zuttiyeh specimen, Sohn and Wolpoff (1990, in press) reconsider such previous approaches, evaluate their conclusions and bases for them, and–like Hublin–afford comparisons with the ZKD-1 samples, and also (in press) set out the distribution in some sub-Saharan African hominids of certain features they consider in common between ZKD-1 and Zuttiyeh. Their principal conclusions are:

> Zuttiyeh, and like other west Asian human populations, exhibits some cranial features considered by them as (East) Asian geographic (clade) features; thus, certain features considered by some investigators as of "modern" aspect, are in fact to be considered "east Asian rather than 'modern,' thereby substituting a clade explanation for a grade one." (Sohn and Wolpoff, in press)
>
> Zuttiyeh, among other such west Asian fossil humans, exhibits the most (about 10) such features, perhaps as a consequence of its more substantial antiquity.
>
> Zuttiyeh, instead of resembling closely one or another west Asian human fossil sample, in fact, shows "much overlap" (among over 70 dimensional values) with those several samples, and the source of such features is demonstrably Asian rather than European (where Neandertal autapomorphies prevail) or African (as few features purportedly linking east Asian (= ZKD- 1) population(s) and Zuttiyeh have a low to negligible

occurrence in a sample (N = 6) of more-or-less "archaic" human cranial remains from the late Pleistocene of sub-Saharan Africa.

It is impossible to concur with the outcome of certain comparisons, the extent of some conclusions, and some attendant inferences from the exercise of Sohn and Wolpoff. There are certain undeniable and general resemblances between the Zuttiyeh specimen and the ZKD-1 sample (and that of Hexian) of *H. erectus*. However, as some other investigators have also noted, these are of a very general nature, and overall plesiomorphic in character. Even with the limited basis of comparison afforded by Zuttiyeh there is a set of morphological divergences, as set out briefly above (and table 5), wherein the latter fails to accord with the well-documented *H. erectus* morphological pattern. These divergences are such that it is surely inappropriate to conclude that "the east Asian features in Zuttiyeh . . . suggest a broader Pan-Asian distribution of regional features than is normally recognized." On the contrary, there are substantive divergences both from penecontemporaneous populations both in eastern (and southern) Asia and, of course, in Europe as well. Contrariwise, there are resemblances to successive Levantine human population samples, as Sohn and Wolpoff (and others before them) have noted, but also significant similarities with another hominid sample, that of Jebel Irhoud, from the Maghreb of northwestern Africa.

The hominid locality at Jebel Irhoud (31°56'N, 8°52'W) represents an infilled karstic sinkhole/cavern formed in pre-Cambrian schists and limestones near a summit toward the eastern end of Jelibet massif (east of Safi, southern Morocco). It is part of a complex karstic system within the massif, of varying Pliocene and Pleistocene ages, recognized in the course of commercial exploitation for baritine (barium ore) which occurs abundantly in such mineralized cavities. This particular cavern, underlain by another (unexplored), contained four meters of infilling, constituting a basal massive rockfall overlain by a succession of silty loams (beds 22-8, upwards) with small rock exfoliates in the basal beds, succeeded by sandy-silty sediments, sometimes with gravels and/or small subangular blocks of local rock (beds 7-1). Human presence is indicated by 13 successive occurrences (beds 21-10, and 8, upwards) of lithic artifacts, five of which also have carbon fragments from fire, as well as some heat-spalled artifacts (stratigraphy of J. Tixier, test excavations 1967, 1969). The initial human remains (adult cranium, juvenile calvaria, juvenile mandible, nos. 1, 2, 3, respectively) were recovered, along with substantial fauna and Mousterian artifacts, without full documentation of accurate geological/stratigraphic provenience other than having occurred "low" in the infilling sequence. A partial juvenile humerus (no. 4) was recovered later under controlled excavations in level 18, associated with Mousterian lithics (217) and fauna (Hublin, Tillier & Tixier 1987). Another human mandible fragment (no. 5) lacking teeth, has recently been recognized among the originally collected fauna. The lithic industry (cf. Balout 1965a, 1965b) is a Mousterian of Levallois facies, particularly rich in sidescrapers, employing mostly flint and quartzite. ESR determinations (Grün and Stringer 1991) on three equid teeth from level 17 gave ages of 90-125 ka (EU) and 105-190 ka (LU), suggesting (due to U-concentrations and other factors) a substantial depositional history, and an age probably within isotopic stage 6.

The associated vertebrate fauna, which is suggestive of a xeric, steppic situation, is diverse, including birds as well as six orders of mammals (table 6). The greatest diversity is among the rodents (N = 8-9 spp.), carnivores (N = 6), and particularly artiodactyls, all bovids (N = 8-9 genera) and as many as 11 species. There are some differences between the initial (Amani 1991; Amani and Geraads 1993) and subsequently excavated (Thomas 1981) faunal collections, the former of which has elements (rodent, an alcelaphine bovid) in common with older, mid-Pleistocene faunal occurrences of the Moroccan Atlantic littoral.

The juvenile specimens (nos. 3, 4) might possibly represent a single individual. The right humeral diaphysis (no. 4), extending from below the surgical neck to the supracondylar area, exhibits features probably plesiomorphic in *Homo*, and often retained in Neandertals and some other archaic Pleistocene human populations (Hublin, Tillier & Tixier 1987). The cortical bone is thickened and the medullary cavity correspondingly reduced. The shaft is rather flattened (mediolaterally). The deltoid impression is narrow, defined between strong linear crests, and the deltoid tuberosity is elevated and set anteriorly. The supraepicondylar crest is very well developed. The subadult mandible (no. 3), preserving most of the left side and part of the right body, has deciduous teeth (dc, dm1 and rt. dm2), erupted M_1, and M_2 still in the crypt, and thus an age of ca. 8-9 years. The full description (Hublin and Tillier 1981) of the specimen documents a distinctive set of features, and consistent divergences from the well-known Neandertal morphological pattern, including the set of the anterior teeth, condylar position and projection, condylar height relative to the coronoid, the angular incisure, small retromolar space, and internal symphysial structure; presumably derived features include the small condyle, posteriorly reduced body height, position of the digastric mm impressions. There is no fullblown bony chin formation, but its several components (symphyseal and lateral tubers, mandibular incurvature subincisivally, and mental fossae) are all indicated. The teeth are notably macrodont, exhibit substantial molar crown wrinkling, slight taurodontism, an incisiform dc, and +5 pattern on M_1.

Irhoud-2 (Hublin 1991) comprises a partial calvarium with the frontal and other vault parts of the left side. It is post-adolescent (fused spheno-occipital synchondrosis), but has partially patent vault sutures, more so ex-

Table 6. Vertebrate taxa from Jebel Irhoud (Morocco).

Taxon	E. Ennouchi collections, early 1960s	J. Tixier excavations 1967, 1969
Reptilia		
Testudo sp.	X	
Aves		
Phasianidae		*Coturnix coturnix*
Corvidae		?*Pica pica*
Falconiformes		*Aegyptidae*, gen./sp. indet.
Struthio		sp. nov.
Insectivora		
Crocidura	*russula*; cf. *tarfayaensis*	
Erinaceus	sp. indet.	
Lagomorpha		
Lepus	sp. indet.	*capensis*
Rodentia		
Hystrix	*cristata*	
Gerbillus	*campestris; grandis*	aff. *campestris*
Meriones	*shawi*	aff. *shawi*
Mus	*spretus*	*musculus*
Paraethomys	*ras*	aff. *filfilae*
Lemniscomys	barbarus	sp. indet.
Eliomys		aff. *quercinus*
Carnivora		
Canis	*aureus;* sp. nov.	*aureus*
Vulpes	*vulpes* cf. *atlantica*	
Hyaena	*hyaena*	
Panthera	*pardus; leo*	*pardus*
Perissodactyla		
Equus	cf. *mauritanicus*; ++ spp. ?	cf. *mauritanicus*
Rhinocerotinae	gen./sp. indet.	*D. hemitoechus*
Artiodactyla		
Taurotragus	cf. *oryx*	? *T.* sp.
Oryx	cf. *gazella*	*dammah*
Rabaticeros	? *R. arambourgi*	
Connochaetes	*taurinus*	
Damaliscus	? *D.* sp.	
Gazella	*atlantica; tingitana;* ?*cuvieri;* ? *rufina*	*cuvieri*
Addax		*nasomaculatus*
Bos	*B.* sp. cf. *B.* cf. *primigenius*	*primigenius*
Caprini	indet. (size of *C. lervia*)	

ternally than internally. Thus, it is presumably a young adult. All the vault bones are quite thin. There is a persistent metopic suture and this metopism may well have effected some aspects of related cranial morphology. It shares some overall resemblances to Irhoud-1 (below), but has some distinctions of its own. The vault is long and broad with low position of maximum breadth (across supramastoid area). The frontal is very broad, has moderate postorbital constriction, lacks a frontal keel, is elevated and has substantial frontal convexity (probably linked to metopism). The supraorbital torus is distinctive in that both lateral segments are arched, exhibit strong development vertically mid-orbitally, and decrease in such thickness both medially and particularly laterally. An oblique supraorbital sulcus (of different expression on each side) traverses the medial and lateral components, and the latter, though thinned, is salient and protrusive. A glabellar depression is between the toral components. A supraorbital notch, either open or nearly closed, is bounded by a substantial tubercle. There is no development of a frontal sinus. The parietal is elongated, perhaps has a prelambdoid depression, has an uncommonly shortened squamal compared to sagittal margin (unlike Irhoud-1). Parietal bosses are evident, though less salient than in Irhoud-1. There is some thickening, but no angular torus in the asterionic region. The vault is rather less flat sagittally than in Irhoud-1 and, in posterior view, the vault has a pentagonal form with subvertical walls. The occipital is not especially convex, little angulation is present between occipital and nuchal planes; an occipital torus is scarcely defined, and largely replaced by thickening of the lower planum and the very distinct superior nuchal lines; there is a weak concavity, but no true supratoral sulcus or fossa. The inion-endinion relationship is obscure. The temporal squama is elevated and arched, higher than in Irhoud-1; both have a long squamal/parietal articulation, and lack any internal crestal thickening. The supramastoid crest is substantial, and bounded above by a broad sulcus. The mastoid process is similar in form, but more robust than in Irhoud-1, is well delineated, and inferiorly directed. The juxtamastoid processes and the digastric notches are variably developed bilaterally. The mandibular fossa is overall like Irhoud-1, but is even deeper set, with sharp anterior margin and postglenoid process. There are suggestions of a substantially developed sphenoidal sinus.

Irhoud-1 (Hublin 1991; also 1993) is a cranium of a rather young adult, judging from largely unfused sutures, which preserves the face (without teeth) and lacks the base. There is rather minimal distortion. The vault is quite long, comparable in length to Qafzeh-6, shorter than European Neandertals, and rather low. Maximum height (vertex) is situated anteriorly, at bregma, rather than posteriorly on the parietal. The parietal upper margin is long, at the upper Neandertal limit (and lower limit of Skhul/Qafzeh group) and only very slightly curved. Maximum cranial breadth is rather elevated, just above the squamous suture, and forward (above external auditory meatus = e.a.m.), rather than more or less posteriorly, behind mastoids in most Neandertals, and above mastoids in Skhul/Qafzeh group. The vault appears rather low (occipital view), but has a non-Neandertal contour in its relative angularity (as in Skhul/Qafzeh), having a more pentagonal form, broadened inferiorly, rather flattened superiorly, the lateral walls quite vertical but subparallel, with distinct, high set parietal bosses.

The frontal rises gradually and fairly steeply, more so than in most Neandertals, as indicated by angular values, and more like the Skhul/Qafzeh group. Its convexity is also more similar to the latter than to various Neandertals. The frontal breadth dimensions are substantial, indicating some reduction (comparable to Zuttiyeh) of frontal constriction, and values overlap ranges of both Neandertals and Skhul/Qafzeh group; in comparison with ZKD-1 the specimen has relatively less elongated, narrowed frontal proportions. There are bilateral, weak frontal bosses, and a distinct, but weak frontal keel. A supratoral sulcus is more marked laterally, there being substantial projection of external orbital processes; medianly, there is a supraglabellar planum above glabella on the lower squama. A supraorbital torus is present, damaged on the left, with a glabellar depression between the lateral segments. There is some differentiation of the toral components, less marked than in Irhoud-2; each toral component has a semicurved form, with (right) a flattening between medial and lateral components, or (left) a more distinct differentiation with some vertical attenuation and flattening of the lateral component. A supraorbital notch is clearly evident. The frontal sinus is largely centrally located, extending superiorly somewhat into the supraglabellar planum and laterally along the medial orbit to the supraorbital notch.

The occipital exhibits only moderate angulation between superior and inferior (nuchal) planums. The squamal (occipital) planum is long, relatively elevated, thus differing in angulation from most Neandertals, and approximating in form the Skhul/Qafzeh condition. The occipital lacks transverse narrowing ("bunning") of Neandertals, having a regularly curved contour from side to side, and similarly no depression sagittally, there being gentle convexity of the superior planum. The (damaged) nuchal planum is somewhat concave (superiorly), and appears overhung by the superior planum. Opisthocranion is in an elevated position, and inion is a bit above the Frankfurt plane; inion and endinion are well separated (18), a more primitive feature. A veritable occipital torus is not developed (as also in Irhoud-2), there being rather the elevation of occipital above nuchal planum, and a corresponding "break in slope," and substantial expression of the superior nuchal lines, which fade out bilaterally. The maximum occipital projection (posteriad) is medial, and there was probably an ovate inial prominence, but no evidence of any supratoral sulcus/fossa, only a flattening of the suprajacent planum.

The vault bones are moderately thick, the frontal exceeding European Neandertals or Skhul/Qafzeh, the parietal within those ranges except for enhanced robusticity, without angular torus in the asterionic region, and the temporal squama also thickened. The endocranial capacity, estimated between 1480 cc and 1305 cc by two investigators, is elevated relative to an *H. erectus* sample, but below European Neandertals or the Skhul/Qafzeh group. The Irhoud-1 endocast (Holloway 1981) is quite undistorted, is platycephalic and preserves very poorly expressed convolutional patterns. The left occipital and (both) right frontal petalia patterns are distinguishable, as in many mid-Pleistocene and later nonmodern hominids, as well as, of course, *H. sapiens.* It exhibits strong development of left inferior frontal convolution (Broca's cap). Saban (1984) documented a rather surprisingly simple middle meningeal vascular pattern in which the (3) main branches show only weak ramification, and lack major anastomoses. By comparison Irhoud-2 exhibits enhanced ramification of the fronto-bregmatic components of the anterior branch and superior anastomoses to the middle (parietal) branch, but no connections between the middle and posterior branches, the latter of which is doubled with interanastomotic connections. The lack of elaboration of ramification and of anastomotic connections between, especially, middle and anterior branches, has been documented as well in (subadult) individuals from Qafzeh (nos. 11, 15) and from Skhul (no. 1), and some of these features are also reflected in juvenile (European) Neandertals.

The temporal squama is elongated, and relatively low (as in most *H. erectus*, and some Neandertals), with an elevated outline and weakly convex superior (sutural) curvature, rather than the angular form of ZKD-1. The supramastoid crest is salient, bounded above by a sulcus, and extends linearly prolonging the zygomatic structure; the latter is incomplete but exhibits substantial lateral divergence from the temporal (as in both Neandertals and Skhul/Qafzeh). There is no sulcus between the zygomatic root and the temporal scale, and the former, superior to the e.a.m., has a more horizontal axis than in Neandertals. The mastoid lacks Neandertal features, is rather small, elevated and not flattened, projects inferior to the juxtamastoid eminence, lacks an anterior tuber(cle), and in overall size approaches Skhul/Qafzeh, not Neandertals; the juxtamastoid eminence is not projecting, and the narrow digastric groove passes obliquely lateroposteriorly, and broadens back of the mastoid root. The mandibular fossa is large, deep, with a salient articular tubercle and well-developed postglenoid process. The temporal facies of the sphenoidal greater wing is of moderate size; there is no sharp infratemporal crest, although there is marked angulation between the temporal and infratemporal facies.

Irhoud-1, fortunately, has the face quite well preserved, but, unfortunately, lacks dentition. The face is very broad, approximating Skhul 5, but short and low. Thus, it is quite unlike Neandertals and closer to the Skhul/Qafzeh pattern. The total face is mesognathic, with midfacial prognathism less than European Neandertals, and similar to Skhul/Qafzeh and some Levantine Neandertals. The subspinal angle is elevated, as in Skhul/Qafzeh, and unlike Neandertals. The interorbital breadth is substantial. The orbits are both broad and high, therefore large, and not dissimilar dimensionally to some Neandertals and larger than Skhul/Qafzeh; but their shape differs from the former in quadrangular conformation and lateroinferior inclination. The nasal aperture is also non-Neandertal, vertically elongate and triangular, with maximum breadth very inferiorly, and having strong upward convergence of the lateral margins. The nasoalveolar clivus is well-defined, lacks a sharp crest, and evidences a slight subnasal gutter; the anterior nasal spine (broken) was probably substantial, and the floor of the nasal cavity is moderately depressed. Nasal bones are displaced, but do exhibit strong transverse curvature, and were probably not particularly elevated (by comparison to Neandertals), judging from the position of the adjacent maxillary processes.

The suborbital region, including malar, is quite well preserved. The orbital face is well-developed, the robust frontal process bears a tubercle/crest for temporal mm fibers, and the lateral facies of the malar frontal process is frontally disposed as in Skhul/Qafzeh (not laterally as in Neandertals). The fronto-orbital processes (damaged) are non-Neandertal as well, and are narrow and well-inclined inferoanteriorly. The malar body is large, robust and frontally disposed with a strong inferior incurvature, and a marked lateral flexure (much as in Qafzeh-6, and unlike Neandertals, even Shanidar-2 and -4, which variably express a few such features). The maxilla is inflected, with horizontal and inframalar incurvatures, and the suborbital face is distinctly concave. The maxillary wall is quite pneumatized, but without prolongation superiorly (frontal process), laterally (zygomatic), or inferiorly (alveoli). The alveolar process is low, close to Qafzeh-9, and unlike the higher process in Neandertals. The palate (damaged) is large, broad, and moderately deep, rather shallower than in Skhul/Qafzeh group.

The Jebel Irhoud specimens, following Hublin (1991), "testify to the presence in North Africa, within the span of isotope stages 5-6, of a grade of *Homo sapiens* a little more ancient and a little more primitive than the people of the Skhul-Qafzeh group," and "present a mosaic of plesiomorphic and modern characters," but do not yet qualify as "anatomically modern," and cannot at all be considered as African Neandertals. There are definitely phenetic links with antecedent/penecontemporaneous human populations in western Asia (Zuttiyeh); and it may even be surmised that there are direct biological relationships farther afield, into equatorial and, perhaps, even southern Africa.

DISCUSSION

Weidenreich (1946, 1947a) produced a scheme (pedigree) "illustrating the ten known, consecutive, evolutionary phases of man and their speciations." Weidenreich was an anatomist who trained and served under Gustav Schwalbe (Strasburg); having devoted over thirty years to human evolutionary studies, and having first-hand familiarity with the fossil record in Europe and Asia, he had a perspective that merited serious consideration. He (1947b) had a progressivist, typological and, some have claimed, an essentially orthogenetic conception of hominid evolution. He considered "that all primate forms recognized as hominids . . . can be regarded as *one species*," and that if "all hominid types and their variations, regardless of time and space, are taken into consideration, their arrangement in a continuous evolutionary line, leading from the most primitive to the most advanced, does not meet with any difficulty." Howells (1951:85) insightfully noted that "Weidenreich's chart of human lines during the Pleistocene calls for an unparalleled parallelism, in fact, for an incredible convergence following an initial radiation, so that what finally became the modern races were the descendants of more diverse forms, who had been staggered in their relative progress until the very end, when all arrived at the tape together, since modern man is skeletally strongly uniform." Some significant aspects of Weidenreich's formulation, of course, subsequently fell away, as a consequence of the acceptance of *Australopithecus* species within Hominidae (now Homininae or Hominini), extensive documentation of *Homo* antiquity, diversification, and presumptive origins in Africa, and demise of the concept of gigantism (*Gigantopithecus* subsumed in Ponginae) as an aspect ("phase") of hominid evolution.

Coon's (1962) emendation and amplification of the Weidenreichian perspective was the first wide-ranging, in-depth effort after the midcentury to evaluate the hominid fossil record within the perspective of extant, somehow descendant, human populations. "All of the evidence available from comparative ethnology, linguistics, and prehistoric archeology indicates a long separation of the principal races of man," Coon wrote. "Fossil men now extinct differed from each other in race and were not members of separate species, except in the sense that one species grew out of another," and "*Homo erectus* then evolved into *Homo sapiens* not once but five times, as each subspecies, living in its own territory, passed a critical threshold from a more brutal to a more sapient state."

Both G. G. Simpson (1963) and E. Mayr (1962) wrote overall positive reactions to this effort. The latter considered that "it seems that the basic framework of Coon's thesis is as well, or better, substantiated than various possible alternatives." Although S. M. Garn (1963) offered "an overall favorable report" on the volume he protested that ". . . the idea of five separate evolutions (at five different times) from *erectus* to *sapiens* does not make equal evolutionary sense." And Dobzhansky (1963), who accepted the volume "as a milestone in the study of fossil man," also asserted that "the possibility that the genetic system of living men, *Homo sapiens*, could have independently arisen five times, even twice, is vanishingly small."

J. B. Birdsell (1963), a previous coauthor (Coon, Garn & Birdsell 1950) and one-time tutee of Coon at Harvard, produced surely the most penetrating and critical appraisal. "In judging the author to have failed in his attempts to trace living populations backward in time to the known samples of early fossil types, it must be considered largely as a consequence of attempting a problem at present impossible to solve," and further, "If Coon wishes to define gene flow as a major force in the evolution of modern races, he is obligated to provide a model which demonstrates, first, that it is indeed feasible in terms of the demands of time and space and, second, that the basic assumptions involved in its construction make it a more probable explanation than the radiation favored by most investigators" (1963:185). The last is in reference to "the position of the majority of students of human evolution" among whom "it is considered that the living races of man, all of which belong to a single species, *Homo sapiens*, have evolved in the last 100,000 years or so from a single ancestral form equivalent to Coon's *Homo erectus* and radiated over the land surfaces of the Old World in Late Pleistocene times, replacing marginal forms of humanity . . . in genetical terms the uniformity of our present populations is stressed in this concept, and modern differences are primarily the consequence of late, local adaptive evolutional processes" (1963:179).

In a series of insightful, and sometimes insufficiently appreciated, papers published over an interval of some two decades, W. W. Howells sought to introduce some methodological rigor, hypothesis testing, and critical aspects of evolutionary (population) biology into evaluations of the Pleistocene hominid fossil record. Almost every major concern evidenced in the more recent, ever proliferating literature on later phases of hominid evolution, including modern human antecedents and origins, were explored in these seminal contributions.[5] These dealt with *Homo erectus* (3, in all), Neandertals (6), and modern populations/races, including modern human origins (5). Howells sought to set straight the unlikely consequences attendant on Coon's approach toward linking the known fossil record with modern human populations, and the latter's (and others) insufficient regard for some evolutionary processes, including isolation, limitations on gene flow, migration (dispersal), and extinctions, and their roles in hominid phylogenesis. I consider that Howells's perspective is directly pertinent to the current polarizations and controversies that characterize studies of modern human origins.

Students of later hominid evolution, including problems attending modern human origins, differ substantially in respect to their acceptance of and support for several basic concepts of evolutionary biology. These include: species concepts; the roles of processes of anagenesis (phyletic evolution) and cladogenesis (speciation); the existence and effect (if any) of (spatial) isolation versus propinquity and contiguity of species populations in respect to gene flow; the existence of endemism; the nature and effects of demographics of past (Pleistocene) human populations; the role of drift (sampling effects) versus adaptation; acceptance (or not) of punctuational versus gradualistic phylogenetic transformation; and appropriateness and applicability of phylogenetic (cladistic) concepts and analysis. There are also notable differences in evaluations of "discrete" morphological traits, particularly of the cranium and dentition, their occurrence, expression, and frequencies, as a potential or real reflection of population affinities. Characterizations as "evolutionists" or "migrationists" only partially or inadequately reflect or subsume some of such different perspectives.

The literature relevant to the issue of modern human origins has increased vastly in the last several decades, in part as a consequence of major symposia devoted to the subject. Important, but usually partisan critical reviews and statements of position include contributions by Aiello (1993, who has not otherwise been directly involved in such studies), Bräuer (1989, 1992), Frayer et al. (1993), Smith, Falsetti, and Donnelly (1989; Simmons and Smith 1991), Stringer and Andrews (1988; and Stringer 1989, 1990, 1992a), and Wolpoff (1989, 1992; also, Wolpoff, Wu & Thorne 1984). In these, and other referenced papers therein, will be found not only consideration of much primary fossil hominid documentation, but also explications of the varying perspectives, precepts and assumptions of a number of primary investigators actively concerned with a modern human origins research program.

There is now a consensus that the *ultimate* roots of *Homo sapiens* are African, either as an antecedent species (*H. ergaster* = *H. erectus*) or, rarely, as the earliest, still primitive representative of *sapiens* proper. Similarly, an extra-African into Eurasia dispersal event, some time within the early Pleistocene, is uniformly acknowledged. From the perspective of some, including myself, the stem and (probably) dispersant species is *H. ergaster*, from which subsequently *H. erectus* derived in eastern and southeastern Asia. The latter is significantly derived, in aspects of craniodental morphology, from its African antecedent. I consider this most probably an anagenetic process resulting in a cladogenetic event, perhaps largely as a consequence of combined founder effect and peripheral isolation circumstances. The earliest, even early, representatives of hominid colonization of Europe thus far remain unknown, but the species *H. erectus* is not identifiable in that record. In either instance, this might be and has been sometimes termed a "Garden of Eden" hypothesis in respect to *initial* origins. Adherents of a multiregional evolution (MRE) model (hypothesis) recognize *no* subsequent cladogenetic events, *no* major extinctions; they stress persistent gene flow between contemporaneous populations to constrain a single polytypic species, having temporal genetic continuity, with a "pattern of central variability and regionally distinct monomorphisms . . . maintained by long-lasting gradients balancing gene flow, usually from the centre towards the edges, against local selection and in some cases drift" (Wolpoff 1989, 1992). Such prolonged, sustained continuity has sometimes been broadly subsumed under an appellation "candelabra model."

Several variants of a so-called "Noah"s ark" (Howells 1976), or center-of-origin model, permit extinction(s), accept isolation and varying degrees of endemism, generally entail one (or more) cladogenetic event(s), and propose population displacement(s)/replacement(s), perhaps (not necessarily) with assimilation or hybridization. Such a model requires an appropriate source and geographic area, distinct from those of "archaic" replaced populations. Thus, "Out of Africa," "African Eve," or "Garden of Eden-II" hypotheses have been considered appropriate appellations. These include Afro-European *sapiens* (AES) (Bräuer 1992) and recent African evolution (RAE) (Stringer and Andrews 1988; Stringer 1990, 1992b, 1993) models. Among the forceful arguments for some form of such an hypothesis is documentation of deep-rooted clades of archaic, but nonetheless derived, *Homo* spp. in both Europe and in eastern/southeastern Asia. For some investigators, including myself, these manifestations reflect differentiation at the evolutionary (phylogenetic) species level, viz., *Homo heidelbergensis-neanderthalensis* and *Homo erectus*, respectively. Most (not all) investigators recognize the substantial, even remarkable, endemism (see Hublin 1990) of European human populations, extending well back into the Middle Pleistocene. The evidence from eastern/ southeastern Asia may very well be viewed from this same perspective. This manifestation is most appropriately considered as a spatiotemporal outcome of hominid evolution in these marginal areas, one of which was periodically sundered from its continental affiliation, rather than reflective of a phase/ grade of hominid evolution. Although morphological stasis has been both claimed and disclaimed in the case of *H. erectus (s.str.)* populations, it is, in fact, the overall persistence, for whatever cause(s), of the fundamental *bauplan* over upwards of 0.75 ma, as exemplified particularly in cranial morphology that has struck the attention of most students of human paleontology.

In Asia, both eastern and southeastern, the presence of another distinctive and seemingly very derived hominid form in the lowermost Pleistocene is quite probable.

The informal appellation, archaic *Homo sapiens,* which increasingly gained currency over the past quarter-century, has served more to obscure than to clarify aspects of hominid phylogeny. It, of course, implies attribution *both* to a particular species (*sapiens*) on the one hand, and simultaneously a grade within said species on the other hand. The latter is usually contrasted with "archaic modern" and "anatomically modern." Thus, although the usage seemingly appears straightforward it is, in fact, informal; no basis for this practice exists within the International Code of Zoological Nomenclature. The term has no formal basis, and as it has been applied almost wholesale to an extraordinary diversity of fossil hominid remains from Eurasia as well as Africa, which often differ substantively one from another, and both spatially and temporally, it is frankly best abandoned altogether. There is, however, nothing to stand in the way of employing local (site, locality) names to refer to recognizably delimitable population samples within formally established taxonomic categories. In this matter, Gingerich (1985:30) remarks: "Simpson (1943), Newell (1956), and others have proposed that successive taxa in evolutionary lineages be designated subspecies, a proposition that confuses temporal and geographic concepts and requires three names to be written when a simple binomial conveys the same information. In practice, subspecies are most often proposed and used by paleontologists unsure that formal designation of new taxa is warranted." This succinct perspective mirrors our own view exactly, and we follow closely its strictures.

The historical roots of the concept of "archaic" *Homo sapiens* are unclear, but are at least implicit in LeGros Clark's writings of thirty years ago. The concept has a strong aspect of grade about it and some authors, unspecified here, have overtly stated this as their basis for employment of the concept and phraseology (adopted from the usual meaning, "of, relating to, or characteristic of a much earlier, often, more primitive period"). Singular characters (apomorphies) adequate to diagnose minimally such morphs are rarely (if ever) specified, and if they exist may well be largely plesiomorphies. The concept has been applied on both intracontinental and intercontinental bases. Further modifiers of "early" and "late" have equally been applied to this adjective, as has also the former to the appellation "anatomically modern" *Homo sapiens* (e.g., among others, Bräuer 1989, 1992; Bräuer and Rimbach 1990). Almost surely the latter appellation, which presumably signifies humans of modern aspect, should be equally considered as redundant and preferably abandoned. Such usages are, at most, unusual, if ever explicit, in vertebrate systematics. Similarly, the expressions "intermediate" ("lying or occurring between two extremes or in a middle position") and "transitional" ("passage from one form, state, style, or place to another") seem equally uninformative except in a (inappropriate) gradistic sense.

This disenchantment and dissatisfaction with terminology appropriate to the subject at hand mirrors concerns expressed by Tattersall (1986, 1992) on species recognition and terminological adequacy in hominid evolutionary studies. Thus, I concur with his observation (1986:169) that "The interpretation of most human fossils subsequent to about 0.5 myr as belonging somehow to *Homo sapiens* is perhaps the most comprehensive smokescreen of all, and could only have been maintained by a zealous refusal to consider characters other than brain size," and the conclusion (1986:170) that "In any group other than Hominidae the presence of several clearly recognizable morphs in the middle to upper Pleistocene would suggest (indeed demonstrate) the involvement of several species." And thus, "there is neither theoretical nor practical justification for continuing to cram virtually all fossil humans of the last half-million years into the species '*Homo sapiens*'" (Tattersall 1992:348). If *Homo neanderthalensis* is the appropriate clade for western Eurasian hominids often subsumed under the "archaic *sapiens*" rubric, what then might be the status of similarly attributed African and Asian counterparts? This matter can only be briefly exposed here and requires more extended treatment in future.

African examples include Salé and Kébibat (Morocco), Bodo (Ethiopia), Ndutu and (perhaps) Eyasi (Tanzania), Kabwe (Zambia), and Elandsfontein (South Africa), among an earlier grouping, and Jebel Irhoud (Morocco), Omo Kibish-2 (Ethiopia), Eliye Springs, and (*perhaps*) Ileret = ER-3884 (Kenya), Ngaloba (Tanzania), and Florisbad (O.F.S., South Africa), among a later group. Distinct species taxa were once proposed for Kabwe (*Homo rhodesiensis* A. S. Woodward 1921) and for Florisbad (*Homo helmei* T. F. Dreyer 1935), the first of which has priority. The cranial morphologies of Ndutu (Clarke 1990) and of Florisbad (Clarke 1985), as well as the setting and context of the latter occurrence (Kuman and Clarke 1986), have now been elucidated and afford the important evidence requisite for the evaluation of phyletic affinities. Hublin (1991; also 1985) has similarly recently afforded comparable descriptive and comparative treatment of the Salé, Kébibat, and Irhoud samples. There appear to be (*at least*) two major morphs represented by these various specimens, which may (or may not) constitute a single clade, sampled from the Middle into the earlier Upper Pleistocene. There is also some basis for the proposal of successive species taxa, the earlier (*H. rhodesiensis*) of which shares some purported resemblances (Rightmire 1990) to the *H. heidelbergensis-neanderthalensis* clade, and the later (*H. helmei*) of which is demonstrably diverse, and can be shown to exhibit extra-African, specifically Asian linkages, both in the west (Zuttiyeh) and in the east (Dali, and to which the subspecific name *daliensis* has been applied by Wu Xinzhi). Although, as indicated previously, I consider the Maba (and Hathnora) and Jinniushan specimens as sampling distinctive populations, Pope (1992:287) has recently admitted that "Jinniushan and Maba may represent immigrating populations from outside of China." The further elucidation of (upper) Middle Pleistocene population relationships within Africa, and between Africa and Asia certainly stands now as one of the central problems in human evolutionary studies.

CONCLUSIONS

If paleoanthropology is to be a viable endeavor there must be conjoint emphasis on evidences of the hominid past derived from the archeological as well as the (human) paleontological records and their contexts. Despite extraordinary enhancement of the available data base, these endeavors persist largely in isolation one from another and their respective goals appear often disparate if not contradictory and antithetical. This situation is particularly characteristic of the debates and controversies surrounding the origin of modern humans and the delineation of their proximate antecedents.

The phylogenetic roots of modern humans are demonstrably in the Middle Pleistocene. The distribution of those antecedent populations appear to lie outside of western and eastern Eurasia, and more probably centered broadly on Africa. Hominid populations of western and eastern Eurasia may now be reasonably considered as specifically distinct, *Homo heidelbergensis-neanderthalensis* and *Homo erectus*, respectively, each characterized by singular autapomorphies and representing partially or largely isolated derivatives of the initial extra-African hominid dispersal in Lower Pleistocene times.

The extension of the concept of *Homo sapiens* deeply into the Middle Pleistocene has been neither progressive nor elucidative. As a consequence, recognizable, definable morphs have been lumped arbitrarily, in spite of their distinctive overall morphological patterns and varying assortments of apomorphies so that the species category has become scarcely definable. Temporal criteria have been employed inappropriately in place of primary biological criteria in assembling and seriating past population samples. Often gradistic, progressivist perspectives have bedeviled such efforts. As a consequence, biological realities have been obscured and the need to examine adaptations and evolutionary processes obfuscated and thwarted. There is serious need for normal procedures in evolutionary biology to prevail.

Various models, as many as four if variants are included, have been espoused in the explication of modern humans. All have substantive drawbacks and limitations, and Smith, Falsetti, and Donnelly (1989:58) are surely correct that "no model explains the available data on the subject unequivocally, whatever their proponents suggest." A far-reaching multiregional, continuity framework has, in fact, the least pragmatic and operational appeal as it stretches the bounds of credulity in respect to population distributions, affinities, and potentialities for gene flow. Continuity in some geographic areas is contraindicated by a diversity of evidence (most of Europe), and is ill-established, even chimerical, as some investigators have demonstrated elsewhere (Australasia). Nonetheless, aspects of continuity, isolation, drift, extinction(s), population displacement(s), introgression, and assimilation must, at times, all have been active factors and processes relevant to the late Pleistocene phases of human evolution. Efforts should increasingly be directed toward the identification and elucidation of such processes in particular situations and in regard to their potential consequences.

There is a measure of congruence between the human paleontological and associated behavioral evidence, as inferred from the archeological record, associated residues and their contexts, in respect to adaptations between and among these extinct populations. Although rather substantial, and convincing evidence attests to the biological nature (morphology, broadly conceived) of early *Homo sapiens* representatives, a lack of congruence of relevant archeological (paleobehavioral parameters) evidence is commonly cited. A central focus of intensified research into modern human origins should be directed towards the enhanced recognition and explication of the archeological diversity manifest within the terminal Middle and initial Upper Pleistocene interval, with the purpose of elucidating the roots of modern human behaviors subsequently fully manifest, almost worldwide, in the late Pleistocene.

ENDNOTES

1. Tuff IB of Olduvai Bed I has an age of 1.80 mya, and is nearly coincident with the proposed age 1.81 mya of several authors of the Pliocene-Pleistocene boundary in southern Italy stratotype section(s).

2. At Gunung Batak (eastern Java), first worked by E. Dubois but known even before, the earliest vertebrate fauna (9 spp., with suggestions of mainland proximity) occurs in continental (fluvial and laharic) sediments *above* an andesitic volcanic breccia affording a K/Ar age of ~1.87 mya (isochron plot), or 1.59-1.76 mya (conventional assessment) (Bandet et al. 1989). The succedent and diverse vertebrate assemblage, that of Kedung Brubus (22 spp.), derives from the eponymous locality.

3. Longgupo Cave, Washna (Wanxian Co., Sichuan) (30°45'N, 108°40'E) contains nearly 30 meters of infilling (Huang Wanpo et al. 1991). It yields a very diverse macro- and microvertebrate fauna (116 spp. of mammals, plus birds), the latter mostly from the mid-lower 7 m of alluvial deposits, below the brecciated upper part of the section. The mammals from the middle part of this section have strong resemblances to those from the Liucheng (Guangxi) Cave, which yielded important gnathic/dental remains of *Gigantopithecus blacki*. Primates include *Gigantopithecus* (dentition) and gnathic/dental remains considered hominid. Two very likely artifacts are reported, one from a hominid-yielding level. A paleomagnetic sampling records a basal normal, and three successive normal intervals with intervening reversed zones. The former are considered to represent Réunion (1, 2) and Olduvai (N) subchrons within the Matuyama (R) chron. This requires confirmation as, perhaps, only the upper Matuyama and

its subchrons and the succedent Brunhes (N) chron might be documented. In any case this important occurrence is as old as, or *potentially* substantially older than the early artifactual localities in the Nihewan Basin. Unfortunately the affinities of the few hominid-referred specimens are difficult to determine.

4. There are, in fact, conflicting isotopic age assessments for the upper (layers 1-3) infilling of ZKD-1. The initial determination (U-S) afforded ages of 230 (+ 30/− 23) kya and 256 (+ 62/− 40) kya (Zhao et al. 1980), comparable with initial ESR results (Huang et al. 1991); these results have been commonly employed as a basis for judging the site's upper age limit and for effecting extraregional correlations, both with the Loess Plateau sequence of events and those of the deep-sea ^{18}O record. The results of Yuan et al. (1991), employing the same method on fossil bone from layer 2, were only slightly older, 256 (+ 47/− 34) kya. However, the equally new results of U-S age assessment of a calcite, from travertine (and considered a closed system) between Layers 1 and 2, afford a very substantially greater age, 420 (+ 110/− 54) kya. The latter results tend to reinforce some criticisms previously voiced by Aigner (1986, 1988) on too ready acceptance of the outcome of dating efforts at ZKD-1, and who has herself strongly favored a "short" chronology for the infilling at this important locality.

5. The outcome of his long-standing concerns and deliberations on such matters are now available in a recent book, *Getting Here* (Howells 1993), which only came to my attention, for a review, after this text was completed.

REFERENCES

Agrawal, D. P., B. S. Kotlia, and S. Kusumgar. 1988. Chronology and significance of the Narmada Formations. *Proceedings of the Indian National Science Academy* 54-A(3): 418-24.

Aguirre, E. 1991. Les premiers peuplements humains de la péninsule ibérique, 143-50. In *Les Premiers Européens*, ed. E. Bonifay and B. Vandermeersch. Paris: Éditions du Comité des Travaux Historiques et Scientifiques.

Aiello, L. C. 1993. The fossil evidence for modern human origins in Africa: A revised view. *American Anthropologist* 95:73-96.

Aigner, J. S. 1986. The age of Zhoukoudian Locality 1: The newly proposed 0^{18} correspondences. *Anthropos* (Brno) 23:157-73.

Aigner, J. S. 1988. Dating the earliest Chinese Pleistocene localities: The newly proposed 0^{18} correspondences, 1032-61. In *The Palaeoenvironment of East Asia From the Mid-Tertiary*. Vol. II, ed. P. Whyte, J. S. Aigner, N. G. Jablonski, G. Taylor, D. Walker and P. Wang. Hong Kong: Centre of Asian Studies.

Amani, F. 1991. *La faune de la grotte à Hominidé du Jebel Irhoud (Maroc)*. Thèse du Doctorat, Faculté des Sciences, Université de Rabat.

Amani, F., and D. Geraads. 1993. Le gisement moustérien du Djebel Irhoud, Maroc: Précisions sur la faune et la biochronologie, et description d'un nouveau reste humain. *Comptes Rendus de l'Academie de Sciences* (Paris) 316:847-52.

An, Z., T. Liu, X. Kan, J. Sun, P. Wang, W. Gao, Y. Zhu, and M. Wei. 1987. Loess-paleosoil sequence and chronology at Lantian Man localities. *Aspects of Loess Research*. Beijing: China Ocean Press.

Badam, G. L. 1989. Observations on the fossil hominid site at Hathnora, Madhya Pradesh, India, 153-71. In *Perspectives in Human Evolution*, ed. A. Sahni and R. Gaur. Delhi: Renaissance Publishing House.

Balout, L. 1965a. Le Moustérien du Maghreb. *Quaternaria* 7:43-58.

Balout, L. 1965b. Données nouvelles sur le problème du Moustérien en Afrique du Nord, 137-46. In *Actas del V Congreso Panafricano de Prehistoria y de Estudio del Cuaternario*. Santa Cruz de Tenerife: Museo Arqueologico.

Bandet, Y., F. Sémah, S. Sartono, and T. Djubiantono. 1989. Premier peuplement par les mammifères d'une région de Java est, à la fin du Pliocéne: âge de la faune du Gunung Butak, prés de Kedungbrubus (Indonésie). *Comptes Rendus de l'Academie de Sciences* (Paris) 308:867-70.

Bartstra, G.-J. 1977. The height of the river terraces in the transverse Solo valley in Java. *Modern Quaternary Research in SE Asia* 3:143-55.

Bartstra, G.-J. 1987. Late *Homo erectus* or Ngandong man of Java. *Palaeohistoria*, Groningen, 29:1-7.

Bartstra, G.-J., S. Soegondho, and A. van der Wijk. 1988. Ngandong man: Age and artifacts. *Journal of Human Evolution* 17:325-37.

Bar-Yosef, O. 1984. Near East, 233-98. In *Neue Forschungen zur Altsteinzeit*. Forschungen zur Allgemeinen und Vergleichenden Archologie, Bd. 4.

Bar-Yosef, O. 1992. Middle Palaeolithic chronology and the transition to the Upper Palaeolithic in southwest Asia, 261-72. In *Continuity or Replacement: Controversies in* Homo sapiens *Evolution*, ed. G. Bräuer and F. H. Smith. Rotterdam: A. A. Balkema.

Bar-Yosef, O., and N. Goren-Inbar. 1993. The lithic assemblages of Ubeidiya. A Lower Paleolithic site in the Jordan Valley. *Qedem*, Jerusalem. 34:266 pages.

Baryshnikov, G. 1989. Les mammifères du Paléolithique inférieur du Caucase. *L'Anthropologie* 93:813-30.

Besançon, J., L. Copeland, F. Hours, S. Muhesen, and P. Sanlaville. 1981. Le Paléolithique d'El Kowm: Rapport préliminaire. *Paléorient* 7:33-55.

Birdsell, J. B. 1963. Review of *The Origin of Races*, by C. S. Coon. *Quarterly Review of Biology* 38:178-85.

Bonifay, E. 1989. Un site du très ancien Paléolithique de plus de 2 M.A. dans le Massif-Central français: Saint-Eble-Le Coupet (Haute-Loire). *Comptes Rendus de l'Academie de Sciences* (Paris) 308:1567-70.

Bonifay, E. 1991. Les premières industries du Sud-Est de la France et du Massif central, 63-80. In *Les Premiers Européens*, ed. E. Bonifay and B. Vandermeersch. Paris: Éditions du Comité des Travaux Historiques et Scientifiques.

Bonifay, E., A. Consigny, and R. Liabent. 1989. Contribution du Massif Central français à la connaissance des premiers peuplements préhistorique de l'Europe. *Comptes Rendus de l'Academie de Sciences* (Paris) 308:1491-96.

Bonifay, M.-F. 1986. Intérêt des études taphonomiques au Pléistocène ancien: Soleilhac et Ceyssaguet. *Bulletin Museum National d'Histoire Naturelle,* Paris 4, 8:269-81.

Bonifay, M.-F. 1991. Archéologie du comportement: Remarques sur l'apport des faunes de grands mammifères au Pléistocène ancien, 111-13. In *Les Premiers Européens*, ed. E. Bonifay and B. Vandermeersch. Paris: Éditions du Comité des Travaux Historiques et Scientifiques.

Bracco, J. P. 1991. Typologie, technologie et matières premières des industries du très ancien Paléolithique en Velay (Massif central, France), 93-100. In *Les Premiers Européens*, ed. E. Bonifay and B. Vandermeersch. Paris: Éditions du Comité des Travaux Historiques et Scientifiques.

Bräuer, G. 1989. The evolution of modern humans: A comparison of the African and non-African evidence, 123-54. In *The Human Revolution: Behavioural and Biological Perspectives on the Origin of Modern Humans*, ed. P. Mellars and C. Stringer. Edinburgh: Edinburgh University Press.

Bräuer, G. 1992. Africa's place in the evolution of *Homo sapiens*, 83-98. In *Continuity or Replacement: Controversies in* Homo sapiens *Evolution*, ed. G. Bräuer and F. H. Smith. Rotterdam: A. A. Balkema.

Bräuer, G., and K. W. Rimbach. 1990. Late archaic and modern *Homo sapiens* from Europe, Africa, and southwest Asia: Craniometric comparisons and phylogenetic implications. *Journal of Human Evolution* 19:787-807.

Campbell, B. G. 1973. New concepts in physical anthropology: Fossil man. *Annual Review of Anthropology* 1:27-54.

Champion, D. E., and M. A. Lanphere. 1988. Evidence for a new geomagnetic reversal from lava flows in Idaho: Discussion of short polarity reversals in the Brunhes and late Matuyama polarity chrons. *Journal of Geophysical Research* 93:11,667-80.
Chen, T., and S. Yuan. 1988. Uranium-series dating of bones and teeth from Chinese Palaeolithic sites. *Archaeometry* 30:59-76.
Chen, T., S. Yuan, S. Gao, and Y. Hu. 1987. Uranium series dating of fossil bones from Hexian and Chaoxian fossil human sites (in Chinese; English abstract). *Acta Anthropologica Sinica* 6:249-54.
Cheng, G., S. Li, and J. Lin. 1977. Discussion on the age of *Homo erectus yuanmoensis* and the event of early Matuyama (in Chinese; English abstract). *Scientia Geologica Sinica* 1:34-43.
Clarke, R. J. 1985. A new reconstruction of the Florisbad cranium, with notes on the site, 301-5. In *Ancestors: The Hard Evidence*, ed. E. Delson. New York: Alan R. Liss.
Clarke, R. J. 1990. The Ndutu cranium and the origin of *Homo sapiens*. *Journal of Human Evolution* 19:699-736.
Coon, C. S. 1962. *The Origin of Races*. New York: Alfred A. Knopf.
Coon, C. S., S. M. Garn, and J. B. Birdsell. 1950. *Races. A Study of the Problems of Race Formation in Man*. Springfield, IL: C. C. Thomas.
Copeland, L., and F. Hours. 1983. Le Yabroudien d'El-Kowm (Syrie) et sa place dans le Paléolithique du Levant. *Paléorient* 9:21-38.
Cosijn, J. 1931. Voorloopige mededeeling omtrent het voorkomen van fossiele be enderen in het heuvel-terrein ten noorden van Djetis en Perning (Midden-Java). *Verh. Geol. Mijnbk. Gen. Geol.* ser. 9:113-19.
Cosijn, J. 1932. Tweede mededeeling over het voorkomen van fossiele be enderen in het heuvelland ten noorden van Djetis en Perning (Java). *Verh. Geol. Mijnbk. Gen. Geol.* ser. 9:135-48.
Dobzhansky, T. 1963. Review of *The Origin of Races*, by C. S. Coon. *Scientific American* 208:169-72.
Dodonov, A. E. 1986. Antropogen Yuzhnogo Tadzhitistana (Anthropogene of South Tajikistan) (in Russian). *Trudy, Geological Institut, Akademia Nauk SSSR* 409:163 pp. Moscow: Nauka.
Dodonov, A. E., Y. R. Melamed, and K. F. Nikiforova, eds. 1977. *International Symposium on the Problem of the Neogene/Quaternary Boundary. Guidebook* (in Russian and English). Moscow: Nauka.
Dzaparidze, V., G. Bosinski, T. Bugianisvili, L. Gabunia, A. Justus, N. Klopotovskaja, E. Kvavadze, D. Lordkipanidze, G. Majsuradze, N. Mgeladze, M. Nioradze, E. Pavlenisvili, H.-U. Schmincke, D. Sologasvili, Tusabramisvili, M. Tvalcrelidze, and A. Vekua. 1992. Der altpaläolithische Fundplatz Dmanisi in Georgen (Kaukasus). *Jahrbuch Römisch-Germanischen Zentralmuseums Mainz* 36(1989):67-116.
Farrand, W. R. 1993. Confrontation of geological stratigraphy and radiometric dates from Upper Pleistocene sites in the Levant. In *Late Quaternary Chronology and Paleoclimates of the Eastern Mediterranean*, ed. O. Bar-Yosef and R. Kra. Madison, WI: Prehistory Press.
Franzen, J. 1985a. What is "*Pithecanthropus dubius* Koenigswald, 1950," 221-26. In *Ancestors: The Hard Evidence*, ed. E. Delson. New York: Alan R. Liss.
Franzen, J. 1985b. Asian australopithecines? 255-63. In *Hominid Evolution: Past, Present and Future*, ed. P. V. Tobias. New York: Alan R. Liss.
Frayer, D. W., M. H. Wolpoff, A. G. Thorne, F. H. Smith, and G. G. Pope. 1993. Theories of modern human origins: The paleontological test. *American Anthropologist* 95:14-50.
Frenzel, B., M. Pécsi, and A. A. Velichko, eds. 1992. *Atlas of Paleoclimates and Paleoenvironments of the Northern Hemisphere. Late Pleistocene-Holocene*. Budapest: Geographical Research Institute, Hungarian Academy of Sciences, and Stuttgart: Gustav Fischer Verlag.
Gao, J. 1975. Australopithecine teeth associated with *Gigantopithecus* (in Chinese; English

abstract). *Vertebrata Palasiatica* 13:81-88.
Garn, S. M. 1963. Review of *The Origin of Races*, by C. S. Coon. *American Sociological Review* 28:637-38.
Garrod, D. A. E. 1956. Acheuléo-Jabroudien et "Pré-Aurignacien" de la grotte du Tabun (Mont Carmel). *Quaternaria* 3:39-59.
Garrod, D. A. E. 1962. The Middle Palaeolithic of the Near East and the problem of Mount Carmel Man. *Journal of the Royal Anthropological Institute* 92:232-59.
Garrod, D. A. E., and D. M. Bate. 1937. *The Stone Age of Mount Carmel.* Vol. 1, *Excavations at the Wady el-Mughara*. Oxford: Clarendon Press.
Geraads, D., and E. Tchernov. 1983. Fémurs humains du Pléistocène moyen de Gesher Benot Ya'acov (Israel). *L'Anthropologie* 87:138-41.
Gerasimov, I. P., and A. A. Velichko, eds. 1982. *Paleogeography of Europe During the Last One Hundred Thousand Years* (Atlas-monograph) (in Russian; English summary). 156 pp., 15 maps. Moscow: Nauka.
Gingerich, P. D. 1985. Species in the fossil record: Concepts, trends, and transitions. *Paleobiology* 11:27-41.
Gisis, I., and O. Bar-Yosef. 1974. New excavations in Zuttiyeh Cave, Wadi Amud, Israel. *Paléorient* 2:175-80.
Givirtzman, G., E. Shachnai, N. Bakler, and S. Ilain. 1985. Stratigraphy of the *kurkar* group (Quaternary) of the coastal plain of Israel. *Geological Survey of Israel, Current Research* 1983/1984:70-82.
Goldberg, P. 1973. Sedimentology, Stratigraphy and Paleoclimatology of Et-Tabun Cave, Mount Carmel, Israel. Ph.D. diss., Geology, University of Michigan, Ann Arbor. Ann Arbor: University Microfilms.
Goren, N. 1981. The lithic assemblages of the site of 'Ubeidiya, Jordan Valley. Ph.D. diss., Hebrew University, Jerusalem.
Goren-Inbar, N. 1992. The Acheulian site of Gesher Benot Ya'aqov–An Asian or an African entity? 67-82. In *The Evolution and Dispersal of Modern Humans in Asia*, ed. T. Akazawa, K. Aoki, and T. Kimura. Tokyo: Hokusen-sha.
Goren-Inbar, N., and S. Belitzky. 1989. Structural position of the Pleistocene Gesher Benot Ya'aqov site in the Dead Sea Rift Zone. *Quaternary Research* 31:371-76.
Goren-Inbar, N., S. Belitzky, Y. Goren, R. Rabinovich, and I. Saragusti. 1992a. Gesher Benot Ya'aqov–The 'bar': An Acheulian assemblage. *Geoarchaeology* 7:27-40.
Goren-Inbar, N., S. Belitzky, K. Verosub, E. Werker, M. Kislev, A. Heimann, I. Carmi, and A. Rosenfeld. 1992b. New discoveries at the Middle Pleistocene Gesher Benot Ya'aqov Acheulian site. *Quaternary Research* 38:117-28.
Goren-Inbar, N., I. Zohar, and D. Ben-Ami. 1991. A new look at old cleavers–Gesher Benot Ya'aqov. *Mitekufat Haeven* 24:7-33.
Grosswald, M. G. 1980. Late Weichselian ice sheet of northern Eurasia. *Quaternary Research* 13:1-32.
Grün, R., and C. B. Stringer. 1991. Electron spin resonance dating and the evolution of modern humans. *Archaeometry* 33:153-99.
Grün, R., C. B. Stringer, and H. P. Schwarcz. 1991. ESR dating of teeth from Garrod's Tabun Cave collection. *Journal of Human Evolution* 20:231-48.
Gu, L. D., Yu P. Zhang, W. Huang, Y. Thang, H. Ji, Y. You, S. Ding, and X. Guan. 1986. *The Cenozoic of Lantian (Shaanxi)* (in Chinese). Beijing: Science Press.
Guseinov, M. M. 1981. *The Azykh Cave* (in Russian and English). Baku: Institute of History, Academy of Science of the Azerbaijan SSR.
Guseinov, M. M. 1985. *Ancient Paleolithic of Azerbaidzhan* (in Russian; English summary). Baku: 'Elm.
Heimann, A., D. Mor, and G. Steinitz. 1987. K-Ar dating of the Ramat Korazim basalts, Dead Sea Rift, 54-56. Israel Geological Society, Annual Meeting, Mispe Ramon.
Hennig, G. J., and F. Hours. 1982. Dates pour le passage entre l'Acheuléen et le Paléolithique Moyen à El Kowm (Syrie). *Paléorient* 8:81-84.

Hill, A., S. Ward, A. Deino, G. H. Curtis, and R. Drake. 1992. Earliest *Homo. Nature* 355:719-22.

Holloway, R. L. 1981. Volumetric and asymmetry determinations on recent hominid endocasts: Spy I and II, Djebel Irhoud I, and the Salè *Homo erectus* specimens, with some notes on Neandertal brain size. *American Journal of Physical Anthropology* 55:385-93.

Howell, F. C. 1978. Hominidae, 154-248. In *Evolution of African Mammals*, ed. V. Maglio and H. Cooke. Cambridge: Harvard University Press.

Howell, F. C. 1991. Foreword, v-viii. In *Koobi Fora Research Project.* Vol. 4, *Hominid Cranial Remains*, B. A. Wood. Oxford: Clarendon Press.

Howell, F. C., P. Haesaerts, and J. de Heinzelin. 1987. Depositional environments, archeological occurrences and hominids from Members E and F of the Shungura Formation (Omo basin, Ethiopia). *Journal of Human Evolution* 16:665-700.

Howells, W. W. 1951. Origin of the human stock. Concluding remarks of the chairman, 79-86. In *Origin and Evolution of Man*. 1950. Cold Spring Harbor Symposia on Quantitative Biology, vol. XV. Cold Spring Harbor: The Biological Laboratory.

Howells, W. W. 1976. Explaining modern man: Evolutionists *versus* migrationists. *Journal of Human Evolution* 5:477-95.

Howells, W. W. 1980. *Homo erectus* – who, when and where: A survey. *Yearbook of Physical Anthropology* 23:1-23.

Howells, W. W. 1993. *Getting Here. The Story of Human Evolution.* Washington, D.C.: Compass Press.

Hrdlicka, A. 1930. The skeletal remains of early man. *Smithsonian Miscellaneous Collections* 83:1-379. Washington, D.C.: Smithsonian Institution.

Huang, P., S. Jin, R. Liang, Z. Lu, Z. Yuan, B. Cai, and Z. Fang. 1991. Study of ESR dating for burying age of the first skull of Beijing Man and chronological scale of the cave deposit in Zhoukoudian site Loc. 1 (in Chinese; English abstract). *Acta Anthropologica Sinica* 10:107-15.

Huang, Wanpo, et al. 1991. *Wushan Hominid Site* (in Chinese; English summary). Beijing: Ocean Press.

Huang, W., D. Fang, and Y. Ye. 1982. Preliminary study of the fossil hominid and fauna from Hexian, Anhui (in Chinese; English abstract). *Vertebrata Palasiatica* 20:248-56.

Hublin, J.-J. 1976. *L'Homme de Galilée*. Mèmoire de D. E. H. Université de Paris VI.

Hublin, J.-J. 1985. Human fossils from the North African Middle Pleistocene and the origin of *Homo sapiens*, 283-88. In *Ancestors: The Hard Evidence*, ed. E. Delson. New York: Alan R. Liss.

Hublin, J.-J. 1990. Les peuplements Paléolithiques de l'Europe: un point de vue paléobiogéographique, 29-37. In *Paléolithique moyen récent et Paléolithique supérieur ancien en Europe*, ed. C. Farizy. Nemours: Mémoires du Musée de Préhistoire d'Ile de France 3.

Hublin, J.-J. 1991. L'emergence des Homo sapiens archaiques: Afrique du nord-ouest et Europe occidentale. 2 vols. Thèse, Doctorat d'Etat es Sciences, Université Bordeaux.

Hublin, J.-J., and A.-M. Tillier. 1981. The Mousterian juvenile mandible from Irhoud (Morocco): A phylogenetic interpretation, 167-85. In *Aspects of Human Evolution*, ed. C. B. Stringer. London: Taylor and Francis.

Hublin, J.-J., A.-M. Tillier, and J. Tixier. 1987. L'humérus d'enfant mousterién (Homo 4) du Jebel Irhoud (Maroc) dans son contexte archéologique. *Bulletins et Memoires. Societe d'Anthropologie de Paris* XXV 4:115-42.

Hyodo, M., W. Sunata, and E. E. Susanto. 1992. A long-term geomagnetic excursion from Plio-Pleistocene sediments in Java. *Journal of Geophysical Research* 97:9323-35.

Islamov, Y. I. 1990. Sel'Oungour, un nouveau site du Paléolithique infèrieur en Asie Centrale. *L'Anthropologie* 94:675-88.

Islamov, Y. I., A. A. Zubov, and V. M. Kharitonov. 1988. Paleolithic site Sel'Ungur in Fergana Valley (in Russian; English abstract). *Voprossi Antropologii* 80:38-49.

Jacob, T. 1981. Solo man and Peking man, 87-104. In *Homo erectus. Papers in Honor of David-*

son Black, ed. B. A. Sigmon and J. S. Cybulski. Toronto: University of Toronto Press.
Jacob, T., and G. H. Curtis. 1971. Preliminary potassium-argon dating of early man in Java. *University of California, Berkeley, Archeological Research Facility, Contributions* 12:50.
Jelinek, A. J. 1981. The Middle Paleolithic in the southern Levant from the perspective of the Tabun Cave, 265-80. In *Préhistoire du Levant*, ed. J. Cauvin and P. Sanlaville. Paris: Centre National de la Recherche Scientifique.
Jelinek, A. J. 1982a. The Middle Palaeolithic in the Southern Levant with comments on the appearance of modern *Homo sapiens*, 57-104. In *The Transition from the Lower to the Middle Paleolithic and the Origin of Modern Man*, ed. A. Ronen. Oxford: British Archaeological Reports International Series S151.
Jelinek, A. J. 1982b. The Tabun Cave and Paleolithic Man in the Levant. *Science* 216:1369-75.
Jelinek, A. J. 1990. The Amudian in the context of the Mugharan Tradition at the Tabun Cave (Mount Carmel), Israel, 81-90. In *The Emergence of Modern Humans: An Archaeological Perspective*, ed. P. Mellars. Ithaca: Cornell University Press.
Jelinek, A. J. 1992. Problems in the chronology of the Middle Paleolithic and the first appearance of Early Modern *Homo sapiens* in southwest Asia, 253-75. In *The Evolution and Dispersal of Modern Humans in Asia*, ed. T. Akazawa, K. Aoki, and T. Kimura. Tokyo: Hokusen-sha.
Jelinek, A. J., W. R. Farrand, G. Haas, A. Horowitz, and P. Goldberg. 1973. New excavations at the Tabun Cave, Mount Carmel, Israel, 1967-1972: A preliminary report. *Paléorient* 1:151-83.
Jenkinson, R. D. S., H. M. Rendell, R. Dennell, M. A. Jah, and S. A. Sutherland. 1989. Upper Siwalik palaeoenvironments and palaeoecology in the Pabbi Hills, Northern Pakistan. *Zeitschrift für Geomorphologie* n.f. 33:417-28.
Jia, L., and W. Huang. 1990. *The Story of Peking Man*. Hong Kong and Oxford: Foreign Language Press and Oxford University Press.
Keith, A. 1927. A report on the Galilee skull, 53-106. In *Researches in Prehistoric Galilee 1925-1926*, F. Turville-Petre. London: Council, British School of Archaeology in Jerusalem.
Keith, A. 1931. *New Discoveries Relating to the Antiquity of Man*. London: Williams and Norgate, Ltd.
Keller, H. M., R. A. Tahirkheli, M. A. Mirza, G. E. Johnson, N. M. Johnson, and N. D. Opdyke. 1977. Magnetic polarity stratigraphy of the Upper Siwalik deposits, Pabbi Hills, Pakistan. *Earth and Planetary Science Letters* 36:187-201.
Kennedy, K. A. R., and J. Chiment. 1991. The fossil hominid from the Narmada valley, India: *Homo erectus* or *Homo sapiens*? *Bulletin of the Indo-Pacific Prehistory Association* 10:42-58. P. Bellwood, ed. Indo-Pacific Prehistory 1990. Vol. 1.
Kennedy, K. A. R., A. Sonakia, J. Chiment, and K. K. Verma. 1991. Is the Narmada hominid an Indian *Homo erectus*. *American Journal of Physical Anthropology* 86:475-96.
Koenigswald, G. H. R. von. 1950. Fossil hominids from the Lower Pleistocene of Java. *Proceedings of the International Geological Congress* (1948) 9:59-61.
Koenigswald, G. H. R. von. 1953, 1954a. The Australopithecinae and *Pithecanthropus*. I, II, III. *Proceedings of the Koninklijke Nederlandse Akademie van Wetenschappen* (B) 56:403-13, 427-38; 57:85-91.
Koenigswald, G. H. R. von. 1954b. *Pithecanthropus, Meganthropus* and the Australopithecinae. *Nature* 173:795-96.
Koenigswald, G. H. R. von. 1957a. Remarks on *Gigantopithecus* and other hominoid remains from southern China. *Proceedings of the Koninklijke Nederlandse Akademie van Wetenschappen* (B) 60:153-59.
Koenigswald, G. H. R. von. 1957b. Erratum: *Hemanthropus* n.g. not *Hemianthropus*. *Proceedings of the Koninklijke Nederlandse Akademie van Wetenschappen* (B) 60:416.
Koenigswald, G. H. R. von. 1968. Java: Prae-Trinil man. *Proceedings of the 8th International Congress of Anthropological and Ethnological Sciences*, Tokyo-Kyoto 1968, 1:104-105.
Koenigswald, G. H. R. von. 1973a. The oldest hominid fossils from Asia and their relation to

human evolution. *Accademia Nazionale dei Lincei*, Rome 182:97-118.
Koenigswald, G. H. R. von. 1973b. *Australopithecus, Meganthropus* and *Ramapithecus*. *Journal of Human Evolution* 2:487-91.
Koenigswald, G. H. R. von. 1980. Neue Einblicke in die Geschichte der Hominiden. *Annalen des Naturhistorischen Museums in Wien* 83:181-95.
Kramer, A. 1989. The evolutionary and taxonomic affinities of the Sangiran mandibles of central Java, Indonesia. Ph.D. diss. University of Michigan, Ann Arbor. Ann Arbor: University Microfilms.
Kramer, A. In press. A critical analysis of claims for the existence of southeast Asian australopithecines. *Journal of Human Evolution*.
Kramer, A., and L. W. Konigsberg. In press. The phyletic position of Sangiran 6 as determined by multivariate analyses. *Courier Forschungsinstitut Senckenberg*.
Krantz, G. S. 1975. An explanation for the diastema of Java *erectus* skull IV, 361-72. In *Paleoanthropology, Morphology and Paleoecology*, ed. R. H. Tuttle. The Hague: Mouton.
Kukla, G. 1987. Loess stratigraphy in central China. *Quaternary Science Reviews* 6:191-219.
Kukla, G., and Z. S. An. 1989. Loess stratigraphy in central China. *Palaeogeography, Palaeoclimatology, Palaeoecology* 72:203-25.
Kukla, G., Z. S. An, J. L. Melice, J. Gavin, and J. L. Xiao. 1990. Magnetic susceptibility record of Chinese loess. *Transactions Royal Society of Edinburgh: Earth Sciences* 81:263-88.
Kuman, K., and R. J. Clarke. 1986. Florisbad–New investigations at a Middle Stone Age hominid site in South Africa. *Geoarchaeology* 1:103-25.
LeGros Clark, W. E. 1964. *The Fossil Evidence for Human Evolution*. Chicago: University of Chicago Press.
Li, C. K., W. Y. Wu, and Z. D. Qiu. 1984. Chinese Neogene: Subdivision and correlation (in Chinese). *Vertebrata Palasiatica* 22:163-78.
Li, P., F. Chien, H. Ma, C. Pu, L. Hsing, and S. Chu. 1977. Preliminary study on the age of Yuanmou man by paleomagnetic technique. *Scientia Sinica* 20:645-64.
Li Tianyuan and D. Etler. 1992. New Middle Pleistocene crania from Yunxian in China. *Nature* 357:404-7.
Li, Y., and B. Wen. 1986. *Guanyindong. A Lower Paleolithic site at Qianxi County, Guizhou Province* (in Chinese; English summary). Beijing: Cultural Relics Publishing House.
Liu, T., and M. Ding. 1983. Discussion on the age of "Yuanmou man" (in Chinese; English abstract). *Acta Anthropologica Sinica* 2:40-48.
Liu, Z. 1985. Sequence of sediments at Locality 1 in Zhoukoudian and correlation with loess stratigraphy in northern China and with the chronology of deep-sea cores. *Quaternary Research* 23:139-53.
Lovejoy, C. O. 1970. The taxonomic status of the '*Meganthropus*' mandibular fragments from the Djetis Beds of Java. *Man* n.s. 5:228-36.
Lü, Z. 1985. *The Excavation and Significance of the Jinniushan Site* (in Chinese). Beijing: Peking University Archaeological Department.
Lü, Z. 1989. On the time of Jinniushan (Gold Ox Mount) man and the position of its evolution (in Chinese). *Liaohai Cult. Relics J.* 1:44-55.
Lü, Z., Y. Huang, P. Li, and Z. Meng. 1989. Yiyuan fossil man (in Chinese). *Acta Anthropologica Sinica* 8:303-13.
Lubine, V. P. 1977. *Mousterian Culture in Caucasus* (in Russian). Leningrad: Nauka.
Lubine, V. P., and G. F. Barychnikov [Baryshnikov]. 1984. L'activité de chasse des plus anciens habitants du Caucase (Acheuléen, Moustérien). *L'Anthropologie* 88:221-29.
Lumley, H. de, and A. Sonakia. 1985. Contexte stratigraphique et archéologique de l'homme de la Narmada, Hathnora, Madhya Pradesh, Inde. *L'Anthropologie* 89:3-12.
Lumley, M.-A. de, and A. Sonakia. 1985. Première découverte d'un *Homo erectus* sur le continent Indien a Hathnora, dans la moyenne vallée de la Narmada. *L'Anthropologie* 89:13-61.
Markova, A. K. 1982. Small mammals from the Paleolithic site of Azykh (in Russian; English abstract). *Paleontologicheskii Sbornik*, Lvov, 19:14-28.

Mayr, E. 1962. Origin of the Human Races. Review of *The Origin of Races*, by C. S. Coon. *Science* 138:420-22.

McCown, T. D., and A. Keith. 1939. *The Stone Age of Mount Carmel*. Vol. II, *The Fossil Human Remains from the Levalloiso-Mousterian*. Oxford: Clarendon Press.

Mercier, N. 1992. Apport des méthodes radionucléaries de datation à l'Europe et du Proche-Orient au cours du Pléistocène moyen et supérieur. Thèse, Docteur en Sciences, Université de Bordeaux.

Mercier, N., H. Valladas, O. Bar-Yosef, B. Vandermeersch, C. Stringer, and J.-L. Joron. 1993. Thermoluminescence date for the Mousterian burial site of Es Skhul, Mt. Carmel. *Journal of Archaeological Science* 20:169-74.

Nikiforova, K. V., and A. E. Dodonov, eds. 1980. *The Neogene-Quaternary boundary* (in Russian; English abstracts). Moscow: Nauka.

Nikiforova, K. V., and E. A. Vangengheim, eds. 1988. *Biostratigrafeya Pozdnego Pliotsena-Rannego Pleystotsena Tadzhikistana. (Biostratigraphy of the late Pliocene-early Pleistocene of Tadzhikistan)* (in Russian). Moscow: Nauka.

Orban-Segenbarth, R., and F. Procureur. 1983. Tooth size of *Meganthropus palaeojavanicus*: An analysis of distances between some fossil hominids and a modern human population. *Journal of Human Evolution* 12:711-20.

Pei, W., and S. Zhang. 1985. A study on lithic artifacts of *Sinanthropus* (in Chinese; English summary). *Paleontologica Sinica* 168, n.s. D., 12:1-277.

Peretto, C. 1991. Les plus anciens gisements préhistoriques du bassin du Pô (Italie septentrionale), 153-59. In *Les Premiers Européens*, ed. E. Bonifay and B. Vandermeersch. Paris: Éditions du Comité des Travaux Historiques et Scientifiques.

Peretto, C., ed. 1992. *I primi abitanti della Valle Padana: Monte Poggiolo, nel quadro della conoscenze Europee*. Milan: Jaca Book.

Picard, L. 1963. The Quaternary in the northern Jordan Valley. *Proceedings of the Israel Academy of Sciences and Humanities* 1:1-34.

Picard, L. 1965. The geological evolution of the Quaternary in the central-northern Jordan Graben, Israel. *Geological Society of America, Special Paper* 84:337-66.

Pope, G. G. 1991. Evolution of the zygomaticomaxillary region in the genus *Homo* and its relevance to the origin of modern humans. *Journal of Human Evolution* 21:189-213.

Pope, G. G. 1992. Craniofacial evidence for the origin of modern humans in China. *Yearbook of Physical Anthropology* 35:243-98.

Pope, G. G., and J. E. Cronin. 1984. The Asian Hominidae. *Journal of Human Evolution* 13:377-96.

Qian, F., G. Zhou, et al. 1991. *Quaternary geology and paleoanthropology of Yuanmou, Yunnan, China* (in Chinese; English summary). Beijing: Science Press.

Qing, F. 1985. On the age of "Yuanmou Man"–A discussion with Liu Tungsheng et al (in Chinese; English abstract). *Acta Anthropologica Sinica* 4:324-32.

Qiu, G. 1990. The Pleistocene human environment of north China (in Chinese; English abstract). *Acta Anthropologica Sinica* 9:340-49.

Ranov, V. A., R. S. Davis, and A. E. Dodonov. In press. Kul'dara and the central Asian Lower Paleolithic. *Quaternary Research*.

Rightmire, G. P. 1990. *The Evolution of* Homo erectus. *Comparative Anatomical Studies of an Extinct Human Species*. Cambridge: Cambridge University Press.

Roe, D. A., ed. 1983. *Adlun in the Stone Age*. 2 vols. Oxford: British Archaeological Reports International Series 159.

Rolland, N. 1992. The Palaeolithic colonization of Europe: An archaeological and biogeographic perspective. *Trabajos de Prehistoria* 49:69-111.

Rollefson, G. 1980. The Paleolithic industries of Ain-el-Assad (Lion's Spring), near Azraq, Eastern Jordan. *Annual of the Department of Antiquities, Jordan* XXIV:129-43; 301-9.

Saban, R. 1984. Anatomie et évolution des veines méningées chez les hommes fossiles. *Comité des Travaux Historiques et Scientifiques. Ministère de l'Éducation Nationale, Section des Sciences, Mémoire* 11:1-289.

Sanlaville, P. 1977. Etude géomorphologique de la région littorale du Liban. Thése de Doctorat d'Etat, Section des Etudes Géographiques I, Publications Université Libanaise. 2 vol. Cartes.

Sartono, S. 1976. Genesis of the Solo terraces. *Modern Quaternary Research in SE Asia* 2:1-21.

Sartono, S. 1982. Sagittal cresting in *Meganthropus palaeojavanicus* von Koenigswald. *Modern Quaternary Research in SE Asia* 7:201-10.

Sartono, S., F. Sémah, K. A. S. Astadiredja, M. Suklen-Darmono, and T. Djubiantono. 1981. The age of *Homo modjokertensis. Modern Quaternary Research in SE Asia* 6:91-101.

Schick, K., N. Toth, Q. Wei, J. D. Clark, and D. Etler. 1991. Archaeological perspectives in the Nihewan basin, China. *Journal of Human Evolution* 21:13-26.

Schulman, N. 1978. The Jordan Rift Valley. *Tenth International Congress on Sedimentology* 2:57-94.

Schwarcz, H. P., W. M. Buhay, R. Grün, H. Valladas, E. Tchernov, O. Bar-Yosef, and B. Vandermeersch. 1989. ESR dating of the Neanderthal site, Kebara cave, Israel. *Journal of Archaeological Science* 16:653-59.

Schwarcz, H. P., P. Goldberg, and B. Blackwell. 1980. Uranium series dating of archaeological sites in Israel. *Israel Journal of Earth-Sciences* 29:157-65.

Schwarcz, H. P., R. Grün, B. Vandermeersch, O. Bar-Yosef, H. Valladas, and E. Tchernov. 1988. ESR dates for the hominid burial site of Qafzeh in Israel. *Journal of Human Evolution* 17:733-37.

Sémah, F. 1983. *Stratigraphie et paléomagnétisme du Pliocène supérieur et du Pléistocène de l'île de Java (Indonésie).* Muséum National d'Histoire Naturelle, Paris, Travaux du Laboratoire de Préhistoîre 20:261.

Sémah, F. 1986. Le peuplement ancien de Java, ébauche d'un cadre chronologique. *L'Anthropologie* 90:359-400.

Sémah, F., A.-M. Sémah, T. Djubiantono, and H. T. Simanjuntak. 1992. Did they also make stone tools? *Journal of Human Evolution* 23:436-39.

Simmons, T. 1990. Comparative Morphometrics of the Frontal Bone in Hominids: Implications for Models of Modern Human Origins. Ph.D. diss., University of Tennessee, Knoxville. Ann Arbor: University Microfilms.

Simmons, T., A. B. Falsetti, and F. H. Smith. 1991. Frontal bone morphometrics of southwest Asian Pleistocene hominids. *Journal of Human Evolution* 20:249-69.

Simmons, T., and F. H. Smith. 1991. Late Pleistocene human population relationships in Africa, Europe, and the Circum-Mediterranean. *Current Anthropology* 32:623-27.

Simpson, G. G. 1963. Review of *The Origin of Races*, by C. S. Coon. *Perspectives in Biology and Medicine* 6:268-72.

Skinner, J. H. 1970. El Masloukh: A Yabrudian site in Lebanon. *Bulletin du Musée de Beyrouth* 23:143-72.

Smith, F. H. 1985. Continuity and change in the origin of modern *Homo sapiens. Zeitschrift für Morphologie und Anthropologie* 75:197-222.

Smith, F. H., A. B. Falsetti, and S. M. Donnelly. 1989. Modern human origins. *Yearbook of Physical Anthropology* 32:35-68.

Sohn, S., and M. H. Wolpoff. 1990. Zuttiyeh: A new look at an old face. *Acta Anthropologica Sinica* 9:359-70.

Sohn, S., and M. H. Wolpoff. In press. The Zuttiyeh face, a view from the east. *American Journal of Physical Anthropology.*

Sondaar, P. Y. 1984. Faunal evolution and the mammalian biostratigraphy of Java. *Courier Forschungsinstitut Senckenberg* 69:219-35.

Stekelis, M. 1960. The Palaeolithic deposits of Jisr Banat Yaqub. *Bulletin of the Research Council of Israel* 9G:61-90.

Stringer, C. B. 1989. Documenting the origin of modern humans, 67-96. In *The Emergence of Modern Humans: Biocultural Adaptations in the Late Pleistocene*, ed. E. Trinkaus. Cambridge: Cambridge University Press.

Stringer, C. B. 1990. The emergence of modern humans. *Scientific American* 263:98-104.

Stringer, C. B. 1992a. Reconstructing recent human evolution. *Philosophical Transactions of the Royal Society, London* 337(B):217-24.

Stringer, C. B. 1992b. Replacement, continuity and the origin of *Homo sapiens*, 9-24. In *Continuity or Replacement: Controversies in* Homo sapiens *Evolution*, ed. G. Bräuer and F. H. Smith. Rotterdam: A. A. Balkema.

Stringer, C. B. 1993. New views on modern human origins, 75-94. In *The Origin and Evolution of Humans and Humanness*, ed. D. T. Rasmussen. Boston, London: Jones & Bartlett.

Stringer, C. B., and P. Andrews. 1988. Genetic and fossil evidence for the origin of modern humans. *Science* 239:1263-68.

Stringer, C. B., R. Grün, H. P. Schwarcz, and P. Goldberg. 1989. ESR dates for the hominid burial site of Es Skhul in Israel. *Nature* 338:756-58.

Tattersall, I. 1986. Species recognition in human paleontology. *Journal of Human Evolution* 15:165-75.

Tattersall, I. 1992. Species concepts and species identification in human evolution. *Journal of Human Evolution* 22:341-49.

Tchernov, E. 1986. *Les Mammifères du Pléistocène inférieur de la vallée du Jordain à Oubeidiyeh.* Paris: Association Paléorient.

Tchernov, E. 1987. The age of the 'Ubeidiya Formation, an early Pleistocene hominid site in the Jordan Valley, Israel. *Israel Journal of Earth-Sciences* 36:3-30.

Tchernov, E. 1988. La biochronologie du site de Ubeidiya (Vallée du Jordain) et les plus anciens hominides du Levant. *L'Anthropologie* 92:839-61.

Tchernov, E. 1992a. Eurasian-African biotic exchanges through the Levantine corridor during the Neogene and Quaternary. *Courier Forschungsinstitut Senckenberg* 153:103-23.

Tchernov, E. 1992b. Biochronology, paleoecology, and dispersal events of hominids in the southern Levant, 149-88. In *The Evolution and Dispersal of Modern Humans in Asia*, ed. T. Akazawa, K. Aoki and T. Kimura. Tokyo: Hokusen-sha.

Tedford, R. H., L. F. Flynn, Qui Zhanxiang, N. D. Opdyke, and W. R. Downs. 1991. Yushe Basin, China: Paleomagnetically calibrated mammalian biostratigraphic standard for the late Neogene of eastern Asia. *Journal of Vertebrate Paleontology* 11:519-26.

Teilhard de Chardin, P., and J. Piveteau. 1930. Les mammifères fossiles de Nihewan (Chine). *Annales de Paléontologie* 19:1-134.

Terra, H. de, and P. Teilhard de Chardin. 1936. Observations on the Upper Siwalik Formation and later Pleistocene deposits in India. *Proceedings of the American Philosophical Society* 76:791-822.

Thomas, H. 1981. La faune de la grotte à Néanderthaliens du Jebel Irhoud (Maroc). *Quaternaria* 23:191-217.

Tobias, P. V. 1966. A member of the genus *Homo* from 'Ubeidiya. *The Lower Pleistocene of the Central Jordan Valley. The Excavations at 'Ubeidiya, 1960-1963*, ed. M. Stekelis. Jerusalem: The Israel Academy of Sciences and Humanities.

Tobias, P. V., and G. H. R. von Koenigswald. 1964. A comparison between the Olduvai hominines and those of Java and some implications for hominid phylogeny. *Nature* 204:515-18.

Trinkaus, E. 1983. *The Shanidar Neandertals*. New York: Academic Press.

Trinkaus, E. 1984. Western Asia, 251-93. In *The Origins of Modern Humans: A World Survey of the Fossil Evidence*, ed. F. H. Smith and F. Spencer. New York: Alan R. Liss.

Trinkaus, E. 1989. Issues concerning human emergence in the later Pleistocene, 1-17. In *The Emergence of Modern Humans: Biological Adaptations in the Later Pleistocene*, ed. E. Trinkaus. Cambridge: Cambridge University Press.

Turville-Petre, F. 1927. *Researches in Prehistoric Galilee, 1925-1926*. London: British School of Archaeology in Jerusalem.

Tyler, D. E. 1991. A taxonomy of Javan hominid mandibles. *Human Evolution* 5:401-20.

Tyler, D. E. 1992. A taxonomy of Javan hominid mandibles (in Chinese and English). *Acta Anthropologica Sinica* 11:285-99.

Valladas, H., J.-L. Joron, G. Valladas, B. Arensburg, O. Bar-Yosef, A. Belfer-Cohen, P.

Goldberg, H. Laville, L. Meignen, Y. Rak, E. Tchernov, A.-M. Tillier, and B. Vandermeersch. 1987. Thermoluminescence dates for the Neanderthal burial site at Kebara in Israel. *Nature* 330:159-60.
Valladas, H., J. L. Reyss, J. Joron, G. Valladas, O. Bar-Yosef, and B. Vandermeersch. 1988. Thermoluminescence dating of Mousterian 'Proto-Cro-Magnon' remains from Israel and the origin of modern man. *Nature* 331:614-16.
van der Plicht, J., A. van der Wijk, and G.-J. Bartstra. 1989. Uranium and thorium in fossil bones: Activity ratios and dating. *Applied Geochemistry* 4:339-42.
Vandermeersch, B. 1981. Les premiére *Homo sapiens* au Proche Orient, 97-100. In *Les Processus de l'Hominisation*, ed. D. Ferembach. Paris: Centre National de la Recherche Scientifique.
Vandermeersch, B. 1989. The evolution of modern humans: Recent evidence from southwest Asia, 155-64. In *The Human Revolution*, ed. P. Mellars and C. Stringer. Edinburgh: Edinburgh University Press.
Velichko, A. A., ed. 1984. *Late Quaternary Environments of the Soviet Union*. Minneapolis: University of Minnesota Press.
Velichko, A. A., G. V. Antonova, E. M. Zelikson, A. K. Markova, et al. 1980. Paleogeography of the Azykh site – The oldest settlement of prehistoric man within the USSR (in Russian). *Akademia Nauk, SSSR, Geographical Series, Izvestia*, Moscow 8:20-35.
Vos, J. de, S. Sartono, S. Hardja-Sasmita, and P. Y. Sondaar. 1982. The fauna from Trinil, type locality of *Homo erectus*: A reinterpretation. *Geologie Mijnbouw* 61:207-11.
Watanabe, N., and D. Kadar, eds. 1985. *Quaternary Geology of the Hominid Fossil Bearing Formations in Java*. Bandung, Indonesia: Geological Research and Development Centre, Special Publication no. 4.
Weidenreich, F. 1943. The skull of *Sinanthropus pekinensis*: A comparative study of a primitive hominid skull. *Palaeontologica Sinica*, n.s. D, No. 10 (whole series 127).
Weidenreich, F. 1946. *Apes, Giants and Man*. Chicago: University of Chicago Press.
Weidenreich, F. 1947a. Facts and speculations concerning the origin of *Homo sapiens*. *American Anthropologist*, n.s. 49:187-203.
Weidenreich, F. 1947b. The trend of human evolution. *Evolution* 1:221-36.
Wolpoff, M. H. 1975. Some aspects of human mandibular evolution, 1-64. In *Determinants of Mandibular Form and Growth*, ed. J. A. McNamara. Ann Arbor: University of Michigan Press.
Wolpoff, M. H. 1980. *Paleoanthropology*. New York: Alfred A. Knopf.
Wolpoff, M. H. 1989. Multiregional evolution: The fossil alternative to Eden, 62-108. In *The Human Revolution: Behavioural and Biological Perspectives on the Origin of Modern Humans*, ed. P. Mellars and C. Stringer. Edinburgh: Edinburgh University Press.
Wolpoff, M. H. 1992. Theories of modern human origins, 25-63. In *Continuity or Replacement: Controversies in* Homo sapiens *Evolution*, ed. G. Bräuer and F. H. Smith. Rotterdam: A. A. Balkema.
Wolpoff, M. H., Wu Xinzhi, and A. G. Thorne. 1984. Modern *Homo sapiens* origins: A general theory of hominid evolution involving the fossil evidence from East Asia, 411-83. In *The Origins of Modern Humans: A World Survey of the Fossil Evidence*, ed. F. H. Smith and F. Spencer. New York: Alan R Liss.
Wood, B. A. 1991. *Koobi Fora Research Project.* Vol. 4, *Hominid Cranial Remains.* Oxford: Clarendon Press.
Wood, B. A. 1992. Origin and evolution of the genus *Homo*. *Nature* 355:783-90.
Wu, M. 1983. *Homo erectus* from Hexian, Anhui found in 1981 (in Chinese; English abstract). *Acta Anthropologia Sinica* 2:109-15.
Wu Rukang. 1964. A newly discovered mandible of *Sinanthropus* type—*Sinanthropus lantianensis*. *Scientia Sinica* 13(5):801-12.
Wu Rukang. 1966. The hominid skull of Lantian, Shensi. *Vertebrata Palasiatica* 10:1-22.
Wu, R., and X. Dong. 1982. Preliminary study of *Homo erectus* remains from Hexian, Anhui (in Chinese; English abstract). *Acta Anthropologica Sinica* 1:1-13.

Wu, R., and X. Dong. 1985. *Homo erectus* in China, 79-89. In *Palaeoanthropology and Palaeolithic Archaeology in the People's Republic of China*, ed. Wu Rukang and J. W. Olsen. Orlando: Academic Press.
Wu, R., and S. Lin. 1983. Peking Man. *Scientific American* 248:86-94.
Wu, R., and R. Peng. 1959. Fossil human skull of early Paleo-anthropic stage found at Mapa, Shaokuan, Kwangtung Province. *Paleovertebrata et Paleoanthropologia* 4:159-64.
Wu, R., M. Ren, X. Zhu, Z. Yang, C. Hu, Z. Kong, Y. Xié, and S. Zhao. 1985. *Multidisciplinary Study of the Peking Man Site at Zhoukoudian* (in Chinese). Beijing: Science Press.
Wu, R., and X. Wu. 1982. Hominid fossil teeth from Xichuan, Henan (in Chinese; English abstract). *Vertebrata Palasiatica* 20:1-9.
Wu, X. 1988. Comparative study of early *Homo sapiens* from China and Europe (in Chinese; English abstract). *Acta Anthropologica Sinica* 7:287-93.
Wu, X. 1989. Early *Homo sapiens* in China, 24-41. In *Early Humankind in China* (in Chinese), ed. R. Wu, X. Wu, and S. Zhang. Beijing: Science Press.
Wu, X., and M. Wu. 1985. Early *Homo sapiens* in China, 91-106. In *Palaeoanthropology and Paleolithic Archaeology in the People's Republic of China*, ed. Wu Rukang and J. W. Olsen. Orlando: Academic Press.
Yokoyama, T., and I. Koizumi. 1989. Marine transgressions on the Pleistocene Pecangan Formation in the Sangiran area, central Java, Indonesia. *Palaeogeography, Palaeoclimatology, Palaeoecology* 72:177-93.
Zhang, Y. 1984. The "Australopithecus" of west Hubei and some early Pleistocene hominids of Indonesia (in Chinese; English abstract). *Acta Anthropologica Sinica* 3:85-92.
Zhang, Y. 1985. *Gigantopithecus* and "*Australopithecus*" in China, 69-78. In *Palaeoanthropology and Palaeolithic Archaeology in the People's Republic of China*, ed. Wu Rukang and J. W. Olsen. Orlando: Academic Press.
Zhang, Y., W. Huang, Y. Tang, H. Ji, Y. You, Y. Tong, S. Ding, X. Huang, and J. Zheng. 1978. Cenozoic stratigraphy of the Lantian region, Shaanxi (in Chinese). *Memoirs of the Institute of Vertebrate Paleontology and Palaeoanthropology*, Beijing, A-14. Beijing: Science Press.
Zhou, G., and X. Zhang, eds. 1984. *Yuanmou Man. A Collection of Photos and Essays on Prehistoric Man and Cultures in Yuanmou Basin, Yunnan Province* (in Chinese). Kunming: Yunnan People's Press.

Index